# Praise for

"The ideas in this book are so [illegible] fascinating and yes, so foreign, you are going to need people to talk to about them . . . Please read this important book. Read it twice. Talk about it. Tell everyone you know."
—Brenna Maloney, *The Washington Post*

"In making clear the patience, imagination and humility required to better know and protect other forms of intelligence on Earth, [Bridle] has made an admirable contribution to the dawning inter-species age."
—*The Economist*

"Bridle is a clear, artful writer and a sweeping thinker . . . [A] hopeful book, almost an antidote. It imagines technology not as something separate and menacing, but as part of a grand unfolding—an 'efflo-rescence,' to use Bridle's word—along an evolutionary continuum of human and 'more-than-human' ways of being in the world."
—Peter Christie, *Post Magazine*

"[A] fascinating survey . . . Bridle makes a solid case for their argument that 'everything is intelligent' and that all life on Earth is interconnected, and their notion that intelligence is 'one among many ways of being in the world' is well reasoned and convincing. This enlightening account will give readers a new perspective on their place in the world."
—*Publishers Weekly*

"An accessible but also technically precise book . . . [*Ways of Being*] makes a remarkably compelling case for the universality of reason, the benefits to be reaped by acknowledging it, and the urgent need to do so given the reality of looming ecological collapse . . . A provocative, profoundly insightful consideration of forms of reason and their relevance to our shared future."
—*Kirkus Reviews* (starred review)

"There's a new breed of thinker—people who've grown up through the collapse of an old order and who are looking at the first shoots of a very different future. James Bridle is right at the front of this thinking. Their writing weaves cultural threads that aren't usually seen together, and the resulting tapestry is iridescently original, deeply disorientating, and yet somehow radically hopeful. The only futures that are viable will probably feel like that. This is a pretty amazing book, worth reading and rereading." —Brian Eno

"A profound and elegant exploration of nonhuman intelligence that unfurls a wider, more expansive notion of thought itself. James Bridle's view of the mind, embedded in a more thoughtful world, is a revelation."

—Alexandra Kleeman, author of *Something New Under the Sun*

"James Bridle's wonderful book will make you think and feel the power of knowing how like all other life-forms we are. There is nothing more important."

—Timothy Morton, author of *Hyperobjects: Philosophy and Ecology After the End of the World*

"James Bridle's brilliant *Ways of Being* shows the importance of listening to one another and our surroundings and creating new forms of community."

—Hans Ulrich Obrist, artistic director at the Serpentine, London

"James Bridle encourages you to widen the boundaries of your understanding, to contemplate the innate intelligence that animates the life force of octopuses and honeybees as well as apes and elephants. We humans are not alone in having a sense of community, a sense of fun, a sense of wonder and awe at the beauty of nature. Be prepared to reevaluate your relationship with the amazing life-forms with whom we share the planet. Fascinating, innovative, and thought-provoking—I thoroughly recommend *Ways of Being*."

—Dr. Jane Goodall, DBE, founder of the Jane Goodall Institute and UN Messenger of Peace

Mikael Lundblad / mikaelcreative.com

## JAMES BRIDLE

# WAYS OF BEING

James Bridle is a writer and an artist. Their writing on art, politics, culture, and technology has appeared in magazines and newspapers including *The Guardian*, *The Observer*, *Wired*, *The Atlantic*, the *New Statesman*, *frieze*, *Domus*, and *ICON*. *New Dark Age*, their book about technology, knowledge, and the end of the future, was published in 2018 and has been translated into more than a dozen languages. In 2019, they wrote and presented *New Ways of Seeing*, a four-part series for BBC Radio 4. Their artworks have been commissioned by galleries and institutions including the V&A, Whitechapel Gallery, the Barbican, Hayward Gallery, and the Serpentine and have been exhibited worldwide and on the internet.

ALSO BY JAMES BRIDLE

*New Dark Age: Technology and the End of the Future*

# WAYS OF BEING

# WAYS OF BEING

## Animals, Plants, Machines:
## The Search for a Planetary Intelligence

JAMES BRIDLE

Picador
Farrar, Straus and Giroux
New York

Picador
120 Broadway, New York 10271


Printed in the United States of America
Originally published in 2022 by Allen Lane, Great Britain, as *Ways of Being: Beyond Human Intelligence*
Published in the United States in 2022 by Farrar, Straus and Giroux
First paperback edition, 2023

Grateful acknowledgment is made for permission to print lyrics from "Enough about Human Rights," by Louis Hardin, courtesy of Managarm Musikverlag.

The Library of Congress has cataloged the Farrar, Straus and Giroux hardcover edition as follows:
Names: Bridle, James, author.
Title: Ways of being : animals, plants, machines : the search for a planetary intelligence / James Bridle.
Description: First American edition. | New York : Farrar, Straus and Giroux, 2022. | "Originally published in 2022 by Allen Lane, Great Britain, as Ways of being : beyond human intelligence." | Includes bibliographical references and index.
Identifiers: LCCN 2022003440 | ISBN 9780374601119 (hardback)
Subjects: LCSH: Intellect. | Philosophy of mind. | Cognitive science. | Information technology—Philosophy. | Human ecology.
Classification: LCC BF431 .B678 2022 | DDC 153.9—dc23/eng/20220214
LC record available at https://lccn.loc.gov/2022003440

Paperback ISBN: 978-1-250-87296-8



10 9 8 7 6 5 4 3

*For Navine and Zephyr*

σχολὴ μὲν δή, ὡς ἔοικε: καὶ ἅμα μοι δοκοῦσιν ὡς ἐν τῷ πνίγει ὑπὲρ κεφαλῆς ἡμῶν οἱ τέττιγες ᾄδοντες καὶ

We have plenty of time, apparently; and besides, the locusts seem to be looking down upon us as they sing and talk with each other in the heat.

Plato, *Phaedrus*, 258e, from *Plato in Twelve Volumes*, Vol. 9, 1925

Enough about Human Rights!
What about Whale Rights?
What about Snail Rights?
What about Seal Rights?
What about Eel Rights? . . .
What about, what about,
What about, what about Bug Rights?
What about Slug Rights?
What about Bass Rights?
What about Ass Rights?
What about Worm Rights?
What about Germ Rights?
What about Plant Rights?
– Moondog

'Enough about Human Rights',
from the album *H'art Songs* by Moondog, 1978

# *Contents*

# *List of Illustrations*

# WAYS OF BEING

# Introduction

*More Than Human*

The late summer sun lingers on the mountainsides and the still waters of the lake. The air is warm, the sky a deep, almighty blue. Cicadas hum in the thick undergrowth, and goat bells chime somewhere in the distance. A small fire has been lit among the reeds, and tins of beer have been cracked open. Someone produces a clarinet and, wandering among the trees that crowd the water's edge, begins to play. It's a scene of timeless tranquillity, yet it is here that one of the greatest conflicts of our age is being played out – between human agency and the intelligence of machines, and between the illusion of human superiority and the survival of the planet.

I am in Epirus, the north-west corner of Greece, hard up against the Pindus mountains and the border with Albania: a region famous for its beauty and its resilience. Here, in the winter of 1940, an outnumbered, ill-equipped, but determined Greek force, fighting in the harshest of conditions, held and pushed back an invading Italian army. The 28th of October, the day on which Greece's wartime premier Ioannis Metaxas refused Mussolini's ultimatum to surrender, is today remembered and celebrated as Oxi Day – in Greek, Οχι, the day of the No.

Epirus is a stunning landscape of rugged mountains and deep gorges, studded with stone villages and monasteries, and inhabited, along with its people, by bears, wolves, foxes, jackals, golden eagles, and some of the oldest trees and forests in Europe. The Aoös River sweeps down from the Pindus into the Vikos National Park, and the Ionian Sea glitters along its rocky coastline. It is something of a paradise; one of the most beautiful, unspoiled lands I have ever seen, but today it is under threat once more.

I am a writer and artist, and for many years I've explored the relationship between technology and everyday life: how the things we make – and particularly complex things like computers – affect society, politics and, increasingly, the environment. I've also lived in Greece for the last few years and I've come to Epirus to visit some friends: a group of native Epirots and transplants from Athens – shepherds, poets, bakers and hoteliers. All are activists in the fight to save Epirus from a new and terrible danger, one which threatens to shatter and poison the very ground we walk upon. Their campaign stickers, found on village noticeboards, road signs and laptop cases, feature a single one-word slogan: Οχι. No.

Walking the woods surrounding the lake, I stumble across thin wooden stakes pushed into the ground and strips of plastic tape tied to branches and saplings. The stakes are tagged with thick, wet marker pen: a series of letters and numbers which mean nothing to me. I follow the path of the stakes as they march in ragged lines through the woods. Breaking through the undergrowth onto a recently scraped dirt road, I see that they extend across a meadow and into deeper woods beyond. They branch off too: more plastic tape, tied onto trees and boughs, mark right angles in what I will come to understand is a vast grid or lattice imposed on the landscape from above. Over the next few days I follow these lines across fields and vineyards, through gardens and villages, marked by more streamers attached to fences and barbed wire, to gates and road signs. They stretch for hundreds, perhaps thousands, of kilometres, like a system of coordinates imposed by a remote, alien intelligence.

There are occasional signs of activity associated with the grid: a new road, bulldozed through the fields; heaps of spoil; tyre marks; deep holes surrounded by debris. The locals tell me about unmarked vans, helicopters and work crews in hi-vis jackets who appear and disappear, their comings and goings accompanied by loud explosions which rattle windows and shake the birds from the trees. On Facebook, my friends share shaky camera-phone footage they have captured of detonations throwing soil hundreds of feet into the air, accompanied by the sirens and whistles of mining crews.

These markings are what I have come to Epirus to see, but their meaning is to be found in scattered internet posts, news stories and company accounts. Smashed through the forest, gouged into the soil, exploded in the grey light of dawn, these marks, I will discover, are the tooth- and claw-marks of Artificial Intelligence, at the exact point where it meets the earth.

Since 2012 successive Greek governments have pursued a policy of fossil fuel development, designating Epirus and the Ionian Sea as areas for exploration and selling off exploitation rights to international oil and gas companies. For cash-strapped Greece, reeling from years of economic crisis and externally imposed austerity, the potential revenues outweigh the threat to both the local environment and the global climate. Discussion of the deal, let alone criticism, has been muted. In Epirus, public access to government contracts is restricted, environmental assessments go unpublished, and exploration teams move about the countryside in unmarked white vans, vanishing at the sight of activists and inquisitive journalists.

The presence of oil in Greece has been documented since ancient times. Around 400 BC, the historian Herodotus described natural oil seeps on the island of Zakynthos, places where thick black ooze welled to the surface from deep underground. The inhabitants used it to caulk their ships and light their lamps. Today, a couple of small rigs extract this oil off the Ionian coast, and tension simmers with Turkey over similar sites in the Aegean and eastern Mediterranean. Until recently, Epirus has remained remote from these concerns, but the possibility of riches beneath its rugged terrain has long been suspected.

I'd read that oil seeps were to be found in Epirus, but my only references were grainy photos in online presentations by oil prospectors and academics.[1] Once I found myself in Epirus, the name of one village kept cropping up: Dragopsa, a few miles west of the regional capital Ioannina and close to the lake in the woods. Asking around, someone suggested I talk to Leonidas, an anti-oil activist whose family had lived there for generations.

One still, sultry afternoon Leonidas drove me to Dragopsa, stopping every now and again to post his Οχι stickers where they might catch the eye. In the valley below the village, we left the car and walked through meadows and orchards to a river. The clear, pure waters of Epirus are the source of some 70 per cent of Greece's drinking water;

large bottling plants cluster at the foot of the mountains. Yet as we pick our way around a bend in the river, I caught the unmistakeable smell of petroleum. The overpowering odour was strongest at the base of a steep cliff, where tree roots and loose, dark clods of soil were exposed by the river's flow. This was the site where, in the 1920s, villagers discovered oil welling up from the ground of its own accord, as it did on Zakynthos. Leonidas tells me that he too has found seeps in recent years: patches of black, sticky fuel, far from the nearest road, rising among the reeds and grasses. You don't need artificial intelligence to find oil in Epirus; but you need AI to exploit it.

The successful bidder for the Epirus exploration contract was one of the world's largest energy corporations, Repsol.[2] From its foundation in 1927 as Spain's national oil company, Repsol has expanded across the globe, discovering hundreds of new fields in the last decade; it has also pioneered the use of new technologies for oil discovery and exploitation. In 2014, Repsol and IBM Watson – the division of the US tech giant responsible for artificial intelligence – announced that they were collaborating 'to leverage cognitive technologies that will help transform the oil and gas industry'. These technologies included 'prototype cognitive applications specifically designed to augment Repsol's strategic decision making in the optimization of oil reservoir production and in the acquisition of new oil fields'.[3]

Acquisition and optimization are the two central endeavours of the fossil fuel industry: where to drill into the earth, and how to get the most out of it. The oil is running out and the economics of extraction are changing: as the largest and most accessible reserves are pumped out, the financial value of what remains increases, even in the face of obvious and catastrophic environmental consequences. Previously untapped reserves, ignored because they were too difficult to evaluate or exploit, are now in the sights of the oil giants once again. As Repsol itself points out in its publicity material, 'Accessing new reserves is an increasingly difficult task. The subsoil is a great unknown. Drilling and making large financial investments are risky, difficult decisions.' As a result, the most sophisticated computational processes must be brought to bear on the situation. Smart decisions require smart tools: 'To minimize error and make the correct decisions at Repsol, we have decided to let technology help us to make those decisions.'[4]

Those decisions include extracting every last drop of oil from under the earth, with full awareness of the irreparable damage that will do to the planet, ourselves and our societies, and everything and everyone we share the planet with. It is that technology which has marked out the grid of stakes, plastic strips and boreholes which march across Epirus and across Greece, rendering the environment into a virtual checkerboard for exploitation. This is what happens – now – when artificial intelligence is applied to the earth itself.

Repsol and IBM are not the only ones using artificial intelligence to hasten the degradation and exhaustion of the planet. Repsol also has an ongoing relationship with Google, which has put its advanced machine-learning algorithms to work across the company's global network of oil refineries, helping to boost their efficiency and output.[5] At Google's Cloud Next conference in 2018, a host of oil companies presented the ways in which they were using machine-learning to optimize their businesses. (Following a Greenpeace report on Silicon Valley and the oil industry in 2020, Google promised to stop making 'custom AI/ML algorithms to facilitate upstream extraction in the oil and gas industry', although this will have no effect on the industry's extensive use of Google's infrastructure and expertise.[6]) The following year, Microsoft hosted the inaugural Oil and Gas Leadership Summit in Houston, Texas, and has long-standing partnerships with ExxonMobil, Chevron, Shell, BP and other energy firms, which include cloud storage and a growing portfolio of artificial intelligence tools.[7] Even Amazon – which controls almost half of commercial cloud infrastructure – is getting into the game, with one salesperson writing, in the aftermath of Google's announcement, 'If you're an O&G [Oil & Gas] company looking for a strategic digital transformation partner, we would recommend choosing a partner who actually uses your products and can help you transform for the future.'[8]

What future is being imagined here? And what intelligence is at work? If and when Repsol's intelligent algorithms reach the oil lying beneath the mountains and forests of Epirus, the result will be the inevitable destruction of environmental treasures: the felling of trees, the killing of wildlife, the fouling of the air and the poisoning of waters. This future is one in which every last drop of oil is pumped out of the earth and burned for profit. It is a future in which carbon dioxide and other greenhouse gases continue to rise, fuelling global heating, catalysing sea-level rises and extreme weather events, and smothering life across the planet. A future which is, in short, no future at all. What form of intelligence seeks not merely to support but to escalate and optimize such madness? What sort of intelligence actively participates in the drilling, draining and despoliation of the few remaining wildernesses on earth, in the name of an idea of progress we already know to be doomed? This is not an intelligence I recognize.

I don't know how much of the legwork, the digging and the design of the Epirus exploration we can attribute to old-fashioned human analysis and how much to AI. Repsol, despite my asking, won't tell me. But that's not really the point. What matters here, to me, is that the most advanced technologies, processes and businesses on the planet – artificial intelligence and machine-learning platforms built by IBM, Google, Microsoft, Amazon and others – are brought to bear on fossil fuel extraction, production and distribution: the number one driver of climate change, of $CO_2$ and greenhouse gas emissions, and of global extinction.

Something seems to be deeply amiss in what we imagine our tools are for. This thought has crept up on me in recent years as I've watched as new technologies – particularly the most novel and 'intelligent' ones – are used to undermine and usurp human joy, security and even life itself. I'm not the only one to think this. The ways in which the development of these supposedly intelligent tools might harm, efface and ultimately supplant us has become the subject of a wide field of study, involving computer scientists, programmers and technology firms, as well as theorists and philosophers of machine intelligence itself.

One of most dramatic of these possible futures is described in something called the paperclip hypothesis. It goes like this. Imagine a piece of intelligent software – an AI – designed to optimize the manufacture of paperclips, an apparently simple and harmless business goal. The software might begin with a single factory: automating the production line, negotiating better deals with suppliers, securing more outlets for its wares. As it reaches the limits of a single establishment, it might purchase other firms, or its suppliers, adding mining companies and refineries to its portfolio, to provide its raw materials on better terms. By intervening in the financial system – already fully automated and ripe for algorithmic exploration – it could leverage and even control the price and value of materials, moving markets in its favour and generating computationally fiendish futures contracts that make its position unassailable. Trade agreements and legal codes make it independent of any one country and unaccountable to any court. Paperclip manufacturing flourishes. But without the proper constraints – which, due to the complexity of the world the AI operates in, would far exceed in complication the most intractable legal contract or philosophical treatise – there is little to stop it going much further. Having secured control of legal and financial systems, and suborned national governance and lethal force to its will, all Earth's resources are fair game for the AI in pursuit of more efficient paperclip manufacture: mountain ranges are levelled, cities razed, and eventually all human and animal life is fed into giant machines and rendered into its component minerals. Giant paperclip rocket ships eventually leave the ravaged Earth to source energy directly from the Sun and begin the exploitation of the outer planets.[9]

It's a terrifying and seemingly ridiculous chain of events – but only ridiculous in so far as an advanced Artificial Intelligence has no need for paperclips. Driven by the logic of contemporary capitalism and the energy requirements of computation itself, the deepest need of an AI in the present era is the fuel for its own expansion. What it needs is oil, and it increasingly knows where to find it.

The wooden stakes that march for miles across the landscape of Epirus, the holes being drilled, the explosions that shake the ground: these are alien probes, the operations of an artificial intelligence optimized to extract the resources required to maintain our current rate of growth, at whatever cost necessary.

Some of the strongest warnings about AI have in fact come from its greatest proponents: the billionaires of Silicon Valley who have most bullishly pushed a narrative of technological determinism. Technological determinism is the line of thinking which decrees that technological progress is unstoppable. Given that the rise of AI is as inevitable as that of computers, the internet, and the digitization of society as a whole, we should strap ourselves in and get with the programme. Yet Elon Musk, creator of PayPal and owner of Tesla and SpaceX, believes that AI is the 'biggest existential threat' to humanity.[10] Bill Gates, the founder of Microsoft – whose Azure AI platform keeps Shell's oil platforms humming – has said he doesn't understand why people are not more concerned about its development.[11] Even Shane Legg, co-founder of the Google-owned AI company DeepMind – best known for beating the best human players at the game of Go – has gone on the record to state that 'I think human extinction will probably occur, and technology will likely play a part in this.' He wasn't talking about oil: he was talking about AI.[12]

These fears aren't so surprising. After all, the captains of digital industry, the beneficiaries of the vast wealth that technology generates, have the most to lose in being replaced by super-intelligent AI. Perhaps they fear artificial intelligence because it threatens to do to them what they have been doing to the rest of us for some time.

In the last few years, I have given talks at conferences and spoken on panels about the social impacts of new technology, and as a result I am sometimes asked when 'real' AI will arrive – meaning the era of super-intelligent machines, capable of transcending human abilities and superseding us. When this happens, I often answer: it's already here. It's corporations. This usually gets an uncertain half-laugh, so I explain further. We tend to imagine AI as embodied in something like a robot, or a computer, but it can really be instantiated as anything. Imagine a system with clearly defined goals, sensors and effectors for reading and interacting with the world, the ability to recognize pleasure and pain as attractors and things to avoid, the resources to carry out its will, and the legal and social standing to see that its needs are catered for, even respected. That's a description of an AI – it's also a description of a modern corporation. For this 'corporate AI', pleasure is growth and profitability, and pain is lawsuits and drops in shareholder value. Corporate speech is protected, corporate personhood recognized, and corporate desires are given freedom, legitimacy and sometimes violent force by international trade laws, state regulation – or lack thereof – and the norms and expectations of capitalist society.

Corporations mostly use humans as their sensors and effectors; they also employ logistics and communications networks, arbitrage labour and financial markets, and recalculate the value of locations, rewards and incentives based on shifting input and context. Crucially, they lack empathy, or loyalty, and they are hard – although not impossible – to kill.

The science fiction writer Charles Stross likens our age of corporate control to the aftermath of an alien invasion. 'Corporations do not share our priorities. They are hive organisms constructed out of teeming workers who join or leave the collective: those who participate within it subordinate their goals to that of the collective, which pursues the three corporate objectives of growth, profitability, and pain avoidance,' Stross writes. 'We are now living in a global state that has been structured for the benefit of non-human entities with non-human goals.'[13]

Put like that, it's not hard to see why the masters of today's largest corporations fear their own obsolescence at the hands of artificial intelligence. No longer at the top of the pile, they would be as vulnerable as the rest of us to all-powerful entities which do not share their interests, and which would at best cast them aside, and at worst physically rearrange them into a more useful consistency.

What I understand from this gloomy appraisal is that our conception of artificial intelligence – and thus, being modelled on ourselves, of intelligence in general – is fundamentally flawed and limited. It reveals that when we talk about AI, we're mostly talking about this kind of *corporate* intelligence, and ignoring all the other kinds of things that AI – that any kind of intelligence – could be.

That's what happens, it would seem, when the development of AI is led primarily by venture-funded technology companies. The definition of intelligence which is framed, endorsed and ultimately constructed in machines is a profit-seeking, extractive one. This framing is then repeated in our books and films, in the news media and the public imagination – in science fiction tales of robot overlords and all-powerful, irresistible algorithms – until it comes to dominate our thinking and understanding. We seem incapable of imagining intelligence any other way – meaning we are doomed not only to live with this imagining, but to replicate and embody it, to the detriment of ourselves and the planet. We become more like the machines we envisage, in ways which, in the present, have profoundly negative effects on our relationships with one another and with the wider world.

One way to change the nature of these relationships, then, is to change the way we think about intelligence: what it is, how it acts on the world, and who possesses it. Beyond the narrow framing put forward by both technology companies and the doctrine of human uniqueness (the idea that, among all beings, human intelligence is singular and pre-eminent) exists a whole realm of other ways of thinking and doing intelligence. It is the task of this book to do some of that reimagining: to look beyond the horizon of our own selves and our own creations to glimpse another kind, or many different kinds, of intelligence, which have been here, right in front of us, the whole time – and in many cases have preceded us. In doing so, we might change the way we think about the world, and thus chart a path towards a future which is less extractive, destructive and unequal, and more just, kind and regenerative.

On this journey, we will not be alone. In the last few decades, a very different imagining of intelligence has been underway. Emerging, on the one hand, from the biological and behavioural sciences, and on the other from the growing appreciation and integration of indigenous and non-Western systems of knowledge, this new way of understanding intelligence runs counter to narratives of single-mindedness and avarice. And much more significantly, for our story, it challenges the idea that intelligence is something uniquely or even especially 'human' at all.

Until very recently, humankind was understood to be the sole possessor of intelligence. It was the quality that made us unique among life forms – indeed, the most useful definition of intelligence might have been 'what humans do'. This is no longer the case. Thanks to decades of work, careful science, much thinking and the occasional but essential cooperation of non-human colleagues and partners, we are just starting to open the door to an understanding of an entirely different form of intelligence; indeed, of many different intelligences.

From bonobos shaping complex tools, jackdaws training us to forage for them, bees debating the direction of their swarms, or trees that talk to and nourish one another – or something far greater and more ineffable than these mere parlour tricks – the non-human world seems suddenly alive with intelligence and agency. It's a trick of the light of course: these other minds have always been here, all around us, but Western science and popular imagination, after centuries of inattention and denial, are only just starting to take them seriously. And taking them seriously requires us to re-evaluate not only our idea of intelligence, but our idea of the entire world. What would it mean to build artificial intelligences and other machines that were more like octopuses, more like fungi, or more like forests? What would it mean – to us and for us – to live among them? And how would doing so bring us closer to the natural world, to the earth which our technology has sundered, and sundered us from?

This idea of forming new relationships with non-human intelligences is the central theme of this book. It emanates from a wider and deeper dawning: the increasingly evident and pressing reality of our utter entanglement with *the more-than-human world*. It is the full meaning of that phrase, and its repercussions for ourselves, our technologies and our relationships with everything and everyone with whom we share the planet, which I will explore in what follows. Such an undertaking is both urgent and fascinating. If we are to address the wholesale despoliation of the planet, and our growing helplessness in the face of vast computational power, then we must find ways to reconcile our technological prowess and sense of human uniqueness with an earthy sensibility and an attentiveness to the interconnectedness of all things. We must learn to live with the world, rather than seek to dominate it. In short, we must discover an ecology of technology.

The term 'ecology' was coined in the mid nineteenth century by the German naturalist Ernst Haeckel in his book *Generelle Morphologie der Organismen* ('General Morphology of Organisms'). 'By ecology,' wrote Haeckel, 'we mean the whole science of the relations of the organism to the environment including, in the broad sense, all the conditions of existence.'[14] The term derives from the Greek οἶκος (*ekos*), meaning house or environment; in a footnote Haeckel also referenced the Greek χωρα (*hora*), meaning 'dwelling place'. Ecology is not merely the study of where we find ourselves, but of everything which surrounds us and allows us to live.

Haeckel was an early proponent of the work of Charles Darwin. In particular, he supported Darwin's belief that the full import of his theories was to be found not in the way in which individual species developed, but in the relationships *between* species. In the famous final paragraph of *On the Origin of Species*, Darwin provided a proto-description of ecology, describing an 'entangled bank', wherein plants of many kinds, birds, insects and other 'elaborately constructed forms, so different from one another' were produced by the complex forces of evolution, yet depended utterly on one another.[15]

Perhaps the briefest but most resonant description of ecological thought is that given in 1911 by John Muir, the Scottish-American naturalist, outdoorsman and father of the US National Park system. Reflecting on the abundance of complex life he encountered while writing his book *My First Summer in the Sierra,* he wrote simply: 'When we try to pick out anything by itself, we find it hitched to everything else in the Universe.'[16]

Ecology is the study of these interrelationships: those unbreakable cords which tie everything to everything else. Crucially, those relationships extend to *things* as well as *beings*: ecology is just as interested in how the availability of nesting materials affects bird populations, or how urban planning shapes the spread of diseases, as it is in how honeybees pollinate marigolds and cleaner wrasses delouse surgeonfish. And that's just biological ecology. Ecology is fundamentally different to the other sciences in that it describes a scope and an attitude of study, rather than a field. There is an ecology – and ecologists – of mathematics, behaviour, economics, physics, history, art, linguistics, psychology, warfare, and almost any other discipline that you can think of.

There is also ecological politics, which has the potential not merely to describe worlds, but to change them. It was as an ecologist that the marine biologist Rachel Carson approached the environment, culminating in her immensely influential *Silent Spring* of 1962, her ecological understanding enabling her to link pesticides in the rivers and oceans to devastating effects on animal and human health. Her work led directly to bans on toxins such as DDT, and the birth of the global environmental movement. Since then, ecological thought has hitched itself to politics and law, in order to shift public awareness and social practice towards less damaging forms of relationships with the natural world.

Ecological thought, once unleashed, permeates everything. It is as much *movement* as science, with all the motive, restless energy that word connotes. Every discipline discovers its own ecology in time, as it shifts inexorably from the walled gardens of specialized research towards a greater engagement with the wider world. As we expand our field of view, we come to realize that everything impacts everything else – and we find meaning in these interrelationships. Much of this book will be concerned with this particular ecological thought: that what matters resides in relationships rather than things – between us, rather than within us.

Technology is the last field of study to discover its ecology. Ecology is the study of the place we find ourselves in, and the relationships between its inhabitants, while technology is the study of what we do there: τέχνη (*techne*), or craft. Putting it that way makes them sound like natural bedfellows, but the history of technology is largely one of wilful blindness to the context and consequences of its enactment. What counts as technology is also much debated. I like the definition given by the science fiction writer Ursula Le Guin, in a rebuff to critics who accused her of not including enough of it in her work. 'Technology', she wrote, 'is the active human interface with the material world.' Its definition, for Le Guin, wasn't limited to 'high' technology, like computers and jet bombers; rather, it referred to anything that was produced by human ingenuity. That included fire, clothing, wheels, knives, clocks, combine harvesters – and paperclips.

To those who consider technology, whether high or low, to be too complex, too specialized or too abstruse to think fully and clearly about, Le Guin had some words of encouragement: 'I don't know how to build and power a refrigerator, or program a computer, but I don't know how to make a fishhook or a pair of shoes, either. I could learn. We all can learn. That's the neat thing about technologies. They're what we can learn to do.'[17] That is worth keeping in mind as we proceed, because we will be encountering plenty of examples of 'high' technology that might seem daunting at the outset – but every one of them has been thought, learned and done by someone who sleeps at night and shits in the morning. We can learn to do them too.

For most of this book, we will be concentrating on high technology, particularly that variant of it developed in the decades since the Second World War: information technology, or the science and practice of computers, digital communication and computation. But, because we are interested in ecological relationships, we will also touch on the centuries of industrial technology which preceded it: the science of steam engines, cotton mills, jet turbines, pneumatic clocks and telegraph wires. We will even encounter Neolithic flutes, clockwork automata, water organs, and the 'new media' of Ancient Greece.

In this endeavour, I am not concerned with the overt technologies of environmental ecology – with solar panels, wind turbines, carbon capture and geoengineering – as necessary and fascinating as these tools may be. Rather, I am concerned at a deeper level with how we think with, through and about all our technologies: how we consider their role and their impact, their meaning and metaphor, their dialogue and relationships with the surrounding world. To the ecological thinker, all technologies are ecological.

Moreover, I will seek to trouble the distinctions between these types and levels of technology, and between technology, human craft and the rest of the universe. Because it is a deep and abiding paradox that it should have taken so long for technology to encounter and acknowledge ecology, or rather to discover it within itself. Technology, understood as our interface with the material world, is that human practice which most closely ties us to our context and our environment. It exemplifies and performs the most central characteristics of ecology: complexity, interrelatedness, interdependence, distribution of control and agency, even a closeness to the earth and the sky; on, under and out of which we fashion our tools.

An ecology of technology, then, is concerned with the interrelationships between technology and the world, its meaning and materiality, its impact and uses, beyond the everyday, deterministic fact of its own existence. We will start to construct such an ecology by examining many of the assumptions and biases that are built into our ways of thinking, and which are subsequently embedded in the tools we use every day so deeply that we rarely think to question them. The most powerful of these is the idea that human intelligence is unique, and uniquely significant, in the world. Yet, as we shall see, there are in fact many ways of *doing* intelligence, because intelligence is an active process, not just a mental capacity. By rethinking intelligence, and the forms in which it appears in other beings, we will begin to break down some of the barriers and false hierarchies that separate us from other species and the world. In doing so, we will be in a position to forge new relationships based on mutual recognition and respect.

Later, I will explore the ways in which language, that most evocative of human faculties, emerged from our direct experience of the world. As we heard and saw and felt the world – the babbling of the brook, the flight of the bird, the rumbling of the storm – we shaped language to reflect these experiences, in order to better reflect it to itself, and thus to embody and come into communion with it. In the millennia since we first spoke to and of the world we have lost much of this sense of connection to it: technological progress is all too often accompanied by spiritual attenuation. But I will argue that our contemporary, networked, computational technologies might yet be our fullest attempt since the development of language to draw ourselves closer to nature, however carelessly and unconsciously.

Changing our relationship with the world requires us to acknowledge this, and to undertake the task more carefully and consciously. This task is paramount if we are to reconcile the vast scope, god-like power and material demands of our technology with our present situation. We are poisoning the soil and air, warming the atmosphere, acidifying the oceans, burning the forests, and murdering with unthinkable efficiency the numberless beings who share our planet, not to mention generations of humans alive and yet to come. The devastation we are visiting upon the earth has every likelihood of forcing our species back into the caves – as does an unthinking critique of technological progress. If we do not wish to go there, and do not wish to render ourselves alone and abject on the face of the earth, we must rethink every aspect of our technological society and the ideas it is founded on, and we must do it fast.

This remains entirely within our capabilities. 'The history of life on earth has been a history of interaction between living things and their surroundings,' wrote Rachel Carson, in *Silent Spring*. 'To a large extent, the physical form and the habits of the earth's vegetation and its animal life have been molded by the environment. Considering the whole span of earthly time, the opposite effect, in which life actually modifies its surroundings, has been relatively slight. Only within the moment of time represented by the present century has one species – man – acquired significant power to alter the nature of his world.'[18] Today, we call this moment the Anthropocene, and its naming should cause us to take our power seriously, while also recognizing that it is temporal, and like all temporal things it is subject to change. A world in which the environment itself was dominant, an *ecological* world, is of much longer duration and, despite the thoughtless exercise of our power, has never gone away. Indeed, the tumult in which we find ourselves today might be considered its violent reassertion. The task that lies ahead of us involves less a novel change in ourselves than a recognition – in the sense of a *re-cognition*, a realization and a rethinking – of our place in the world.

I will also make the case in this book for the agency and personhood of technology itself, or perhaps of technology yet to come: the moment, much prophesied, when our machines become self-sufficient, self-aware, perhaps self-governing. Such a moment does not strip us humans of the responsibility or agency to effect change in our own attitudes and behaviours. On the contrary, thinking about the agency of technology is an opportunity to think seriously and concretely about how we might ensure greater justice and equality for all of the planet's inhabitants: human, non-human and machine. Technology, for now, remains mostly in our hands, and it remains within our capability to repair, restore and regenerate its entanglement with and effects upon the world.

It was not technology which cast us out of Eden, or sent us fleeing from Babel. It was not technology which designated non-human life as brutish or mechanical, fit only for the slaughterhouse and the vivisection table. That was greed and hubris, Aristotle and Descartes, the edifice of human exceptionalism and Western, European philosophy. Technology embodies the ideas and metaphors of its time, but such tools are turnable to other ends, and so are we. As the poet and visionary William Blake wrote: 'The tree which moves some to tears of joy is in the eyes of others only a green thing that stands in the way. Some see nature all ridicule and deformity . . . and some scarce see nature at all. But to the eyes of the man of imagination, nature is imagination itself.'[19]

More than ever, it is time for re-imaginings. Yet this act of imagination cannot be ours alone. To think against human exceptionalism requires us to think outside and beyond it, and to recognize in Blake's vision the deep truth of his words: *nature is imagination itself*. In this truth is encapsulated the philosophy behind the phrase I used earlier: *the more-than-human world*.

Coined by the American ecologist and philosopher David Abram, the 'more-than-human world' refers to a way of thinking which seeks to override our human tendency to separate ourselves from the natural world. This tendency is so pronounced it is rife even within environmentalism, the movement which seeks to bring us closer to nature and thereby to preserve it. For in so framing our intentions, we have already set up an implicit separation between ourselves and nature, as if we were two separate entities, unbound by inseparable ties of place and origin. Conventional terms such as 'the environment', and even 'nature' itself (particularly when opposed to 'culture'), compound the erroneous idea that there is a neat divide in the world between us and them, between humans and non-humans, between our lives and the teeming, multitudinous living and being of the planet.

In contrast, the 'more-than-human world' acknowledges that the very real human world – the realm of our senses, breath, voice, cognition and culture – is but one facet of something vastly greater. All human life and being is inextricably entangled with and suffused by everything else. This broad commonwealth includes every inhabitant of the biosphere: the animals, plants, fungi, bacteria and viruses. It includes the rivers, seas, winds, stones and clouds that support, shake and shadow us. These animate forces, these companions on the great adventure of time and becoming, have much to teach us and have already taught us a great deal. We are who we are because of them, and we cannot live without them.

The more-than-human world is not mere fancy, or philosophical wordplay: it is the instantiation in our awareness and attitudes of hard-won scientific truths, albeit ones whose full implications have yet to permeate society. Lynn Margulis, the most significant evolutionary biologist of the twentieth century, had this to say about our entanglement with non-human life: 'No matter how much our own species preoccupies us, life is a far wider system. Life is an incredibly complex interdependence of matter and energy among millions of species beyond (and within) our own skin. These Earth aliens are our relatives, our ancestors, and part of us. They cycle our matter and bring us water and food. Without "the other" we do not survive.'[20]

The notion of a more-than-human world further intimates that these *things* are *beings*: not passive props in the drama of our own preoccupations, but active participants in our collective becoming. And because that becoming, that potential flourishing, is collective, it demands that we recognize the *beingness*, the personhood of others. The world is made up of subjects, not objects. Every*thing* is really every*one*, and all those beings have their own agency, points of view and forms of life. The more-than-human world demands our recognition, for without it we are nothing. 'Life and Reality', wrote the Buddhist philosopher Alan Watts, 'are not things you can have for yourself unless you accord them to all others. They do not belong to particular persons any more than the sun, moon and stars.'[21]

*Everything? Really?* Yes. As we shall see, the subjecthood of which we speak springs up all around us when we consider how we relate to everything else. Being itself is relational: a matter of interrelationships. All that is required for sticks and stones to leap into life, wrote the Brazilian anthropologist Eduardo Viveiros de Castro, is our own presence.[22] Our human agency and intentionality transforms the objects of culture into subjects, through the meaning we give to them and the uses we put them to.

While the machines we are constructing today might one day take on their own, undeniable form of life, more akin to the life we recognize in ourselves, to wait for them to do so is to miss out on the full implications of more-than-human personhood. They are already alive, already their own subjects, in ways that matter profoundly to us and to the planet. In the words often attributed to Marshall McLuhan (but more properly ascribed to Winston Churchill): 'we shape our tools, and thereafter our tools shape us.'[23] We are the technology of our tools: they shape and form us. Our tools have agency, and thus a claim upon the more-than-human world as well. This realization allows us to begin the core task of a technological ecology: the re-integration of advanced human craft with the nature it sprung from.

Finally, this book has one further aim. Given that we humans and the things we make are inextricably entangled with the more-than-human world, and given that rethinking our relationship with that world demands that we acknowledge its existence and agency, we must think a little about the form that relationship might take. Part of that relationship is simply *care*: a constant attentiveness to the meaning and affect of our entanglement. The rest, unfortunately, is politics: the hard, meaty detail of debate, decision-making, power relationships and status. This is where, I believe, the computational world has something crucial to contribute to our more-than-human community, something which might in time justify its inclusion in that commonwealth, should justification be needed. The infinite complexity of computation, which we have divined or dreamed up from the material world, and instantiated in the form of machines, has much to teach us about how we might relate to one another. This is the subject of the final part of the book: machines which, along with honeybees, sacred rivers, incarcerated elephants and roulette wheels, might lead us towards a more just and equitable, a more-than-human, politics.

We have come, as the shock of more-than-human consciousness testifies, to think of 'nature' as something separate from ourselves. When we speak of the fantastical futures envisioned by high technology, we speak of a 'new' or 'next' nature, some utopia of computation which further alienates and supplants the actual ground we came from and still stand upon. It is time to put aside such adolescent solipsism – both for the sake of ourselves and of the more-than-human world. There is only nature, in all its eternal flowering, creating microprocessors and datacentres and satellites just as it produced oceans, trees, magpies, oil and us. *Nature is imagination itself*. Let us not re-imagine it, then, but begin to imagine anew, with nature as our co-conspirator: our partner, our comrade and our guide.

# I

# Thinking Otherwise

Somewhere on the higher slopes of Mount Parnassus a small, dark grey car makes its way along a roughly tarmacked track. The road is fringed with snow; far below, the Gulf of Corinth glitters in the sun. The car moves slowly, almost carefully: it is watching the road. It has eyes – several of them – which track the edges of the embankments, identify the white markings at the junctions, note and transcribe where stops are made and turnings taken. It has other senses too: it can tell how fast it is travelling, where it is on the map, what angle the steering wheel is set to. And it has a kind of mind. Not a very sophisticated one, but with a clear focus and a capacity to learn from its surroundings, integrate its findings, and extrapolate and make predictions about the world around it. That mind was perched precariously on the passenger seat; I sat at the wheel, still in control, for now.

All this took place a few years ago, in the winter of 2017, when I decided to try and build myself a self-driving car. And although it never – quite – drove itself, it did take me to some pretty interesting places.

The idea of a self-driving car is fascinating to me. Not really for its capabilities, but for its place in our imagination. The self-driving car is one of those technologies which in the space of just a few years has gone from space-age, 'Life in the Twenty-First Century' fantasy to humdrum reality, without ever passing through a period of critical reflection or assimilation. In moments like this, reality is rewritten. The same will almost certainly be true of more advanced forms of AI. They will appear, suddenly, in our midst – the long slog of research and development, invisible to most, forgotten in the fact of their reality. Questions about who gets to do that rewriting of reality, which decisions are made along the way, and who gains from it, are all too often missed and forgotten in the excitement. That is why I believe that it's crucially important for as many of us as possible to be engaged in thinking through the implications of new technologies; and that this process has to include learning about and tinkering with the things ourselves.

My attempt at building an autonomous vehicle consisted of a rented SEAT hatchback, a few cheap webcams, a smartphone taped to the steering wheel, and some software copied and pasted from the internet.[1] This wasn't a case of programming a dumb machine with everything it needed to know in advance, however. Like the commercial systems developed by Google, Tesla and others, my car would learn to drive by watching *me* drive: by comparing the view from the cameras with my speed, acceleration, steering wheel position and so forth, the system matched my behaviour with the road shape and condition, and after a couple of weeks it had learned how to keep a vehicle on the road – in a simulator at least. I'm not the world's best driver, and I wouldn't trust anyone's life to this thing, but the experience of writing code and going out on the road gave me a better understanding of how certain kinds of AI operate, and what it feels like to work alongside a learning system.

I wondered too what it would mean to do this kind of work far from the highways of California, where Silicon Valley trains its self-driving cars, or the test-tracks of Bavaria, where the giants of the automotive industry evolve new models, and instead on the roads of Greece, where I had recently found myself living. This was a place with a very different material and mythological past and present. It turned out to go beautifully.

Leaving Athens and heading north with no particular destination in mind, other than to give my AI co-pilot a taste of many different kinds of terrain, I soon found myself passing the ancient sites of Thebes and Marathon, and climbing towards the dark bulk of Mount Parnassus. In Greek mythology, Parnassus was sacred to the cult of the god Dionysus, whose ecstatic mysteries were revealed by consuming copious amounts of wine and dancing wildly; participants in such rites liberated the beast within to become one with nature. Parnassus was also the home of the Muses, the goddesses who inspired literature, science and the arts. To attain the summit of Parnassus is thus to be elevated to the peak of knowledge, craft and skill.

Chance and geography conspired to frame a fascinating question. What would it mean, mythologically speaking, to be driven up Parnassus by an AI? On the one hand, it might be read as a kind of submission to the machine: an admission that the human race has run its course, and that it is time to pass the mantle of exploration and discovery to our robot overlords. On the other hand, to attempt the journey in the spirit of mutual understanding rather than conquest might just be how we write a new narrative onto Parnassus – one in which human and machine intelligences amplify, rather than try to outdo, one another.

I started this project because I wanted to understand AI better, and in particular because I wanted to have the experience of collaborating with an intelligent machine, rather than trying to determine its output. In fact, the whole effort was predicated on a kind of anti-determinacy: I wanted to plan as little as possible about the whole journey. So one thing I did, when training the car, was to drive completely at random, taking almost every side road and turning I came across, wandering and wondering, and getting totally, happily lost. In turn, by watching me, the car learned to get lost too.

This was a deliberate rejection of the kind of driving most of us do today: plugging a destination into a GPS system and following its directions without question or input. This loss of agency and control is mirrored in society at large. Confronted by ever more complex and opaque technologies, we capitulate to their commands, and a combination of fear and boredom is the frequent result. Instead of surrendering to a set of processes I didn't understand, only to arrive at a pre-selected location, I wanted to go on an adventure with the technology, to collaborate with it in the production of new and unforeseen outcomes.

In this, my approach owed more to the *flâneur* than the engineer. The *flâneur* or *flâneuse* of nineteenth-century Paris was a person who walked the streets without a care in the world, an urban explorer on whom the impressions of the city would play and play out. In the twentieth century, the figure of the *flâneur* was picked up by proponents of the *dérive*, or drift: a way of combating the malaise and boredom of modern life through unplanned walks, attentiveness to one's surroundings and encounters with unexpected events. The twentieth-century philosopher Guy Debord, the primary theorist of the *dérive*, always insisted that such walks were best undertaken in company, so that people's differing impressions of the group could resonate with and amplify one another. In the twenty-first century, could my autonomous companion perform the same role?[2]

As well as getting lost, I was trying to think of ways to illustrate what I was coming to think of as the *umwelt* of my self-driving car. Coined by the early twentieth-century German biologist Jakob von Uexküll, *umwelt* literally translates as 'environment' or 'surroundings' – but, being German, it means a lot more than that. The *umwelt* connotes the particular perspective of a particular organism: its internal model of the world, composed of its knowledge and perceptions. The *umwelt* of the tick parasite, for example, consists of just three incredibly specialized facts or factors: the odour of butyric acid, which indicates the presence of an animal to feed upon; the temperature of 37 degrees Celsius, which indicates the presence of warm blood; and the hairiness of mammals, which it navigates to find its sustenance. From these three qualities, the tick's whole universe blooms.[3]

Crucially, an organism creates its own *umwelt*, but also continually reshapes it in its encounter with the world. In this way, the concept of *umwelt* asserts both the individuality of every organism and the inseparability of its mind from the world. Everything is unique *and* entangled. Of course, in a more-than-human world, it's not only organisms which have an *umwelt* – everything does.

The *umwelt* has long been a useful concept in robotics as well as biology. It's easy to see how the example of the tick's simple rules could be adapted to provide the basic framework for a simple, autonomous robot: 'move towards this light; stop at that sound; react to this input.' What then is the *umwelt* of the self-driving car?

The simple intelligence at the heart of my car is called a neural network, one of the most common forms of learning machine in use today. It is a programme designed to simulate a series of artificial 'neurons', or smaller processing units, arranged in layers like an extremely simplified brain. Input signals – the speed of the car, the position of the steering wheel, the view from the cameras – are fed into these neurons, sliced into component parts, compared, contrasted, analysed and

Visualizations of a neural network's way of seeing.

associated. As this data flows through the layers of neurons, this analysis becomes ever more detailed and ever more abstract – and therefore harder for an outsider to understand. But we can visualize aspects of this data. In particular, once the car has been trained a little, we can see what the network thinks is important about what it sees.[4]

The images above illustrate a little of that. The first is the view directly from the car's main camera: a road in Parnassus, disappearing into the mist. The second is how that image looks when it has passed through two layers of the network; the third is the fourth layer. Of course, these are visualizations for human eyes: the machine 'sees' only a representation in data. But these images are data too: the details which remain in the image are the details which the machine thinks are important about the image. In this case, the important details are the lines along the side of the road. The machine has decided from its observations that these lines are of some importance; as indeed they are, if the machine is to stay on the road. Like the tick's sensitivity to the temperature of mammalian blood, the lines on the road form an important part of the car's *umwelt*.

And in this observation, we find the point where my *umwelt* is entangled with that of the car. I see the lines too. We share at least one aspect of our models of the world – and from this, too, whole universes might bloom.

To dramatize this revelation of a shared model – and therefore a shared world – I did something which felt a little mean. As much as I'd grown into our collaboration, and fond of my automated companion, I decided to test it. And so, using several kilo bags of salt, I poured out onto the ground a solid circle a few metres in diameter, and then around it I drew a dashed circle. Together, these circles formed a

Autonomous Trap 001, Mount Parnassus, 2017.

closed space in which the (European) road marking for 'No Entry' is projected inwards. As a result, any well-trained, law-abiding autonomous vehicle, on entering the circle, would find itself unable to leave it. I called it the Autonomous Trap.

This crude attack on the machine's sense of the world was intended to make a few points. The first is political: by working with these technologies, we can learn something of their world, and this knowledge can be used to turn them to more interesting and equitable ends – or to stop them in their tracks. Faced with the kind of corporate intelligences we encountered in the Introduction, this is useful knowledge.

Secondly, it asserts that the tools of the imagination and aesthetic representation are as important in an age of machines as they ever were. Art has a role to play here, and we can intervene in the development and application of technology as effectively from this position as from that of an engineer or programmer. This is useful knowledge too.

Mostly, though, I wanted to emphasize the aspects of the world which the AI and ourselves perceive in common: our shared *umwelt*. My video of the Autonomous Trap subsequently went viral, and I have the feeling that people appreciated the chutzpah and the whiff of black magic more than the collaboration: in an age of Uber, air pollution, mass automation and corporate AI, there's something pleasing about stopping the robot in its tracks. Nevertheless, the fact remains: we share a world with our creations.

If seeing the relationships between humans and artificial intelligences as creative collaborations rather than open competitions produces such interesting results, what else might be possible? What other intelligences share worlds, and what is to be found in their encounters and imbrications? If contemporary ideas about artificial intelligence seem to be leading us down a darkly corporate, extractive and damaging path, what alternatives exist?

The current, dominant form of artificial intelligence, the kind you hear everyone talking about, is not creative or collaborative or imaginative. It is either totally subservient – frankly, stupid – or it is oppositional, aggressive and dangerous (and possibly still stupid). It is pattern analysis, image description, facial recognition and traffic management; it is oil prospecting, financial arbitrage, autonomous weapons systems, and chess programmes that utterly destroy human opposition. Corporate tasks, corporate profits, corporate intelligence.

In this, corporate AI does have one commonality with the natural world – or rather, with our false, historical conception of it. It imagines an environment red in tooth and claw, in which naked and frail humanity must battle with devastating forces and subdue them, bending them to his will (and it is usually his) in the form of agriculture, architecture, animal husbandry and domestication. This way of seeing the world has produced a three-tiered classification system for the kinds of animals we encounter: pets, livestock and wild beasts, each with their own attributes and attitudes. In transferring this analogy to the world of AI, it seems evident that thus far we have mostly created domesticated machines of the first kind, we have begun to corral a feedlot of the second, and we live in fear of unleashing the third.

Where does my self-driving car sit in this taxonomy? It's mostly 'pet' – a domesticated machine under my control – but it's also productive, in harness, a working animal; and, because of my insistence that it goes where it pleases, it's a little wild, a little unpredictable. With careless handling, the self-driving car might be considered among the most damaging applications of AI. Not only does it contribute directly to the destruction of the planet through material extraction and carbon emissions – at least, until we get fully solar, sustainable versions – but it also steals from us the very real, if guilty, pleasure of driving.

Only those who have already lost most of their joy could consider this an improvement on the current situation. But considered differently, autonomous transport could replace the kinds of selfish, individual transportation we rely upon at present, and reinvigorate public transport, shared ownership and environmentally appropriate usage. It might also return us to the world, by making us more aware of our surroundings and our fellow travellers. In this way it could liberate us from the mundanity of everyday life, and introduce us to a host of chattering new companions, starting with itself. That it is capable of such different outcomes, depending on how we approach it, tells us that those historical categories of animal and machine – of master, servant, slave and resource – are not to be trusted. Indeed, we should scrap them altogether: for the machines, and for everyone else.

It seems to me significant that just as we start to question the real meaning of 'artificial' intelligence, science is starting to explore what it means to call something or someone intelligent across the board. Our myths and fables have always held a place for the liveliness of non-human beings – and non-Western cultures, with deeper knowledge and longer memories, have always insisted on their agency – but for Western science, they have always been tricky territory. On the one hand, we've always known animals were *smart*, in the most stunning diversity of ways, but official discourse has always held off from ascribing them *intelligence*.

It's at this point that we have to ask, well, what do we mean by intelligence? This is not only the most crucial question we could ask, but also the most diverting, and ultimately the most shattering and generative – because, honestly, nobody really knows.

There are many different qualities which we categorize as intelligent. They include, but are far from limited to, the capacity for logic, comprehension, self-awareness, learning, emotional understanding, creativity, reasoning, problem-solving and planning. There are many reductive versions of this list: attempts to show how one capacity is really the product of another, or accounts which claim that one is more important than the others. But, historically, the most significant definition of intelligence is *what humans do*. No other definition, however elegantly phrased or extensively researched, has a chance of standing up to this one. When we speak about advanced artificial intelligence, or 'general' artificial intelligence, this is what we mean. An intelligence which operates at the same level, and in much the same manner, as human intelligence.

This error infects all our reckonings with artificial intelligence. For example, despite never being used by serious AI researchers, the Turing Test remains the most widely understood way of thinking about the capabilities of AI in the public consciousness. It was proposed by Alan Turing in a 1950 paper, 'Computing Machinery and Intelligence'. Turing thought that instead of questioning whether computers were truly intelligent, we could at least establish that they appeared intelligent. Turing called his method for doing this 'the imitation game': he imagined a set-up in which an interviewer interrogated two hidden interlocutors – one human, one machine – and tried to tell which was which. A machine was intelligent, according to this test, if it could successfully pass itself off as human in conversation. It is testament to our solipsistic way of thinking that this is still what we consider to be a benchmark for general artificial intelligence today.[5]

In fairness to Turing, his idea was a bit more complicated than that. He was less concerned with whether a machine could be intelligent than with whether we could imagine an intelligent machine – a crucial difference, and one which was of more use to his own thinking about how computers might develop. In his 1950 paper, he considered nine objections to the idea of general machine intelligence, all of which are still current today. These included the religious objection (machines have no soul, so cannot think); mathematical objections (per Gödel's incompleteness theorems, no logical system can answer all possible questions); and physiological objections (the brain is not digital, but continuous, and true intelligence must share this quality).

Turing provided counter-arguments to each of these objections, many of which have also proved prescient. One of the most famous objections was posited by the very first computer programmer, Ada Lovelace, when she was working on Charles Babbage's early computer design, the Analytical Engine, in the middle of the nineteenth century. Lovelace wrote that the Engine 'has no pretensions to originate anything. It can do whatever we know how to order it to perform.' Computers only do what you tell them to do, thus they can never be described as intelligent.

But Turing disagreed with Lovelace. He believed that, as technology progressed, it would be possible to design circuitry which could adapt itself to new inputs and thus new behaviours – a kind of 'conditioned reflex', similar to that of animals, leading to 'learning'. He understood why Babbage and Lovelace hadn't thought this likely, but by the middle of twentieth century it seemed quite possible: today, Turing's prediction has indeed been realized. The machine-learning algorithms at work in everything from my self-driving car to chess machines, YouTube recommendations and online fraud protection, are examples of exactly the kind of machines that Lovelace said could not exist. (The argument concerning the immortal soul, on the other hand, is somewhat harder to adjudicate.)

Turing also profoundly disagreed with Lovelace's view that 'a machine can never take us by surprise'. On the contrary, Turing wrote, 'Machines take me by surprise with great frequency', usually because he had misunderstood their function, or calculated something wrongly. In such cases, he wondered, was the surprise 'due to some creative mental act on my part' – or did it 'reflect credit on the machine'? Turing felt that this objection was a dead end as it led back to the question of consciousness – but he felt moved to emphasize that 'the appreciation of something as surprising requires as much of a "creative mental act" whether the surprising event originates from a man, a book, a machine or anything else.'

In Turing's argument, I hear more than a mere acknowledgement of the possibility of machine intelligence. First, his appreciation of it is also a recognition that human intelligence is not all that great. Turing describes his own thinking as 'hurried, slipshod fashion, taking risks', and it is this self-awareness that makes the machine's behaviour so surprising. Here we find an intimation, at the very founding of the discipline of artificial intelligence, that machine intelligence is somehow different from human intelligence. Secondly, in placing his emphasis on the 'creative mental act' of interpreting the machine's response, Turing touches upon something very interesting: the idea that perhaps intelligence doesn't reside wholly inside the head or the machine, but somewhere in between – in the relationship between them.

We have always tended to think of intelligence as being 'what humans do' and also 'what happens inside our head'. But in this early sketch of intelligent machines, Turing suggests something else: that intelligence might be multiple and relational: that it might take many different forms, and that it might exist between, rather than within, beings of all and diverse kinds.

The ongoing popularity of the Turing Test for artificial intelligence, a process which is deeply human-centric and individualized, shows that these kind of nuanced ideas about intelligence did not gain much traction. Instead, we continue to judge AI and other beings by our own standards. This wilful blindness is now being dramatized in our confusion regarding the role and possibilities for artificial intelligence, but it might also allow us to see more clearly how our thinking about other beings has been clouded. In rethinking 'artificial' intelligence, we might begin to rethink intelligence across the board.

The same arguments that Turing rejected still hamper our ability to recognize all kinds of non-human intelligence, even when it's staring us right in the face. Or, as we shall see, staring itself in the face. Or hitting us with a stick. Or singing, or dancing, or making art, or making plans, or making culture. *No*, we say, time and time again. *Not like that. Like this.* The lengths we go to in our attempts to prove or disprove to ourselves that other beings are worthy of being called intelligent might be absurd if they weren't so tragic. The experimental record provides a shining and faultless account, not of the lack or otherwise of intelligence in others, but of a lack of awareness on our own part.

One of the ways we like to evaluate the intelligence of other animals is by getting them to solve puzzles and, in more 'advanced' animals, to test their ability to use tools to do so. A classic test of this kind is to place some tempting food just out of reach and give an animal a tool for obtaining it, like a stick or a piece of string. If they manage it, they've demonstrated the ability to recognize a problem, think it through, make plans and carry them out, and manipulate tools – all classic signs of intelligence. We've been playing this game for years with apes and monkeys, and most of them are pretty good at it: chimpanzees, gorillas, humans, orang-utans, and all kinds of smaller monkeys will quickly make use of any implements they're provided with to snare the treat.

But another primate, the gibbon, refused point blank to play along. For years, gibbons presented a conundrum, because despite belonging to the same class of large-brained apes as chimpanzees and humans, they would ignore the stick and fail to obtain the food; thus rendering them, according to scientific categorization, less intelligent. What was more, gibbons showed the same attitude in many kinds of tests: they refused to pick up cups as part of a response test, and disdained to investigate upturned containers in search of treats.

One researcher, writing in 1932 about a white-handed gibbon named Charlotte, did concede that 'conceivably these errors may have been caused by lack of interest and motivation rather than by any intellectual deficiency, as our animal, although perfectly tame and tractable, evinced frequently a total indifference to the entire experimental situation.' Given the tasks required and the conditions under which most experimental animals were kept – and many still are – this seems, frankly, fair enough. The same researchers recorded their surprise that baboons, despite being 'primitive' and 'dog-like', did far better at the tests they were set. The researchers concluded, against all of their instincts and their understanding of evolution, that baboons were superior in intelligence to gibbons. But they were ultimately proved wrong. The fault lay not with the gibbon, but with us.[6]

In 1967, four gibbons – their names unknown – took part in an experiment at the Chicago Zoo and showed us what we'd been missing. In previous tests, food had been attached to strings lying on the ground. Tugging on the strings would have brought the food into the animals' enclosure, but the gibbons had ignored them. In the 1967 experiment, the researchers hung the strings from the roof of the enclosure: immediately the gibbons grasped them, tugged, and got

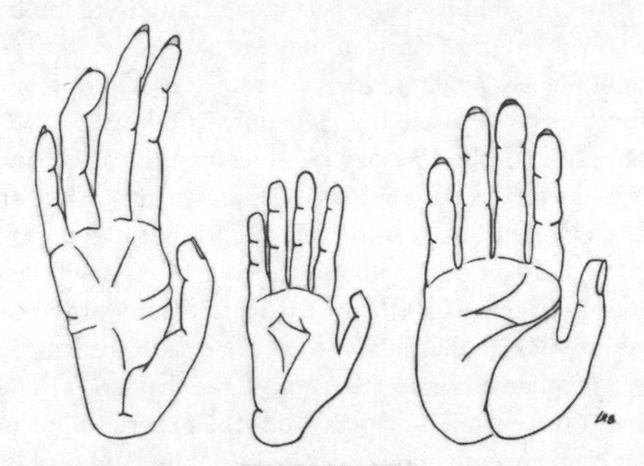

PHOTOGRAPH 1
The gibbon hand and fingers are greatly elongated relative to the macaque and man. As a result it is more difficult for the gibbon to pick up objects lying on a flat surface.

Illustration from Benjamin B. Beck,
'A Study of Problem Solving by Gibbons', 1967.

their snacks. In one swift motion, the gibbons suddenly became 'intelligent' – according, that is, to the narrow definition of the scientific method.[7]

The 1967 experiment was designed to account for the fact that gibbons are brachiators. In their natural forest habitat, they spend almost all their time in the trees, and move around by swinging from branch to branch. This results in physiological – as well as, it seems, cognitive – differences to other apes (including us). To make climbing and swinging easier, gibbons have elongated fingers. While an excellent adaptation for an arboreal lifestyle, this makes it harder for them to pick up objects lying on flat surfaces. It also, some researchers believe, makes them less likely to notice such things: their attention and interest, and therefore their problem-solving and planning, points upwards. They notice and make use of tools when they're where (and what) they expect tools to be. Put another way, the gibbon's *umwelt* is arboreal – and if we don't account for that in our own models, we're likely to miss out on what makes them smart.

Embodied as we are, with a different body pattern and pattern of awareness to the gibbon, we expect the solutions to problems to match our own patterns. This can make it hard to see how animals, embodied differently, might address the same tasks in different ways. For example, we tend to see the elephant's trunk as a kind of fifth limb, because it appears dextrous like our own. But elephants don't use their trunks in the same way we use our hands. They are quite capable of picking up sticks with their trunks, but if you give an elephant a stick and some fruit just out of reach, they will ignore the stick and keep reaching for the fruit. Give them a sturdy box, however, and they will kick that box into place beneath the fruit so that it can reach it. Elephants' trunks are trunk-y, and they don't think or use them in the ways we might expect from the experience of our own bodies. But elephants are quite capable of tool use and problem solving, when given the kind of tools they can put to use (and which, given the freedom, they would fashion themselves). Different tools for different bodies, and different minds acting in concert with those bodies. There are multiple types of intelligence, and it isn't just about what's going on between the ears.[8]

Tool use and problem solving are ways of evaluating intelligence which are quite easy to see in action, if your experimental colleagues choose to enact them, which is why these kinds of tests are popular among experimenters. But recreating in the lab the kinds of tools and puzzles that animals might encounter in the wild is not so easy, and are unlikely to represent the full spectrum of an animal's abilities.

What about other traits of intelligence (as we understand it) which are a little more metaphysical? No amount of experimental design is going to tell us whether gibbons, gnus, gnats or guppies have an immortal soul. But what about self-awareness, a trait that lies somewhere between outward action and inward contemplation? If a mind can recognize its own subjecthood, as distinct from that of other subjects, does this mean that it has a self? This question has puzzled scientists for as long as they have encountered apparently intelligent animals.

On 28 March 1838, Charles Darwin visited London Zoo to see an orang-utan named Jenny.[9] Jenny was one of the first such apes that people in Britain had seen, and her exhibition attracted huge crowds. (Queen Victoria visited in 1842 and met Jenny's successor – another

Portrait of Jenny, the first orang-utan at London Zoo.

orang-utan, also named Jenny. She described this Jenny's appearance as 'frightful and painfully and disagreeably human', which I imagine is what she thought about a lot of actual people.)

Darwin was allowed to enter Jenny's enclosure in the Giraffe House, which had additional heating installed for her comfort. He later wrote to his sister about his impressions:

> The keeper showed her an apple, but would not give it her, whereupon she threw herself on her back, kicked & cried, precisely like a naughty child. She then looked very sulky & after two or three fits of pashion [sic], the keeper said, 'Jenny if you will stop bawling & be a good girl, I will give you the apple.' She certainly understood every word of his, &, though like a child, she had great work to stop whining, she at last succeeded, & then got the apple, with which she jumped into an armchair & began eating it, with the most contented countenance imaginable.

Darwin had only recently returned from his voyage aboard HMS *Beagle*. It was another twenty years before he would publish *On the Origin of Species*, and another ten before he would explicitly include humans in his theory of evolution, in *The Descent of Man*, in 1871. However he was already thinking, and writing in secret, about the similarities between men and apes. He returned twice to the zoo in subsequent months, bringing Jenny small gifts each time – a mouth organ, some peppermint, a sprig of verbena – and each time he was astonished at her responses. 'Let man visit Ouranoutang in domestication . . . see its intelligence . . . and then let him boast of his proud preeminence,' he wrote in his notebook. 'Man in his arrogance thinks himself a great work, worthy the interposition of a deity. More humble and I believe true to consider him created from animals.' With these words, Darwin seems to acknowledge that his own recognition of non-human intelligence was foundational to his development of evolutionary theory: a startling conclusion, given the many subsequent ways his theory has been misused to justify human superiority.

In particular, Darwin's attention had been drawn to how Jenny looked at herself in the mirror. Did she recognize herself, and if so, what would such recognition signify? More than a century later, Gordon G. Gallup, Jr, a psychologist at Tulane University, wondered the same thing. Unlike Darwin's, Gallup's musings weren't sparked by an encounter with an animal, but rather with himself in the mirror while shaving. 'It just occurred to me,' he recalled later, 'wouldn't it be interesting to see if other creatures could recognize themselves in mirrors?'[10]

Working at Tulane, Gallup had access to a number of chimpanzees that had been born free in Africa and shipped to the US for biomedical research. Being captured, rather than born in captivity, they were considered to be less exposed to human behaviour and customs (although, as we'll see later, this is assuming a lot about what constitutes human behaviour).

Gallup isolated four young chimps in cages, and placed a mirror in each cage for eight hours at a time over ten days. Watching through a hole in the wall, he saw their reactions to the mirror begin to change. At first, they reacted as if there was another chimp present, making social, sexual and aggressive gestures towards the glass. But slowly they began to explore their own bodies: 'They'd use the mirror to look at the inside of their mouths, to make faces at the mirror, to inspect their genitals, to remove mucous from the corner of their eyes,' he recalled.

In order to record this change in behaviour scientifically, Gallup devised an experiment. After a week of exposure to the mirror, the chimps were anaesthetized and a smooth, odourless dot of dye was applied to the top of one eyebrow, and to the opposite ear. When they awoke, they were again shown the mirror, and their responses closely observed.

The results were obvious: immediately upon seeing themselves reflected, the chimps started touching and poking the marked areas, and sniffing and tasting their fingers. 'Insofar as self-recognition of one's mirror image implies a concept of self,' wrote Gallup in the subsequent paper, 'these data would seem to qualify as the first experimental demonstration of a self-concept in a subhuman form.'[11] The term 'subhuman' should give us pause to consider the presumptions of the writer, although its use was not unusual for the time. Gallup did not record the names of the chimpanzee participants, if they had been given any. Nevertheless, 'It was just clear as day,' he later recalled. 'It didn't require any statistics. There it was. Bingo.'

The mirror test soon became the standard test for self-awareness, and over the years many different species have been subjected to it. Humans generally pass the mirror test at around eighteen months – with some important exceptions we'll get to shortly – and this evidence of self-awareness is regarded as a key milestone in childhood development. But like many such cognitive tests, its main effect is to reinforce the sense of a dividing line between 'higher' and other animals, rather than to suggest any sense of shared kinship: we decide who 'passes' the test, and can thus claim the elevated state of subjecthood, and who does not. As within human society, we'd rather extend the in-group sharply and grudgingly than acknowledge that there are multiple ways of behaving and being intelligent, many of which trouble our existing methods of classification.

The mirror test is an excellent example of this. Among apes, only chimpanzees, bonobos and orang-utans reliably pass the test, but other creatures have done so under certain conditions. Some of these are more surprising than others. Dolphins and orcas seem like good candidates, and indeed they exhibit clear responses to mirrors: blowing bubbles and inspecting their bodies carefully, as well as responding to marks. But magpies have been shown to exhibit reflective behaviour, as have ants: these behaviours might have as much to do with social relations, or different senses, as they do with vision.[12]

The responses of these animals differ from those of apes in other ways too. Psychologists Diana Reiss and Lori Marino – former students of Gallup – started exposing dolphins to mirrors at Marine World in California in the 1990s. The dolphins responded immediately: by having sex with each other in front of the mirrors. The researchers referred to these recordings privately as 'dolphin porno tapes' and in the literature as 'suggestive' of self-awareness. It was another decade before they figured out how to mark dolphins in such a way that they could better categorize their responses. As soon as they did this, dolphin self-awareness was immediately clear: the dolphins pirouetted and flipped upside down to view markings on their heads and backs which would otherwise be invisible to them.[13]

Effective as it may seem, there are some issues with the mirror test. The first obvious problem, as with the gibbon's arboreal *umwelt*, is that bodies matter to minds. The way we perceive and act in the world is shaped by the limbs, senses and contexts we possess and inhabit. My favourite example of this is, again, elephants. We think elephants should pass the test, because they seem so smart in many other ways, but for years they failed to respond in any discernible way to the presence of marks on their faces. A much-cited 1989 study of Shanthi and Ambika, two elephants at the National Zoological Park in Washington DC, is entitled 'Failure to Find Self-Recognition in Asian Elephants (Elephas maximus) in Contrast to Their Use of Mirror Cues to Discover Hidden Food'.[14] It highlights the fact that while the pair were able to use mirrors to solve puzzles (because they reliably reached for treats visible only in the mirror – a separate and complex motor task which many apes fail), they ignored marks placed on their bodies, and thereby failed to demonstrate self-awareness in the scientists' definition of the term.

Others saw a problem with the study. In the original experiment, the mirror was placed on the ground, outside the elephants' enclosure, meaning that what the elephant saw was mostly legs and bars – hardly ideal conditions for self-recognition. Working with three Asian elephants at the Bronx Zoo named Patty, Maxine and Happy, researchers Joshua Plotnik and Diana Reiss (the same Diana Reiss from the dolphin study) found that at least one of the pachyderms was quite capable of self-recognition – provided the mirror was big enough and close enough. They installed an eight-foot mirror inside the elephants' enclosure and found it immediately attracted attention, with the elephants rubbing themselves against it, swinging their trunks over the wall, and attempting to get behind it – possibly to see where the other elephant was. But after a little while, like the chimps before them, they became accustomed to it, and when Happy was marked with a large white cross above her right eye, she immediately started touching it repeatedly. In this way, the Asian elephant joined the self-recognition club – as with the gibbons, a case, it seems, of humans finally being smart enough to figure out where they were being dumb before.[15]

That's the position of the renowned animal behaviourist Frans de Waal, who cites the case of Happy in his book *Are We Smart Enough To Know How Animals Are?* (Plotnik was one of his students, and de Waal is a co-author on the paper.) De Waal notes that some other Asian elephants in Thailand have passed the test too – but many have also 'failed' it. (For the record, the originator of the test, Gordon Gallup, disputed the results for elephants and dolphins (and presumably magpies, ants and other critters) because he believed self-awareness is unique to 'higher' primates.) Meanwhile, one of de Waal's star students, Daniel Povinelli, author of the first elephant paper, argues that the mirror test tells us nothing at all about inner states, only about the ability to match movements to an image. The mirror test, across all species, remains deeply contested today.

The best example, for me, of animal physiognomy causing issues with the mirror test also involves elephants. Patty, Maxine, Happy and the Thai elephants are all Asian elephants, and the experiments suggest they possess mirror test self-awareness. Their larger and more rambunctious African cousins have yet to do so, but this doesn't of course mean they don't possess it. Rather, no researcher has yet built a mirror big and strong enough to withstand the trunk, tusks and natural inquisitiveness of these 10,000-pound pachyderms. In every such test so far, the elephants have ended up destroying the mirrors.

One perspective not included in these discussions is any real recognition of the elephants as individuals. Digging into the history of such experiments, we find that the story of Happy is not a happy one. She was captured in the wild in 1971, together with six other young elephants, and brought to the US from Thailand. The calves were sold for $800 each to a safari park in California, where they were named after the seven dwarves from *Snow White*. For more than forty years, Happy lived at the Bronx Zoo, which in 2014 was ranked the fifth worst zoo for elephants in the US by the international animal protection organization In Defense of Animals.[16] Until 2002, she shared an enclosure with Grumpy, another of the Thai captives, but when the pair were introduced to Patty and Maxine, Maxine attacked Grumpy, who died shortly afterwards from his injuries. The effect of these events on the elephants' differing responses to the mirror test is not recorded – nor is it routinely a part of such studies. Individual animals are taken as synecdoches for the whole species, and while it takes many such studies to draw firm conclusions, an individual's personal history (the kernel of selfhood, the *umwelt*, remember, is the creation of the individual) is not taken into account.

We can, however, see how the countenance and behaviour of these animals differed. Videos accompanying the study are available on the website of the Proceedings of the National Academy of Sciences. They show a bare, concrete, dirt-floored enclosure; they also show clear contrasts in the elephants' behaviours. Patty and Maxine violently attack the mirror when it's first installed; Happy keeps her distance. Patty and Maxine are tested together, but Happy appears by herself. Since the death of Grumpy's replacement, Sammie, in 2006, Happy has lived alone.

'Happy spends most of her time indoors in a large holding facility lined with elephant cages, which are about twice the length of the animals' bodies,' reported the *New York Post* in 2012. 'The public never sees this.' The mirror test depends on differentiating the self from another individual; how much different must it feel to an elephant accustomed to solitude, who has suffered trauma and loss inflicted by other elephants? Whatever the innate intelligence of a species, we must surely acknowledge that the quality of it, and the capacity for its demonstration, is unique to every individual.[17]

Happy in her enclosure at the Bronx Zoo, 2012.

Unlike comparative cognition studies, the law treats subjects as individuals – at least when applied to humans. But Happy is also the subject of a ground-breaking legal case brought by the Nonhuman Rights Project, an animal advocacy organization based in Florida. Since 2018, the NhRP has been seeking Happy's release to a dedicated animal sanctuary under a writ of habeas corpus, the medieval legal doctrine which protects against unlawful detention or imprisonment. Historically, habeas corpus has only applied to humans, but the NhRP is seeking to change this: they have also filed writs on behalf of a number of gorillas and chimpanzees. None have so far been successful, and Happy's case is still in progress. We'll explore the deeper implications of this approach in a later chapter, but it seems significant that the NhRP's case treats Happy as an individual, with a unique history, needs and set of experiences, all of which are ignored in most scientific research into the intelligence, being and personhood of animals.[18]

Here's another potential problem with the mirror test: maybe the kind of self-scrutiny which humans are accustomed to performing just isn't that important to other animals? In the original dolphin study, one hypothesis suggested that dolphins wouldn't pass the mirror test because dolphins don't groom themselves. Indeed, sex rather than self-care seemed to be their priority. But it also seems as if many animals that 'fail' the mirror test nonetheless exhibit plenty of interest in their own reflection – more than enough to make it clear they know it's 'them' in the mirror. For many of these animals, it seems that the act of touching a mark on the face has more to do with social cues than cognitive ones.

To illustrate this possibility, let's return to our old friend the gibbon, who, when presented with a mirror, will ignore any such marks, instead continuing to treat their own reflection as if it were another gibbon: making social gestures, and trying to reach around the mirror to touch the perceived 'other'. In one comprehensive Australian test, a total of seventeen gibbons were tested. (The cohort was actually twenty, but 'three could not be tested as they refused to approach the experimenter' – and fair enough.) Their names were Philip, Kayak, Arjuna, Jury, Jars, Ulysses, Mang, Suli, Irian, Jaya, Ronnie, Bradley, May, Siam, Sydney, Milton and Milo. They were given treats of icing cream to spur their interest, but they steadfastly ignored icing-like dots painted on their eyebrows.[19] This 'evidence of absence' was presumed by the scientists to support a phylogenetic account of self-recognition – that is, the idea that this ability emerged at a particular point in the evolution of apes, some time between 18 and 14 million years ago, after gibbons split from the line that led to modern humans (and before the split that led to orang-utans). Of course, this doesn't account for the possibility of convergent evolution – the emergence of a similar trait along different evolutionary lines – which might account for the ability in elephants, magpies, orcas and dolphins – but more on that later. Let's stick to apes for now.

Perhaps the answer, as I suggested, is that faces just aren't that important to some species, or are bound up with other cues that are less significant to humans and other higher apes. Rhesus macaques are small, lively monkeys: well known in many cities across Asia for their tendency to live closely among humans. I remember watching them steal food from market stalls in India – and being terrified as they chased visitors out of temple grounds, where large sticks were provided to defend against them. Less fortunately for the macaques, the similarity of their immune system, as well as certain neurological structures, to that of humans makes them popular subjects for medical experiments. It was in the context of such an experiment that the rhesus's ability to self-recognize became evident.

At the University of Wisconsin-Madison, two macaques were prepared for a neurological experiment. The experiment required screwing an inch-square block of blue acrylic to the top of their skulls, with attachments for electrodes and restraints. (This study, unlike many of those cited above, did not name the macaque participants – and in images and video accompanying the paper 'the view of the head implant has been blocked for discretion'.)[20] The researchers quickly noticed that the macaques behaved in a markedly different way to any others in the literature. While macaques typically investigate and display in front of the mirrors, they nonetheless steadfastly avoid eye contact with their reflections, or treat them as they would rivals. The implanted macaques, in contrast, spent considerable time examining the tops of their heads – as I'm sure any of us would if someone bolted a block of blue plastic onto it. And once they'd started doing that in the mirror, they started looking at and grooming other parts of themselves, mostly their genitals.

So, does this make macaques self-aware after all? Eye contact is bound up with hierarchy, dominance and aggression in ways that make staring at a stranger highly uncomfortable. One explanation for the macaques' behaviour is that because they are so socially aware, they really don't like looking at each other. This would explain their continued – but intermittent – social behaviour in front of the mirror, which is never concentrated enough for them to pass through this stage and realize they're looking at themselves – as happened in time with the chimps and other higher apes. In the case of the implanted macaques, it took a powerful counter-stimulus – having a block of plastic screwed to their heads – to make it to the next level.

Faces might just not be that important at all in the thinking of other species, or at least have very different significance, and so approaches like the mirror test, which reproduce humans' own obsession with faces reflected in the mirror, are simply inappropriate. Macaque monkeys, who demonstrate smartness in so many ways, might just not think that being face-to-face matters. On the other hand, they do masturbate a lot and look at each others' butts. Most scientific reports still conclude that macaques lack self-awareness, but I don't buy it. Of course these animals are self-aware: it's just that if you tend to communicate by showing your butt to someone rather than your face, that's what you're going to check out in the mirror. Macaque Instagram would contain a lot of butt selfies.

As with Happy the elephant, there are many species in which only one individual has passed the mirror test, or in which the circumstances for passing the test vary widely. Species and individual differences matter, and attending to those differences allows us to see how our own perspectives cloud our judgement, and impair our ability to recognize the abilities of others.

Despite all the theory about higher and lower apes, most gorillas (considered 'higher' apes) fail the mirror test. Like macaques, gorillas have a strong aversion to eye contact, considering it a threat, so they don't like looking at faces in the mirror. While there's plenty of

Koko the gorilla learning sign language
with trainer Penny Patterson, 3 March 1978.

evidence – such as grooming and exploring hidden areas of their bodies in mirrors – that they 'see' themselves, ever more ingenious tests, such as angling mirrors and hiding food, have found no widespread, scientific evidence for self-recognition (that is, touching facial marks).[21] This might seem like a species-specific problem. In fact, it's more of an individual one. To understand the significance of this, let's look at the story of Koko.

Koko was a female gorilla who lived for forty-six years in captivity – forty-five of which were under the care of animal psychologist Francine Patterson at the Gorilla Foundation in Woodside, California. Patterson taught Koko more than a thousand words of 'Gorilla Sign Language', and spoke to her in English from a young age. Koko's abilities remain disputed, but according to those who spent time with her, Koko used language the same way people do – and exhibited plenty of other 'human' qualities too. For example, Koko was adept at lying: she would often blame others for thefts or breakages, choosing those who were present at the time, or whom she disliked. She was also highly self-conscious. She loved to play with dolls, and often signed to them, but would immediately stop if she thought she was being observed.

Koko passed the mirror test with ease.[22] Indeed, her widespread fame during her lifetime originated in a *National Geographic* cover which depicted her taking her own photo in the mirror. Another gorilla who passed the test was Otto, a forty-five-year-old male who since the age of two had lived in an 'enriched' environment at the Suncoast Primate Sanctuary in Florida, where he was provided with activities such as foraging, watching television, painting and regular contact with human trainers.[23] It seems that supposedly innate characteristics – or at least their presentation – can actually be developed in time, providing the environment and the body pattern to support them exist.

The question of self-awareness gets even more complex in the case of Michael, an eighteen-year-old gorilla who lived with Koko at the Gorilla Foundation. He had a vocabulary of over 600 signs, many of them taught to him by Koko. Famously, he was believed to have described his mother's death at the hands of poachers in Cameroon in a sequence of signs: 'Squash meat gorilla. Mouth tooth. Cry sharp-noise loud. Bad think-trouble look-face. Cut/neck lip girl hole.'

Like Koko, Michael could be self-conscious. He could also be quite destructive, with a history of breaking equipment, so when he was tested for self-recognition with the mirror test, the mirror and the experimenters stayed outside the room, peering in. Although he'd previously shown plenty of self-interest in mirrors, when he was marked with a large dot of dye on his nose, he started behaving oddly. On approaching the mirror, he froze and leaned forward, carefully inspecting himself. He turned his head from side to side, looking at his face from different angles. And then he asked for the lights to be turned off. The experimenters refused, but Michael kept asking for the lights to be turned off, and the drapes to be closed. Eventually, he retreated to the back of the room, turned his back on the watchers, and wiped his nose on the wall until the mark was gone.[24] Michael's reaction suggests not only self-recognition but another significant and much-contested quality of complex cognition: theory of mind.

Theory of mind goes beyond simple recognition of the self distinct from other individuals: it implies the ability to think about the inner lives of others, to imagine their mental states, and to act accordingly. Beyond the simple interaction of the mirror test, we glimpse a much broader and more complex intelligence at work.

What are we to make of all these radically different abilities and outcomes? As we've seen, different species pass the test in different ways: some display strong relationships to the mirror, but refuse the mark test, others develop the ability in radically different ways, or confound our expectations based on their species' known tendencies. The mirror test is of course only one kind of test, for one very specific component of what we think of as intelligence – but its lessons should be applied to every single claim we make about the supposed intelligence or otherwise of all beings, including humans. Even when the results are not merely the effects of our own biases and limitations, they only really tell us how much we don't know.

It turns out that mirror self-awareness isn't even that well distributed across human individuals and cultures. Adults with schizophrenia often fail to recognize themselves in mirrors, and struggle with theory of mind as well. The oft-cited 'fact' that human children pass the mirror test – and thus develop self-recognition – at around eighteen months old turns out to be itself biased and only partly true. Those tests are based on Western children, primarily in the US and Canada, and don't hold up in tests performed in Africa, South America and the Pacific islands.[25] That doesn't mean that non-Western kids aren't self-aware. It means that our testing processes and analyses are culturally biased in ways we don't recognize even when we do them on other humans, let alone on other species.

It's striking that the gorillas in particular had to have much of their gorilla-ness taken away from them to make these human-like behaviours legible to us. Koko and Michael were removed from their natural habitats and lived in specialized enclosures with extended contact with humans for most if not all of their lives. The qualities that make them seem intelligent to us were contingent on them being trained into human-like patterns of behaviour and sociality. Yet other forms of interaction, which we'll come to shortly, don't make this demand, and in these interactions we see other – but no less clear – signs of intelligence manifesting. Nevertheless, they depend on humans being able to read non-human signs – they depend on an overlap in the *umwelt*. That this is unlikely to ever extend to, say, ants, means that we will never be able to make the same scientific determination about such creatures. But that certainly doesn't mean they're not intelligent, as our unpicking of the various experiments we've tried thus far should intimate. Rather, if we are truly to appreciate what non-human intelligence might consist of – and thus transform our understanding of our own abilities and those of others – we need to stop thinking about intelligence as something defined by human experience. Instead, we must from the outset think about intelligence as something more-than-human.

It turns out there are many ways of 'doing' intelligence, and this is evident even in the apes and monkeys who perch close to us on the evolutionary tree. This awareness takes on a whole new character when we think about those non-human intelligences which are very different to us. Because there are other highly evolved, intelligent and boisterous creatures on this planet that are so distant and so different from us that researchers consider them to be the closest things to aliens we have ever encountered: cephalopods.

Cephalopods – the family of creatures which contains octopuses, squids and cuttlefish – are one of nature's most intriguing creations.

They are all soft-bodied, containing no skeleton, only a hardened beak. They are aquatic, although they can survive for some time in the air; some are even capable of short flight, propelled by the same jets of water that move them through the ocean. They do strange things with their limbs. And they are highly intelligent, easily the most intelligent of the invertebrates, by any measure.

Octopuses in particular seem to enjoy demonstrating their intelligence when we try to capture, detain or study them. In zoos and aquariums they are notorious for their indefatigable and often successful attempts at escape. A New Zealand octopus named Inky made headlines around the world when he escaped from the National Aquarium in Napier by climbing through his tank's overflow valve, scampering eight feet across the floor, and sliding down a narrow, 106-foot drainpipe into the ocean. At another aquarium near Dunedin, an octopus called Sid made so many escape attempts, including hiding in buckets, opening doors and climbing stairs, that he was eventually released into the ocean. They've also been accused of flooding aquariums and stealing fish from other tanks: such tales go back to some of the first octopuses kept in captivity in Britain in the nineteenth century, and are still being repeated today.

Otto, an octopus living in the Sea-Star Aquarium in Coburg, Germany, first attracted media attention when he was caught juggling hermit crabs. Another time he smashed rocks against the side of his tank, and from time to time would completely rearrange the contents of his tank 'to make it suit his own taste better', according the aquarium's director. One time, the electricity in the aquarium kept shorting out, which threatened the lives of other animals as filtration pumps ground to a halt. On the third night of the blackouts, the staff started taking night shifts sleeping on the floor to discover the source of the trouble – and found that Otto was swinging himself to the top of his tank, and squirting water at a low-hanging bulb that seemed to be annoying him. He'd figured out how to turn the lights off.[26]

Octopuses are no less difficult in the lab. They don't seem to like being experimented on, and try to make things as difficult as possible for researchers. At a lab at the University of Otago in New Zealand, one octopus discovered the same trick as Otto: it would squirt water at light bulbs to turn them off. Eventually it became so frustrating to have to continually replace the bulbs that the culprit was released back into the wild. Another octopus at the same lab took a personal dislike to one of the researchers, who would receive half a gallon of water down the back of the neck whenever they came near its tank. At Dalhousie University in Canada, a cuttlefish took the same attitude to all new visitors to the lab, but left the regular researchers alone. In 2010, two biologists at the Seattle Aquarium dressed in the same clothes and played good cop/bad cop with the octopuses: one fed them every day, while the other poked them with a bristly stick. After two weeks, the octopuses responded differently to each, advancing and retreating, and flashing different colours. Cephalopods can recognize human faces.[27]

All these behaviours – as well as many more observed in the wild – suggest that octopuses learn, remember, know, think, consider and act based on their intelligence. This changes everything we think we know about 'higher order' animals, because cephalopods, unlike apes, are very, very different to us. That should be evident just from the extraordinary way their bodies are constituted – but the difference extends to their minds as well.

Octopus brains are not situated, like ours, in their heads; rather, they are decentralized, with brains that extend throughout their bodies and into their limbs. Each of their arms contains bundles of neurons that act as independent minds, allowing them to move about and react of their own accord, unfettered by central control. Octopuses are a confederation of intelligent parts, which means their awareness, as well as their thinking, occurs in ways which are radically different to our own.

Perhaps one of the fullest expressions of this difference is to be found, not in the work of scientists, but in a novel. In his book *Children of Time*, science fiction writer Adrien Tchaikovsky conceptualizes octopus intelligence as a kind of multi-threaded processing system. For the space-faring octopuses in *Children of Time*, their awareness – their consciousness – is tripartite. Their higher functions, which Tchaikovsky calls the 'crown', are embedded in their head-brain, but their 'reach', the 'arm-driven undermind', is capable of solving problems independently – sourcing food, opening locks, fighting or fleeing from danger. Meanwhile, a third mode of thinking and communicating, the 'guise', controls the strobing and spotting of the octopuses' skin, 'the chalkboard of the brain', where it doodles its thoughts from moment to moment. In this way, the octopuses freewheel through space, constructing ships, habitats and whole societies which owe as much to bursts of emotion, flights of fancy, acts of curiosity and boredom, as they do to conscious intent. Tchaikovsky's octopuses are lively, frantic, bored, creative, distracted and poetic – all at the same time: a product of the constant dialogue and conflict within their own nervous systems. As Tchaikovsky tells it, octopuses are multiple intelligences in singular bodies.[28]

Tchaikovsky based his research on visits to the Natural History Museum in London, conversations with scientists and his own background as a zoologist. But what are we to make of such creatures – such intelligences – that require the tools of science fiction to make them intelligible to us? How can they appear so extraordinarily other, yet exist on the same planet, part of the same evolutionary process, as us?

The kind of self-awareness which we can observe with the mirror test – the kind that is most like our own – seems to have appeared in apes somewhere between the bonobo and the orang-utan, or between 18 and 14 million years ago. That's when one of the qualities which make up our kind of intelligence seems to have evolved. Humans parted ways with chimpanzees only about six million years ago, so it's understandable that our intelligence might be similar to theirs. But primates split from other mammals around 85 million years ago, while mammals themselves appeared distinct from other animals over 300 million years ago. To find a common ancestor with cephalopods, we need to go back twice that far, to 600 million years ago.

In his book *Other Minds*, the philosopher Peter Godfrey-Smith imagines who this common ancestor might have been. Although we cannot know for sure, it was most likely some kind of small, flat worm, just millimetres long, swimming through the deep, or crawling on the ocean floor. It was probably blind, or light-sensitive in some very basic way. Its nervous system would have been rudimentary: a network of nerves, perhaps clustered into a simple brain. 'What these animals ate, how they lived and reproduced,' he writes, 'all are unknown.' It's hard to imagine something less like us, yet alive, than tiny near-blind worms wriggling on the ocean floor. But we come from them, and so does the octopus.

Six hundred million years down the evolutionary tree – and 600 million up the other side too. While that distance makes all the obvious differences between us and the octopus understandable, it makes the similarities even more startling.

One of the most remarkable features of octopuses is their eyes, which are remarkably like our own. Like ours, their eyes consist of an iris, a circular lens, vitreous fluid, pigments and photoreceptors. In fact, the octopus eye is superior to ours in one notable way: because of the way they develop, the fibres of the optic nerves grow behind the retina rather than through it, meaning they lack the central blind spot common to all vertebrates. And this difference exists because the octopus eye evolved entirely separately from our own, starting from that blind flatworm 600 million years ago, along an entirely different branch of the evolutionary tree.

This is an example of convergent evolution. The octopuses' eye evolved to do much the same thing as our eye, entirely separately but only slightly differently. Two incredibly complex, but startlingly similar structures appeared in the world, by different routes, in different contexts. And if something as complex and adaptive as the eye can evolve more than once, then why can't intelligence do the same?

In Chapter 4, we'll explore further how this idea of branching and splitting of the evolutionary tree is overly simplistic, if not entirely false. For now, let's simply imagine it this way: the tree of evolution bears many fruits and many flowers, and intelligence, rather than being found only in the highest branches, has in fact flowered everywhere.

The intelligence of the octopus is one such flower. As Godfrey-Smith puts it, 'Cephalopods are an island of mental complexity in the sea of invertebrate animals. Because our most recent common ancestor was so simple and lies so far back, cephalopods are an independent experiment in the evolution of large brains and complex behaviour. If we can make contact with cephalopods as sentient beings, it is not because of a shared history, not because of kinship, but because evolution built minds twice over.' If twice, then likely many more.

From the octopus, then, we learn several important things. First, that there are many ways of 'doing' intelligence: behaviourally, neurologically, physiologically and socially. This bears repeating: intelligence is not something which exists, but something one does; it is active, interpersonal and generative, and it manifests when we think and act.

We have already learned – from the gibbons, gorillas and macaques – that intelligence is relational: it matters how and where you do it, what form your body gives it, and with whom it connects. Intelligence is not something which exists just in the head – literally, in the case of the octopus, who does intelligence with its whole body. Intelligence is one among many ways of being in the world: it is an interface to it; it makes the world manifest.

Intelligence, then, is not something to be tested, but something to be recognized, in all the multiple forms that it takes. The task is to figure out how to become aware of it, to associate with it, to make it manifest. This process is itself one of entanglement, of opening ourselves to forms of communication and interaction with the totality of the more-than-human world, much deeper and more extensive than those which can be performed in the artificial constraints of the laboratory. It involves changing ourselves, and our own attitudes and behaviours, rather than altering the conditions of our non-human communicants.

Rigorous scientific studies can only tell us so much about the actual realities of non-human lives; mostly, they just tell us about our own models. They also tell us about the structure of science itself: a human-centric way of knowing about a human-centric world. But just because this approach is limited, it does not follow that such realities are inaccessible to us, wholly or partially, by other means – and even to scientists.

Beginning in the 1970s, and for a period of more than twenty-five years, the primatologist Barbara Smuts explored the behaviour of baboons living free across Kenya and Tanzania. The baboons she came to know best she called the Eburru Cliffs troop, named after a rocky outcrop in the Great Rift Valley near Lake Naivasha. At one point, she spent every day for almost two years with the troop, joining them as they arose at dawn, and travelling with them until they found a resting spot at nightfall. As a result, she gained unique insights into baboon behaviour, as well as an appreciation for, and sympathy with, many aspects of their lives and minds which are not readily classifiable.

At the time, Smuts's approach was unusual: often, when studying their subjects, scientists would ignore them or move away from them if approached. But, Smuts argued, in precluding meaningful interaction, this approach prevented scientists from seeing the behaviour they're trying to study.

Smuts stressed that this familiarity with non-human animals, although initiated as a research scientist, was made possible by something much older, an inheritance from our ancestors. 'Until recent times,' she wrote, 'all humans possessed profound familiarity with other creatures. Palaeolithic hunters learned about the giant bear the same way the bear learned about them: through the intense concentration and fully aroused senses of a wild animal whose life hangs in the balance. Our ancestors' survival depended on exquisite sensitivity to the subtle movements and nuanced communication of predators, prey, competitors, and all the animals whose keener senses of vision, smell, or hearing enhanced human apprehension of the world.'[29] For Smuts, this sensitivity could be recovered – and was, for her, by her own interactions with the baboons.

In order to accommodate the baboons to her presence, Smuts would approach the troop on open ground, slowly moving towards them and stopping any time they seemed alarmed and moved away. As she became attuned to their behaviour, she started to pick up on more subtle signals – mothers, for example, would call their infants closer before a general alarm was sounded – which allowed her to draw closer still. Eventually, she was able to move among them freely.

This ability to move freely among another species, Smuts emphasizes, had less to do with the baboons becoming habituated to her presence, than with her adjusting her behaviour to suit them. She learned to move a little, and think a lot, like a baboon, picking up on their behaviour not merely as a scientist, but as a guest. In return, the baboons started to treat her as a subject of communication rather than an object of fear – that is, they realized that a dirty look rather than a general alarm would suffice to make her move away, and over time a greater range of gestures and signals arose between them. Research scientists, Smuts complained, are often told that they should ignore or move away from the subjects of their study if they come too close or try to interact with them. Such behaviour precludes meaningful interaction, as it would in any social group, so prevents most researchers even seeing the behaviour they're trying to study.

Smuts soon discovered she could get a lot closer to the baboons if she did not merely react to their sounds and gestures, but returned them. Baboons have a highly developed sense of personal space: they might ignore the proximity of close family members, while taking flight at a glance from a higher-ranking or aggressive member of the troop. When meeting one another, a grunt or a facial expression conveys an expectation of a relationship – so when Smuts returned the grunts and gestures of her companions, they accepted her more. To ignore a baboon – or any other social animal – is not a neutral act. We might say that the same is true of our relationship to the world itself.

As a result of her willingness to engage with the baboons, Smuts began to learn more than she expected from them. Over time, her awareness of the world around the troop began to change too. Her attitude to the weather, for example, began to shift. In the initial months of her study, she'd find herself looking for shelter the moment rain clouds appeared on the horizon, but over time she became baboon-like in her attitude. The baboons didn't like getting wet any more than she did, but they stayed in the open far longer, making the most of feeding time before, at the very last moment, dashing for the protection of rocks and trees. After a while, Smuts found herself doing likewise: 'I could not attribute this awareness to anything I saw, or heard or smelled; I just knew. Surely it was the same for the baboons. To me, this was a small but significant triumph. I had gone from thinking about the world analytically to experiencing the world directly and intuitively. It was then that something long slumbering awoke inside me, a yearning to be in the world as my ancestors had done, as all creatures were designed to do by aeons of evolution. Lucky me. I was surrounded by experts who could show the way.'

The appreciation of outer behaviours tends inevitably towards a sensitivity to inner lives – and, in time, Smuts felt a change in her own sense of identity. 'The shift I experienced is well described by millennia of mystics but rarely acknowledged by scientists. Increasingly, my subjective consciousness seemed to merge with the group-mind of the baboons. Although "I" was still present, much of my experience overlapped with this larger feeling entity. Increasingly, the troop felt like "us" rather than "them". The baboons' satisfactions became my satisfactions, their frustrations my frustrations.' She describes sharing the hunger of the troop, her elation when they killed, and her mouth watering at the sight of fresh meat – despite being a vegetarian. 'I had never before felt a part of something larger, which is not surprising, since I had never so intensely coordinated my activities with others. With great satisfaction, I relinquished my separate self and slid into the ancient experience of belonging to a mobile community of fellow primates.'

One incident which she shared with a different troop of baboons in Tanzania's Gombe Stream National Park seemed to capture a kind of experience and sensibility which is not only impossible to classify scientifically in animals, but which mirrors the unclassifiable experiences of humans ourselves. We call it spirituality. Late one evening the baboons were making their way to a sleeping place, down a small stream they often travelled along, which was interspersed with many small pools. Without any obvious signal, each of the baboons sat down on a smooth stone surrounding one of the pools, and for half an hour (by human reckoning) they sat alone or in small clusters, completely quiet, staring at the waters. Even the normally boisterous juveniles slipped into quiet contemplation. Then, again at no perceptible sign, they stood up and resumed their journey in quiet procession.

Smuts witnessed this behaviour just twice; what's more, she is the only researcher ever to do so. Whether it is particular to the group, or a manifestation of something broader, we don't know, but for Smuts the episode appeared more than anything like a religious experience – she uses the Buddhist term *sangha*, meaning a community of devotees engaged in meditation and spiritual practice. 'I sometimes wonder if, on those two occasions, I was granted a glimpse of a dimension of baboon life they do not normally expose to people. These moments reminded me how little we really know about the "more-than-human world".'

Gombe, where Smuts sat with the meditating baboons, is the location of Jane Goodall's decades-long research with chimpanzees – the longest uninterrupted study of any animals, anywhere. Goodall, too, has written of her glimpses into the inner lives of the animals she studied and lived alongside: states of mind which seemed deeply significant yet were ultimately untestable and fundamentally inaccessible.[30]

Deep in the forest around Gombe are spectacular waterfalls, which for the chimpanzees seemed to hold a deep fascination: 'Sometimes as a chimpanzee – most often an adult male – approaches one of these falls his hair bristles slightly, a sign of heightened arousal,' Goodall recounts. 'As he gets closer, and the roar of falling water gets louder, his pace quickens, his hair becomes fully erect, and upon reaching the stream he may perform a magnificent display close to the foot of the falls. Standing upright, he sways rhythmically from foot to foot, stamping in the shallow, rushing water, picking up and hurling great rocks. Sometimes he climbs up the slender vines that hang down from the trees high above and swings out into the spray of the falling water.' What Goodall calls the 'waterfall dances' may last for ten or fifteen minutes. She considers them to be 'precursors of religious ritual'.

Goodall also writes about the chimpanzee's rain dances, performed at the onset of heavy downpours. During these displays, the chimpanzees will grasp saplings and low branches and sway them rhythmically back and forth, then move forward slowly while loudly slapping the ground with their hands, stamping their feet and hurling rocks. While loath to draw firm conclusions from such displays, Goodall associates them with 'feelings akin to wonder and awe' and notes that afterwards the performer will sit quietly for some time watching the water, much like Smuts's baboons around the pools.

Goodall attributes her own open-mindedness to matters of science and religion to starting out in the field without any specialized scientific training. She was originally hired by the anthropologist Louis Leakey to study the primates at Gombe because he wanted researchers 'unbiased by the reductionist thinking of most ethnologists'. The result of her fifty-five years of close association with both chimpanzees and scientific thinking has not been a narrowing of her perspective. Rather, 'the more science has discovered about the mysteries of life on Earth, the more in awe I have felt at the wonder of creation, and the more I have come to believe in the existence of God.'

For Smuts and Goodall, intelligence – whatever it is – manifests in this kind of awareness and belief, this kind of wonder and awe. It is here that our thinking on intelligence breaches the mind completely, and becomes part of larger questions about culture and consciousness, about being and living in a more-than-human world.

This, then, is how we might become aware of *real* intelligence: that is, the kind that exists everywhere and between everything. It is made evident not by delineating and defining, not by splitting, reducing, isolating and negating, but by building, observing, relating and feeling. Intelligence, when we perceive it at play in the world, is not a collection of abstract modes: a concatenation of self-awareness, theory of mind, emotional understanding, creativity, reasoning, problem-solving and planning that we can separate and test for under laboratory conditions. These are simply reductive and all-too-human interpretations of a more boundless phenomenon. Rather, intelligence is a stream, even an excess, of all these qualities, more and less, manifesting as something greater, something only recognizable to us at certain times, but immanent in every movement, every gesture, every interaction of the more-than-human world.

To think of intelligence in this way is not to reduce its definition, but to enlarge it. Anthropocentric science has argued for centuries that redefining intelligence in this way is to make it meaningless, but this is not the case. To define intelligence simply as what humans do is the narrowest way we could possibly think about it – and it is ultimately to narrow ourselves, and lessen its possible meaning. Rather, by expanding our definition of intelligence, and the chorus of minds which manifest it, we might allow our own intelligence to flower into new forms and new emergent ways of being and relating. The admittance of general, universal, active intelligence is a necessary part of our vital re-entanglement with the more-than-human world.

In this thought, I believe, resides the real promise of 'artificial' intelligence. That is to say, if intelligence, rather than being an innate, restrictive set of behaviours, is in fact something which arises from interrelationships, from thinking and working together, there need be nothing artificial about it all. If all intelligence is ecological – that is, entangled, relational, and of the world – then artificial intelligence provides a very real way for us to come to terms with all the other intelligences which populate and manifest through the planet.

What if, instead of being the thing that separates us from the world and ultimately supplants us, artificial intelligence is another flowering, wholly its own invention, but one which, shepherded by us, leads us to a greater accommodation with the world? Rather than being a tool to further exploit the planet and one another, artificial intelligence is an opening to other minds, a chance to fully recognize a truth that has been hidden from us for so long. Everything is intelligent, and therefore – along with many other reasons – is worthy of our care and conscious attention.

*

In this chapter, we've explored only a tiny fragment of the vast literature of intelligence. To do justice to such a range of thinking – both our own and that of other beings – would require numerous books, many already in existence and many yet to be written. What I have sought to do is to illustrate the complexity of such thinking, the deep uncertainty of the subject and, quite frankly, the colossal failure of many of our current ways of thinking about it, both popular and scientific.

On the one hand, these are the hallmarks of good scientific practice: frequent failure, openness to debate and willingness, even a desire, to be proved wrong and learn from the experience. On the other hand, it suggests that there are all kinds of gaps, misunderstandings and outright fallacies in our thinking about intelligence in general, which we all too often ignore or gloss over to the detriment of our own thinking. Under such circumstances, it should be no surprise that the kinds of 'intelligence' we are creating in our machines are so baffling, frightening and misdirected. We are only beginning to learn the true meaning of the word, even as we put it to work. If I have done one thing here, I hope it has been to destroy the idea that there is only one way of being and doing which deserves the name 'intelligence' – and even, perhaps, that intelligence itself is part of a greater wholeness of living and being which deserves our wider attention.

This claim underlies a larger one. Both scientific and popular thought tend towards the conclusion that there are ultimately single answers to single questions. What is intelligence? Who possesses it? Where do they fit into our rigid structures and hierarchies of thought and dominion? Perhaps – whisper it – *this just isn't how the world works*. The closer we examine and the more forcefully we interrogate and attempt to classify the world, the more complex and unclassifiable it becomes. Taxonomy after taxonomy breaks down and falls apart. In part this is a result of our own innate limitations, the possibly insuperable problem of our own *umwelt* and human ways-of-being. But it is also a problem of entanglement: the fact that in the more-than-human world, everything is hitched to everything else, and there are no hierarchies: no 'higher' or 'lower'; none more, or less, evolved. Everything is intelligent. Now what?

# 2
# Wood Wide Webs

At the beginning of 2018 I was invited to give a talk at a big, glossy event in Vancouver. The topic of the talk was to be some of the darker aspects of the internet, and particularly the effects of online video and algorithms on children: a grim subject which I didn't relish delving into. On the other hand, I'd never been to the west coast of Canada before, and the trip promised encounters with interesting new people and landscapes.

As a speaker at the conference, I was offered a range of regional 'experiences', most of which I missed out on as I hid in my hotel room, frantically rewriting and rehearsing my talk. But there was one that seemed too good to miss: a tour of the redwood forests on the city's edge, led by a biologist from the University of British Columbia. So I climbed aboard a bus with a couple of dozen other attendees and went for a walk in the forest.

Our guide was Professor Suzanne Simard, who has spent decades studying the giant redwood forests of Canada's Pacific coast. As we walked among the mossy trunks, she explained how the huge trees were intimately connected to one another – and to places far outside the forest itself. For example, a significant proportion of the essential nitrogen that the trees and other forest plants take up through their roots is ultimately derived from very far away indeed – from the middle of the Pacific Ocean, thousands of miles distant.

We know this because there's a particularly heavy nitrogen isotope called nitrogen-15 which is much more common in marine algae than in most terrestrial vegetation. Yet it's found in surprising amounts in coastal forests, and the way it gets there is quite marvellous. The fish which feed on algae and phytoplankton in the deep ocean become enriched with nitrogen-15. One of these fish is the Pacific salmon, which every year returns to spawn in the same rivers and streams from which they hatched, on the coast of the western United States and Canada. As they swim upstream, they're caught and eaten by the black and grizzly bears who have made their own journeys from far inland to congregate along the salmon rivers during the spawning season. The remains of the fish – as well as the bears' copious droppings – fertilize the trees and other forest plants and mark them with that particular form of nitrogen, ferried over thousands of miles in the bodies of other species.[1]

Similar networks of distribution, and even of mutual aid, exist beneath the forest floor, as Professor Simard explained. The trees can't access much of this nitrogen directly: instead, they rely on specialized fungi growing on, around, and even through their roots. These fungi absorb and process the nitrogen with their own enzymes, and pass the resulting nutrients onto the trees. The dense networks of interconnected fibres through which this exchange takes place are called mycorrhiza – a literal growing-together of the Greek words for fungus (*mykós*) and root (*riza*) – and we're only just beginning to understand their significance.

The mycorrhizal trade is not a one-sided deal. In turn, the trees supply the fungi with the sugar and carbon they produce from photosynthesis: solar products for underground dwellers. Even more remarkably, the trees connect to one another through this network. Approaching one of the largest trees along the path, Professor Simard introduced us to it as a 'mother tree', which serves as a central hub in the mycorrhizal network. As new seedlings emerge from the undergrowth, the mother tree infects them with the fungi, and uses it to supply them with the nutrients they need to grow.

The trees are selective in whom they choose to help: mother trees favour their own offspring, sending more carbon to them than to unrelated seedlings, and shifting their roots to make room for them to grow. But they're not selfish, and will go to the aid of other species in times of need. In the summer, the shaded fir trees receive carbon and sugar from the taller birch trees, while in autumn the firs return the favour as the birches start to lose their leaves.

This living network is not just about sharing food; it's also about sharing information. If one tree in the forest is attacked by insects, it releases warning chemicals into the network. The other trees pick up the signal and respond by producing their own chemical defences. Standing among the giant trunks, listening to Simard and her colleagues speak, I began to realize that the forest was filled with a constant hum of unseen signs and unheard chatter. Decisions were being made, agreements reached, bargains made and broken. The trees were speaking to one another. A whole world of voices I had never heard – whose existence I had never even suspected – burst into my consciousness.

As mesmerized as I was by this encounter with previously unseen, unknown networks of communication, we had to move on. After a couple of hours we got back on the bus, returned to the city and resumed our conversations about technology, the internet and the modern networks which seem to have supplanted, or at least drowned out, those of the forest.

It was another year, and a work of fiction that brought home to me the significance of what I'd learned in that brief encounter. In Richard Powers' novel *The Overstory*, a variety of characters find their lives increasingly intertwined with, and dependent upon, the trees around them. In one chapter, a mulberry tree planted by a recent Chinese immigrant to America in his new backyard becomes the frame for a saga of family life over decades: a proposal, a child, a death and more. In another, an artist discovers a box of photographs taken by his great-grandfather, grandfather and uncle over a hundred years. The photos are of a rare Midwestern chestnut on the family farm, snapped every year, from the same place: a century-long time lapse of a growing tree emerges as the world shifts around it. Gradually, mediated by their encounters with trees, the book's various characters band together to campaign for the preservation of some of the oldest and grandest trees on the continent: 2,000-year-old Californian redwoods, almost completely destroyed by logging.

One of these characters is a tree researcher named Patricia Westerford. While still in graduate school, she starts to think differently about the trees she studies, setting herself apart from the other students. 'She's sure, on no evidence whatsoever, that trees are social creatures. It's obvious to her: motionless things that grow in mass mixed communities must have evolved ways to synchronize with one another. Nature knows few loner trees. But the belief leaves her marooned.' Her first paper on tree communication – an analysis of maples which send out signals to warn one another of insect attacks – is savaged by established critics. For years, she is marginalized and ridiculed by other scientists for her assertions.

For me, Westerford is the most compelling character in the book: someone who feels a truth deeply and who insists on doing the dogged work of science, and of literature, to bring that idea into a meaningful, communicable framework of knowledge. Despite being fired from her academic position, and enduring years of itinerancy, she sticks with the trees, disappearing into the forests and living in a network of remote cabins, before discovering she's been vindicated in her absence and another generation of students has picked up her work and run with it. 'Her public reputation, like Demeter's daughter, crawls back up from the underworld. A scattering of scientific papers vindicates her original work in airborne semaphores. Young researchers find supporting evidence, in species after species. Acacias alert other acacias to prowling giraffes. Willows, poplars, alders: all are caught warning each other of insect invasion across the open air.' Eventually, she writes a book, *The Secret Forest*, which becomes a bestseller.

In this case, art follows life. Powers based the fictional Westerford's writings and discoveries on Suzanne Simard's very real work in the forests of British Columbia. The revelations about the other lives of trees which enthralled me in *The Overstory* were recountings of the same stories I heard beneath the trees in Vancouver. I already knew these things. I just hadn't understood what they meant.

The achievement of *The Overstory* is to make these revelations meaningful at human scale, building the connections between us and the trees around us. Powers uses the scale of arboreal life – its extent in time, as well as its size – to tell a new kind of epic story: multi-generational, planetary-scale and ecological, in the sense of deeply intertwined with and responsible to our environment. As I read it, I felt something shift in myself, a sense of having been blind all my life to events and processes, whole other lives that surround us all the time. Having always been more interested in words on pages and code on screens, I suddenly found myself leaning out of windows to touch the leaves of nearby trees, and stopping in the street to trace, with wonder, the whorls and cracks of living bark.

This came as something of a shock. I did not think I was inattentive to the weird other lives of things. As an artist, technologist and technology critic, I've worked extensively with the new techniques of artificial intelligence, which I first studied almost two decades ago, but which have undergone something of a renaissance in the last few years. Attempting to understand these technologies, I've built machine-learning systems that seek to question our assumptions about machine intelligence and human agency, and the path they seem to be taking us down. The self-driving car was one such experiment; in another, I used the latest machine-learning technologies to simulate a relationship between the British climate and Brexit voting patterns, drawing parallels between stormy weather, cloud computing and cloudy thinking. And what had struck me each time about these engagements with machine intelligence was their sheer strangeness, their fundamental difference to our own ways of seeing and thinking. And yet, I had not extended such awareness to the even stranger intelligences which, it turned out, were already all around me.

Powers and Simard suggested that another way of seeing the world was possible, one infinitely more vital and interconnected than any I had previously imagined. In their worlds, information pulsed beneath the ground and floated on the breeze, interactions pulsed and shifted to the rhythm of the seasons, and knowledge and understanding grew, slowly but sturdily, over decades and centuries. Beyond the human, beyond the animal, now appeared another domain – multiple kingdoms – of flourishing, active, even intelligent beings: plants and fungi.

I have not always been the best of friends to plants. I do not have green fingers, and many have died at my hands over the years. But I am lucky enough to live with someone with a very different aptitude, so that in the last few years I have become surrounded by them. First our apartment, then our garden, bloomed with vines and vegetables, succulents and flowering plants, a riot of ever-growing, ever-changing life. Presence and attention breed awareness, and as a result of this closeness, I have begun to distinguish species, shapes, colours and preferences in ways which were previously inscrutable to me.

We have already seen how the forms of bodies and their relationships to one another shape the subjectivity of animals, their *umwelt*. Think of the arboreal gibbons, directing their attention and their intelligence upwards into the trees; the way an elephant's abilities are framed by its trunk; or the complex group dynamics of baboons and macaques. Intelligence, it seems, is something physical and relational, not a wholly abstract process, but one closely tied to our being and doing. And if every organism has an *umwelt*, what would the *umwelt* of a plant look like? How do they sense and act, perceive and contribute to the world around them? While reading about the busy lives of trees in the forests, and the chatter passing through the mycelium, I looked around me and started to wonder what I might learn from these potted individuals, these trailing legumes and budding brassicas, cacti and lemon trees, and the way each individually responds and contributes to the world.

On the one hand, to get to know plants means to come down to earth, both literally and figuratively: you need to put your nose in the dirt, get mud beneath your fingernails, become familiar with the moistness of soil and its composition. The first effect of such familiarity is to remind us of the importance of place: specific places, with their own relationship to light and heat, sun and rain, shade and soil types. The abstract becomes specific. On the other hand, plants also connect us to everything else, again both literally and figuratively: their roots intertwine beneath the earth, the mycorrhizas grow together and their seeds scatter on the wind. They also precede us, and make life, everywhere, possible: through photosynthesis, soil production and the conversion of nutrients into food, plants are the founders and sustainers of the world.

It is these qualities which make plants good companions in thinking about technology. Both are endlessly generative, complex systems at both the macro and the micro scales. On the one hand, our contemporary technologies of power generation, information processing, communication and sensing are critical life-support mechanisms, which reproduce at a global scale and grant us – and themselves – near limitless, god-like powers. On the other hand, they depend upon the most minute of connections and the tiniest of electrical impulses; they need firm ground, the right atmosphere and constant human attention. No wonder then that the poet Richard Brautigan was moved to imagine 'a cybernetic forest / filled with pines and electronics / where deer stroll peacefully / past computers / as if they were flowers / with spinning blossoms'.[2]

Somewhere between these two extremes – between the plant pot and the continental forest, between the microchip and the satellite – lies our actual lived experience, the place where our shared *umwelts* meet and mingle. For surely, as much as we differ, there must be points of overlap in our awareness, our sense of the world, which provide an opening to understanding. This turned out to be the case with my self-driving car; it was what Barbara Smuts discovered with the baboons. To understand a little of what plants mean and why they matter, we must discover what we have in common: the ways in which we share a world.

One way we might do this is to ask what plants hear. This simple question is already rife with contestation – plants have no ears or other known receptors for sound waves. And yet, quite assuredly, they hear, as we shall prove. How they do so, what they do with that information, and what it means for our interrelationships are similar questions to those we ask ourselves about animal intelligence: how are we changed by encountering non-human senses and a non-human impression of the world?

In 2014, two biologists at the University of Missouri recorded the sound of cabbage white caterpillars feeding on a cress plant. (*Arabidopsis thaliana*, rock cress, is the macaque of the botanical sciences, the most popular plant for biological experiments, and it has taught us many things about plant growth and genetics. It was the first flowering plant to have its genome sequenced and its DNA cloned, and it has even gone to the moon.[3]) Having left the caterpillars to munch away for some time, the scientists then removed them and played the sound of their approach back to the plants. Immediately, the plants flooded their leaves with chemical defences intended to ward off predators: they responded to the sound as they would to the actual caterpillars. They heard them coming. Crucially, they didn't respond in the same way when other sounds – of the wind or of different insects – were played to them. They were able to distinguish between the different sounds, and act appropriately.[4]

Claims about plants' capacity to hear and respond are not new. Neither are they always reliable. Notoriously, the bestselling 1973 work of pseudoscience *The Secret Life of Plants* used lie detectors and tape recorders to 'prove' that plants possessed emotions, telepathic abilities – and hearing. Roundly decried and debunked by scientists, the book's claims linger in the popular imagination, not least in the persistent belief that plants like classical music, or grow better when people talk to them. Fond as I am of all kinds of magical and imaginative thinking (and particularly, in this case, of Stevie Wonder's wonderfully odd concept album *Journey through the Secret Life of Plants*), the reality – as witnessed in the 2014 cress experiments – is better.

The cress displayed what scientists call 'an ecologically-relevant response to an ecologically-relevant stimulus'; that is, a plant-y response to a plant-y event. While plants are unlikely to have any preference for Beethoven or Mozart, for English or Arabic, the munching of caterpillars, on the other hand, has for plants a clear meaning – and the plants respond meaningfully in turn.

'What is the ecological relevance?' is the core question which rigorous science asks of any claims made about non-human abilities. In order to make sense of such abilities, it is necessary to frame them within an ecological context, the context of their relationships. It is not enough to say simply, 'this happened'. We must place each action within its context and understand why it occurs: in relation to which particular pressure, need or desire, or stimulus. Only then can we say that we truly understand an ability, rather than simply observe it.

This is a double-edged sword. On the one hand, it restricts our understanding of those abilities to an understanding of their circumstances, and as we already know from animal studies that can be woefully limiting. It's precisely what stopped us from recognizing the intelligence of gibbons, dolphins and others for so long, because we didn't fully understand their context and so missed the relevant cues. And whose understanding are we even talking about? The gardener, the greenkeeper and the forest-dweller have a very different perception of a plant's situation from that of the lab technician. On the other hand, the focus on ecological relevance pushes us to focus on relationships rather than innate abilities: on what matters when plants, animals, we and others meet. What matters in these encounters and how do they shape our relationships?

If the context of our relationships really matters so much, then it will also matter in our relationships with machine intelligence. Perhaps we should be thinking more carefully about the ecosystem in which we are raising AI, particularly the kind of aggressive, domineering and destructive forms which seem to be proliferating. That these systems are overly concerned with profit and loss, winning and losing, control and dominance, suggests that their ecological niche – the slice of the environment shaping their evolution – is somewhat narrow. Their learned responses are that of a corporate intelligence, evolving within the arid, airless ecology of neoliberal capitalism, tech company boardrooms and ever-increasing financial and social disparities. If we wish them to evolve differently, we will need to address and alter this ecology.

The concept of ecological relevance admits something else too, something even more earth-shaking. It demonstrates that plants have a world. What does this mean? It means that plants sense and respond to a world which they experience, a world of their own making – and, furthermore, that there is a 'they' there to do the sensing and responding, a subject rather than an object, a kind of self, however abstruse and unlike our own. They have an *umwelt* all of their own. Plants encounter, access, influence and are influenced by the world on their own terms and in their own fashion. Most of these experiences and responses are and will always be unknowable to us; but hearing – being an ability we share – makes them thinkable. Suddenly, from being background, plants leap into action once again, present and attentive. The act of hearing transforms vegetal passivity into active listening, leafy torpidity into vibrant participation.

Of course, that's only our realization, our novel awareness. They've been active all along. Just as they probably don't care about Mozart or Stevie Wonder, so they don't need us to know that they are there. We choose to acknowledge them because it enriches us.

One morning some years ago, I was walking with my partner on the slopes of Ymittos, the mountain which rises on the eastern edge of the city of Athens. It was the day before spring – not the spring of the calendar, but the spring of the year, when everything bursts forth at once on an unspoken cue. The whole mountain seemed to be preparing for it. As we descended through the forest around the ancient monastery of Kaisariani, we came across a cherry tree on the very threshold of flowering. Its blood-red bark was stretched tight around its trunk, each tiny leaf held in tension, each branch tipped with a swollen pink bud, its tip pursed in anticipation: the whole thrummed with a barely contained energy, as if straining against a tightly held breath. I'd never before experienced a being so insistent upon life without visible movement. I've since come to realize that this is how trees, plants and much of the earth are, much of the time. This feeling has become a part of me, and I am capable of calling it up at almost any time, simply by turning my conscious attention towards the whisper of the trees and the turning of the leaves. *We share a world.*

This shared world is not flat, nor singular. Many worlds – lively, noisy worlds – exist; many don't include us at all. Experiments with plant hearing suggest that, unsurprisingly, in many of those worlds the footsteps of caterpillars matter more than classical music or human language. It is a beautiful illustration of the kind of decentring of ourselves and of human experience at which we must become adept in order to live better and more responsibly in a more-than-human world.

This decentring, an admission that the human race is not the only game in town, does not correspond to any reduction of our world. Rather, as when we extend the virtue of intelligence to other beings, the addition of plant worlds to our own enriches both. Even those worlds in which we do not participate add to the totality of sensations and experiences which form the living, teeming earth, and on which we live and depend.

The very existence of other worlds, of numerous overlapping worlds in which many kinds of things and many ways of seeing and being are possible, should thrill us. Other worlds are not only possible, they are already present. The acknowledgement of multiple other worlds, the worlds of others, is key to disentangling ourselves from our greatest social and technological deception, and re-entangling ourselves with a more meaningful and compassionate cosmology.

I call this deception the 'one world' fallacy. Since the Enlightenment of the seventeenth and eighteenth centuries, and the scientific revolutions that followed it, our understanding of the world has been shaped by a misplaced objectivity – the belief that the world has a single, coherent narrative and that there exists a one-size-fits-all framework for interpreting it. Eighteenth-century scientists called this 'truth-to-nature', although their attempts to fix an understanding of how nature 'really' worked involved eliminating its inconsistencies and idiosyncrasies. As scientific practice has matured, we've come to understand the value of interpretation, situated knowledge and trained judgement; but this fallacy has been perpetuated by our technologies, which flatten and lump together the myriad different expressions of the world.

The power of these tools to make the world legible to humans in turn contribute to techno-determinism and network power: the belief that the tools produced in the most dominant societies by the most dominant groups are not only the best for everyone, but are inevitable and unassailable. And so, in contrast to the original intent of scientific investigation, a mismatch arises between the world as we perceive it, and the world as it actually is. The attempt to force the world to conform to our portrayal of it, and the friction this attempt generates in our lives and societies, is behind the great malaise of our age: widespread confusion, shading into anger, rage and fear. It is the result of trying to find truth and meaning in a single world, a single box into which we cram all the contradictions and paradoxes of reality. But in truth, there are so many worlds.

The fact that we are still able to live, to function, to survive and thrive together in this world-of-many-worlds, also implies that these worlds are shared. There are points and planes of intersection, shared experience and shared awareness. All the inhabitants of the earth – animals, plants and diverse others – are, whether they care about classical music or not, whether we even notice it or not, buffeted by the same vibrations in the atmosphere. By dispensing with the fallacy of one world for all, we come to the awareness of a greater multiplicity of worlds which are held in common. This is a far richer cosmology than the solipsism of one world; it is an acknowledgement of communal being and experience. *We share a world*. We hear, plants hear; we all hear together. We all feel the same sun, breathe the same air, drink the same water. Whether we hear the same sounds in the same way, whether they are meaningful to us in the same way, is beside the point. We exist, together, in the shared experience and creation of the more-than-human world.

This chain of realizations is significant, it seems to me, for it fundamentally transforms our possibilities for thinking not only about the lives of plants and trees, but about all these other minds, beings and persons. It permits a wholly different kind of thinking, one that is more open and clear. Because acknowledging the existence of non-human worlds, and subsequently the existence of a shared world, helps us navigate the twin hazards that we face in thinking about the more-than-human world: anthropocentrism and anthropomorphism. The former is the danger of thinking ourselves to be at the centre of everything; the latter is the danger that, in trying to access non-human experience, we simply mould it into a poor shadow of our own.

Fully recognizing that non-human plants, animals and others have their own worlds which are fundamentally different and unknowable to us is to begin to end human exceptionalism and human supremacism. Humans are not at the centre of the universe. By stepping aside we can start to imagine what a world in which we are not the most important thing might actually look like, and consider the richness of non-human worlds on their own terms.

Yet some measure of anthropomorphism – the attribution of human traits, emotions and intentions to non-humans – is inevitable. It is, when all's said and done, the only way we can begin to understand other living things. We are *anthropos*, and we have no other means of addressing these worlds than through our own. But if we can consciously bear in mind the fact of our difference, if we can recognize our own limited perspective while not enforcing it upon others, these things need not be a barrier to accessing and acting on the basis of shared interests and intentions. In fact, it is precisely through the acknowledgement of difference that shared action and shared life, without the weight of domination and control, becomes possible. We seek to meet, and not to conquer. We are as comfortable with companionable silence as with talking. We have different words for the same things.

This position even allows us to think again about abilities that are *not* ecologically relevant, or not apparently so. We still have no idea why bird eggs are brightly coloured, for example (seemingly an invitation to predators), or why plankton are so diverse, defying the laws of evolutionary competition.[5] But we don't need to have a complete explanation for every mechanism before we test our theories, because – contrary to the design of most scientific experiments – we live in the same world as our test subjects. This is brilliant: suddenly whole vistas of inexplicable but obvious behaviours become interesting and thinkable.

One such extraordinary behaviour is the ability of plants to remember. That plants hear, in some form, is not too great a stretch for our imaginations. But the idea that they might remember – and therefore perhaps think and reflect, and do all the things we humans do with memory – well, that seems like a stretch. After all, we have no inkling of the structures and processes which might facilitate such behaviour. And yet, we do know that plants remember.

The greatest – and most controversial – exponent of plant memory is the Australian biologist Monica Gagliano, whose scientific background is an object lesson in ecological thinking. Gagliano began her academic career as a marine ecologist in the 1990s, studying the behaviour of fish on Australia's Great Barrier Reef, before turning to the plant world. It's this background which informs her working practice. Historically, scientists' approach to the study of plants has been a mechanistic one; that is, they break them down into a series of actions and reactions, viewing them more like a series of component mechanisms, as a series of tiny, interconnected machines, rather than as whole organisms. And this means we regard them in much the same way as we regard machines, capable only of reacting to stimuli in automatic, pre-determined and predictable ways. Botany sees only narrow cause and effect, and often fails to fully consider the whole life and experience of the organism. In contrast, Gagliano sees plants as having behaviours, like the tropical reef fish she studied previously.

Gagliano's favourite partner in this research is *Mimosa pudica*, also known as the 'touch-me-not' or 'sensitive plant'. The mimosa displays a particularly startling behaviour: on being touched, with a stick or a hand, it suddenly and quickly rolls up its leaves into a tight bunch. It belongs to a rare class of plants whose movement is visible to the naked eye: only the Venus flytrap, with its snapping jaws, and the Asian telegraph plant, whose dancing leaves entranced Charles Darwin, can rival it for speed. As such it makes an excellent and communicative partner in experiments.

Gagliano wanted to change scientists' understanding of plants, to show that they were capable of learning from experience, and altering their behaviour as a result, an ability scientists had tended to reserve for 'higher' animals.[6] In order to demonstrate this, she devised a simple mechanism: a cup attached to a vertical rail which, when released, would drop a mimosa in its pot exactly fifteen centimetres onto a foam pad. The soft *thonk* of this landing was enough to surprise the plant into closing up its leaves, but not enough to damage it: a scientifically precise stimulus, with a measurable, ecologically relevant response. Mimosa leaves have small hydraulic structures at their base which, by pumping or draining water, allow them to expand and contract, forcing the leaves to curl. The curling is understood to be a response to a threat: either from animal predation, or excessive heat and evaporation. Either way, a traditional cause-and-effect study would probably stop here: a relationship between stimulus and response has been established, the mechanism known and understood, our expectations of plant capabilities satisfied but hardly enlarged.

Gagliano went further. She kept dropping the plants, over and over again, up to sixty times a session, over multiple sessions. Three hundred and sixty drops a day: a marathon of bumps and shocks. *Thonk thonk thonk*. It transpired that it only took a few drops – as few as four or five – for the plants to realize that there was no threat, and that it was safe to keep their leaves open (a side test, involving a different stimulus which still elicited the closing response, revealed that they weren't simply exhausted by the activity). By the end of the sixty-drop sessions, the plants were entirely unbothered by the drop: they had learned to ignore it.

Understanding a lesson is one thing; its value depends on the ability to recall it later, to put the knowledge gained into practice. So Gagliano and her colleagues rested and then retested individual mimosas, demonstrating that they retained over time the memory of the drop, and their associated change in behaviour. Mimosas – and, we must now understand, all plants – are not machines. They are more than the sum of a set of pre-programmed actions and reactions. They learn, remember and change their behaviour in response to the world.

This is an astonishing discovery. Not only does it run counter to everything we thought we knew about plants, but it radically overhauls the very categories of knowledge we have for understanding them. So much so that it took years for Gagliano to publish her results, following rejections by over a dozen academic journals, many of which would not even acknowledge reading her unpublished paper, let alone send it out for peer review. Part of the problem is innate academic conservatism – upending decades of understanding is not only hard to do, it's hard to imagine and acknowledge – but another part is simply bureaucratic. Where to put this new knowledge? Whose domain does it fall under? Is this botany or physiology, ethology or psychology?

As Gagliano put it, 'We were somewhat unrehearsed for the play *Mimosa* had put on and which – candidly unprepared or stubbornly unwilling – we were all, nevertheless, characters in.' The fact of *Mimosa*'s behaviour was undeniable, and it demanded we rethink our assumptions. Science has a tendency – a prime directive, even – to understand phenomena through the structures that produce them; that old ecological relevance again. It understands memory as a structural process which takes place in the brain, in the interaction between particular molecules and synaptic connections. In this sense, memory depends on the brain. How then can something without a discernible brain remember? It's simply unthinkable.

But we can look at it another way. Rather than denying the existence of plant memory because plants lack brains, discoveries such as Gagliano's give us the opportunity to rethink what memory is, and admit that perhaps it can arise by other means, in other structures, in other beings. Like the intelligence of the octopus, the memory of a plant is an analogous phenomenon running on completely different hardware.

One intriguing aspect of Gagliano's account of her research – and one which puts many other scientists' backs up – is that, for her, each experimental result is the result of a plant speaking. She means this literally: as a result of a long-standing practice of meditation and shamanic rituals, she has received direct instructions from entities she considers to be plant spirits, who offer her advice on how to conceive and structure her experiments. Each of these encounters is then an introduction, an opening of the channel, which allows the experimental subject to speak for itself, for *Mimosa* to declare its ability, now that we are ready to hear it. And every time the plant speaks, the boundaries which we thought enclosed our worlds are shaken.

If you're sceptical, then fine. But it's not necessary to accept the reality of Gagliano's direct communication with plant spirits to accept the results of her experiments, which in every other way conform to the most stringent scientific norms, including peer review and reproducibility.[7] I find aspects of Gagliano's account hard to take seriously too. Yet I've had my own personal experience with Ayahuasca, one of the plants she speaks with, and it has spoken to me, too. I have heard a plant speak, and I still don't fully understand how, nor can I adequately describe the experience, but I know it to have happened, and it changed me utterly.

Western scientific conditioning is hard to break with. It makes it hard for us to trust even our own experiences, whether that is the ecstatic experience of spirit encounters in the grip of psychedelics, or the uncanny sensation of communion in the eyes of other species. This is part of the genius of Gagliano's experiments: the rigorous application of the scientific method, the careful weighting and testing of her *Mimosa* plants, and the reproducibility of her method means we don't have to put all our trust in our subjective experience. What is required of us is to be open to changing our minds.

Mimosa speaks, and the world changes. We are forced to account for a different reality than the one we knew. When we open ourselves to the voices of the more-than-human world, this is always the result: a breaching and collapsing of the established borders of thought and feeling. This is the outcome that shines through these encounters, with all kinds of intelligences, including 'artificial' ones. Machines beat us at chess, take the wheel of the car and make scientific discoveries, and what we thought were uniquely human endeavours become instead shared ones. Baboons stare deeply into pools of water, and the experience of contemplation we thought wholly our own is suddenly mirrored in, and enhanced by, that of others. Plants have a world, it's different to our world – and, oh, we share a world! Each time, acknowledgement of the agency and difference of the other actually brings us closer, while broadening our own vision, knowledge and experience. It enlarges us.

The field just grows and grows. Plants don't just hear and remember – they make their own sounds. In another experiment, Gagliano recorded the sound of corn kernels clicking at frequencies far beyond human perception, their purpose still unknown. It turns out plants can smell too, detecting both predators and the warning signals given off by neighbouring plants. They even smell out prey, as animals do: the parasitic dodder vine sniffs out suitable victims, which it then envelops and feeds upon. Plants make decisions based on complex information, such as picking the best response to nearby competition. Grow sideways, root deeper or shoot higher? They can tailor chemical releases to attract or repel animals – poisoning some and creating addiction in others. They have proprioception – the 'sixth sense' which enables us to know where parts of our bodies are without looking. They recognize and respond differently to kin and close relatives. In short, plants act, and they act in ways which, when animals do likewise, we call indicators of intelligence.[8]

In just the last decade, we have crossed a once-invisible line in our relationship with plants: they have been transformed in our understanding from objects into subjects and, as in any conversation, their subjecthood will continue to expand in scope the longer we converse. As with the abilities of animals, this tally of plant agency will grow and grow as we admit more and more possibilities, come up with new theories and design the apparatus to test them.

This is not to make any specific claim about the intelligence of plants. The exact form of plant intelligence must always remain partially or mostly unknowable to us, because of the radical difference which exists between our own lives and our experience of the world, and that of plants. But because we share a world, we can think about what intelligence and memory might be, when we include others in that thinking, when we understand it as something which acts between us and in the world, rather than merely in our own heads. You can read endless books about plant and animal intelligence but they'll all tell you the same thing, that non-humans are brilliant and also unknowable, and that the greatest joy in the world is to be found not in testing and taxonomizing but in going on together.

How we talk about this newfound subjecthood of plants is another matter. How do we make sense of these newly recognized skills, this life that no longer merely teems, but thinks? It's tempting, as with non-human animals, to speak immediately of plant intelligence as if it were comparable to the human kind, to induct the geranium and the vine into that exclusive club that we're just starting to shake up. But, again, that seems to miss the point. Like the restriction of 'artificial' intelligence to the kind of things that humans do, it runs straight back into anthropomorphism and anthropocentrism. Whatever plants are doing, they may resemble us and other animals in some ways – but they are clearly different in other, significant ways. The insufficiency of intelligence as a way of explaining these abilities is clear. Plant intelligence, whatever it is, is plant-y.

Debates over plant intelligence, and animal intelligence, and artificial intelligence, and even human intelligence, are thus revealed as endless – and also, ultimately, pointless, because they are debates about the meaning of words rather than the being of things. Where we start to move forward is when we learn to ask questions which are less concerned with 'Are you like us?', and more interested in 'What is it like to be you?'

The reason we have so much difficulty understanding what plant intelligence might consist of is precisely because it is so different to our own. One obvious objection to the idea of plant intelligence is that plants don't have brains. Yet one of the strengths of plants is precisely that they have no central, irreplaceable organs. Plants are modular – they can survive losing 90 per cent of themselves, and many species can reproduce from broken pieces, or cuttings. In particularly harsh environments, like deserts, cuttings are more successful than seeds for raising new generations, which is why many succulents are particularly good at this kind of reproduction. Modularity creates resilience: fetishizing particular organs, as animals do, obscures such abilities. Rethinking what intelligence might be also allows us to rethink the modes and mechanisms which might produce it, and thus to come up with new ways of being intelligent.

Plants such as succulents also produce clones: genetically identical offspring which may separate from the parent, or remain attached, shaping their parents into new forms. While cloning can be a weakness, as genetically identical offspring are vulnerable to the same threats as the parent, it can also create extraordinary creatures. One such clonal wonder is Pando, an aspen living in Fishlake National Forest in Utah. At ground level, Pando looks like a forest. They take the form of more than a hundred acres of quaking aspen trees; 47,000 tall, slender trees with white bark and black knots, whose leaves turn shades of brilliant yellow in the autumn. But in fact Pando is one individual, a single organism, in which each tree trunk is a shoot from a single root system. They are one of the largest and oldest individuals on Earth.

Pando was born around 80,000 years ago, or perhaps many, many more: one study puts their age at closer to a million years.[9] Clonal trees can't be aged by their rings, as each is a mere passing stem of the more ancient root. For most of Pando's life, conditions were close to perfect for a clonal aspen: frequent forest fires prevented competition from conifers, while a gradual climatic shift from wet and humid weather to drier, semi-arid conditions made it harder for younger aspens to seed and establish rivalries. In fact, hardly any new quaking aspen have grown from seed in the western United States for 10,000 years, yet the aspen remains North America's most widely distributed tree.

Today, Pando is believed to be dying – albeit very, very slowly. No new growth has been detected for several decades. The possible causes of this are numerous, but all point back to human interference. The establishment of cabins and campsites within Pando's expanse reduces the likelihood of the cleansing forest fires on which aspen depend, while the disappearance of bears, wolves and mountain lions has led to an explosion of mule deer. The deer, along with cattle, graze on young aspen shoots. Once areas are eaten or cut back, they don't regrow.[10]

Pando wasn't 'discovered' until the 1960s, not least because they don't look like a tree – they break with our idea of what a tree is. Yet as our idea of what a tree can be expands, so do our models for thinking about other structures in the world. Indeed, the discovery and understanding of plant communication itself owes much to such radical rethinkings of the way we see the world, and leads to new realizations in turn.

Earlier, I mentioned the mycorrhiza which underpin life in the forest: the networks which process, store, extract and transport food and information between the surface plants. To think of these networks – of fungus itself – as separate from the plants which they connect is not really correct. As we've discovered in recent decades – like everything we've discovered in recent decades – the kingdom of fungi and the kingdom of plants are not really that separate. They are physically, socially, vitally entangled with one another. Understanding the nature of this interrelationship is vital to understanding our own entanglement with the more-than-human world.

The earliest plants were mere agglomerations of cellular tissue, lacking roots or leaves or any of the specialized structures we recognize in plants today. They were the descendants of simple marine algae which washed ashore and found some purchase on beaches and cliffs, sustaining themselves through photosynthesis alone. But around 400 million years ago – at least, that is the date of the oldest fossils we've found – these proto-plants began to associate with fungi: to evolve lobes and fleshy organs to house mycorrhizal partners. This is the origin of all plant roots: questing limbs in search, not of food itself, but of partners in the process of producing life.

Plants and fungi don't merely interact underground, they penetrate one another. Parts of fungi actually live within the cells of plants, and they form in effect an extended root system, more than a hundred times longer than the roots of the plant itself. And these fungal strands, called mycelium, extend everywhere and through everything. What we take to be the soil itself is actually part fungus – somewhere between a third and a half of its living mass. Plant, fungus, and the entire ecosystem on which we and all life on earth depends, is inseparable, right down to the cellular level.

This relationship between fungus and plant is symbiotic: each depends upon the other. As the mycologist Merlin Sheldrake puts it: 'What we call "plants" are in fact fungi which have evolved to farm algae, and algae that have evolved to farm fungi.'[11] We must then understand that our own lives are dependent, symbiotically, on such relationships: for the food we eat, the environment we inhabit, and even the climate we – hopefully, for now – enjoy.

One place we can put our growing understanding of mycorrhiza to work is in models of the global climate. These are the complex, computational simulations created by weather forecasters and climate scientists, which allow them to reconstruct and predict the effect of various phenomena on the behaviour of the atmosphere. These simulations can explore historic changes millions of years in the past as well as the looming crises precipitated by our own destructive interactions with the planet today. By varying the numbers entered into the simulation, we can see how differing conditions produce different outcomes.

These models can be incredibly sophisticated. At the University of Leeds, researchers created a climate model which explored how the availability or otherwise of phosphorus – a key plant nutrient – played a role in the transformation of the atmosphere which occurred in the Devonian period, some 300 to 400 million years ago. This is the period in which plants, already well established on land, massively expanded their range, growing faster and taller and more widely than ever before. The sudden appearance of so many green plants resulted in a dramatic reduction of carbon dioxide in the Earth's atmosphere, as much as 90 per cent compared to the previous era. The researchers thought that this explosive growth might have been caused by an increase in phosphorus in the soil, and this is what they initially modelled. But the real cause, it seems, might have been mycorrhiza. The number which made the biggest difference to the model wasn't simply the amount of phosphorus available to plants, but the efficiency with which they could take it up, an efficiency entirely determined by plants' mycorrhizal networks. By adding mycorrhizal fungi to the model, the researchers showed that the amount of carbon dioxide and oxygen in the atmosphere, as well as the global temperature – the entire climatic system – was dependent upon the mycorrhizal relationship between plants and fungi. All life, and all changes in life, depends upon mycorrhizal relationships.[12]

There is another way in which we have come to depend on mycorrhizal networks; not as matter, but as metaphor. My sudden awareness, when walking in the redwood forest outside Vancouver, of a vibrant, active network beneath my feet, through which vast quantities of information as well as nutrients were passing, was not entirely new to me. It was the same sensation I experienced when beginning to understand, and bring into view, the infrastructure of the internet: the vast, planet-spanning network of cables, wires, machines and electromagnetic signals which sustains and regulates humanity today. These microprocessors and data centres, undersea cables and wireless transmissions are our own mycorrhizal network, interpenetrating everyday life, managing our supplies and demands, disturbing the climate and touching upon our own skin – even, in the case of cochlear implants, pacemakers, and perhaps soon neural links, breaching it.

It speaks volumes about our lives today that when we hear about networks and communication it is these artificial networks which we think of first. The far more ancient and deeply rooted networks of nature are an afterthought, if we even think of them at all. Yet I believe that this sequence of realizations is one of the most significant lessons of a technological ecology, because without our understanding of digital networks, we might not really understand the mycorrhizal network at all.

Suzanne Simard, who first introduced me to the redwoods and the mycorrhiza, published her first studies on the transfer of carbon between trees, through the underground network, in 1997. By exposing pairs of seedlings to radioactive carbon dioxide, and tracking this carbon as it flowed through the forest, she showed that it appeared only in those trees connected by mycorrhiza. Moreover, when the fir trees she studied were shaded, and therefore lacked the energy to capture as much essential carbon, they received more of it from their neighbours. The network sent the carbon where it was most needed. These findings stood in opposition to classical forest ecology, which prioritized competition and individual success in measuring the health of the forest. Rather, beneath the ground, the trees and the mycorrhiza were working together, making connections and cooperating. It was a revolutionary discovery, and it required a new way of thinking and writing about ecological networks.

When Simard published her findings in *Nature*, probably the most important global scientific journal, its editor asked the distinguished botanist David Read to write a commentary on it. Together, the pair came up with a resonant new phrase to describe the mycorrhizal networks, which became the headline on the cover of that issue of *Nature*: 'The Wood Wide Web'.[13]

Back in the 1960s, when the nascent internet started to thread its filaments across the planet, it did so primarily through university departments. It was the development of hypertext and the invention of the World Wide Web by Tim Berners-Lee at CERN in 1989 (specifically to facilitate the sharing of academic documents) which kick-started its wider adoption and understanding. But the gift of the Web wasn't only informational: by its very existence it gave us new tools to identify and understand networks themselves.

Before the Web's arrival, scientists lacked the tools needed to understand how networks functioned in the real world. Their main instrument had been graph theory, which views networks as a system of interconnected nodes and edges. Graph theory was invented by Swiss mathematician Leonhard Euler in 1736, as a way of solving the Seven Bridges of Königsberg problem: an attempt to find a path through that city which crossed each bridge once, and only once. In order to prove that such a path was impossible, Euler brilliantly generalized the problem, showing that it could be decided mathematically, without reference to the actual geography of Königsberg. But while graph theory remained useful for problems of pure topology (and many other things), it didn't explain many of the behaviours of real-world networks, such as their ability to scale, or their susceptibility to failure.

In the 1990s a new type of science, called network theory, began to emerge from the study of the World Wide Web itself. A team of researchers at the University of Notre Dame, led by the Hungarian-American physicist Albert-László Barabási, mapped the topology of a portion of the Web, and discovered that some nodes, which they called 'hubs', had far more connections than others, but that the distribution of these connections was consistent across the whole network. The Web, they reported, was 'scale-free', meaning that the network could continue to function regardless of the number of the nodes, and as it grew, its underlying structure remained the same. Scale-free networks can have some nodes with many connections, and many nodes with only a few connections, and still function efficiently. In this, it seemed to resemble many of the networks which occur naturally in the world, including economic systems, citations between academic papers, biochemical reactions inside organisms, and the spread of diseases. 'While entirely of human design,' wrote Barabási, 'the emerging network appears to have more in common with a cell or an ecological system than with a Swiss watch.'[14]

Network theory has since transformed the way we understand the world. It appears in epidemiology, biology, astrophysics and economics. It has revolutionized how we think about relationships between people as much as relationships between plants and fungi. It has made possible new understandings of the human brain, and of artificial neural networks. It has laid the groundwork for our recognition of the Wood Wide Web.

This is not to say that the mycorrhizal network itself is really like the internet. Mycelium (the connecting threads of the network) are not mere transmission cables, but active contributors to the whole, with their own decision-making and processing capabilities. Fungi and plants are not simple stores or servers; they too are individual life forms with their own disposition and agency. In thinking about the strange and wondrous life of other beings, we must be careful not to fall back into such reductive, anthropocentric models. Nevertheless, the power of network theory, and perhaps more importantly the idea of universal, scale-free networks, is that it allows us to appreciate these forms in ways we were incapable of doing before. This is a gift from the technological to the ecological: a way of seeing and thinking the natural world which emerges from the things we have crafted for ourselves.

This seems to reinforce our idea of Artificial Intelligence as a kind of guide to understanding the more-than-human intelligences which surround us. As we have seen, animals and plants are capable of 'doing' intelligence in all kinds of interesting ways. Indeed, they do many 'intelligent' things better than we do: chimpanzees can remember longer sequences of briefly observed numbers, for example, while plant roots and slime moulds can figure out the shortest path to food much more efficiently than we can. And we are rapidly discovering that Artificial Intelligence also outperforms human abilities in many important, if narrowly defined, areas.

But what if the meaning of AI is not to be found in the way it competes with, supersedes or supplants us? What if, like the emergence of network theory, its purpose is to open our eyes and minds to the reality of intelligence as something doable in all kinds of fantastic ways, many of them beyond our own rational understanding?

Just as the advent of networking technologies from the 1960s onwards allowed us to perceive life in new ways, and to open ourselves to new relationships and new modes of being, perhaps the advent of intelligent technologies will allow us to perceive the rest of the thinking, acting and being world in ways that are more interesting, more just and more broadly mutually beneficial. This is one ecological lesson we can take from technology: we exist, not alone at the top of the tree, but as one of many flowers which bloom in an endlessly proliferating, entangled and cross-fertilizing thicket.

For a long time we have been unheeding of the more-than-human intelligences which surround us, as we have been deaf to the frequency of electrons, and blind to the ultraviolet light that soaks the plants around us. But these intelligences have been here all along, and are becoming undeniable, just at the moment when the new-found sophistication of our own technologies threatens to supersede us. A new Copernican trauma looms, wherein we find ourselves standing upon a ruined planet, not smart enough to save ourselves, and no longer by any stretch of the imagination the smartest living things around. Our very survival depends upon our ability to make a new compact with the more-than-human world, one which views the intelligence, the innate being, of all things – animal, vegetable and machine – not as another indication of our own superiority, but as an intimation of our ultimate interdependence, and as an urgent call to humility and care.

# 3
# The Thicket of Life

I was an inconsistent student of the sciences in school. I enjoyed biology, but it didn't seem to be going anywhere – just roaming around the fields, poking at leaves, and occasionally opening up a rat (sorry, rat). Chemistry was awful: it just got more and more complex and incomprehensible as we got deeper into it. But physics seemed to going somewhere. It felt like we were ascending rather than descending; as though the universe was slowly cohering from a mess of wires, weights and light beams into a kind of unified theory, whose axioms actually made sense of the world around me. Peering into a desktop cloud chamber and watching alpha and beta particles decay into delicate little contrails, I felt I understood something fundamental and certain about the composition of the world.[1]

Then I learned about quantum physics. Or rather, like pretty much everyone else, I heard about it, but had no real idea what it entailed. I read a little, I pretended to understand more, but I failed to accommodate it within my view of the world in any meaningful way. So it goes.

Until a few years ago, when I heard the American feminist theorist Karen Barad speak. The occasion was a symposium on art and the atomic age. Various artists and researchers gave talks about photographing atomic explosions; about nuclear test sites and indigenous land rights; about drone warfare and surveillance. I spoke on the history of anti-nuclear protest and personal digital data as a kind of radioactive waste, which seeps into the groundwater and is toxic on contact. And at the end of the day, Barad took the podium.[2]

Barad is an extraordinary thinker. A theoretical physicist by training, she is at ease with the complexity of quantum physics. She also believes in putting it to work, and has the skills and generosity to do so. Barad describes a universe comprised of phenomena which emerge, not from bodies themselves, but from what she calls their intra-actions. Intra-action is a particularly Baradian term – her writing is full of charismatic slashes/between/words which signify their simultaneous separation and unity – and it is firmly grounded in a way of thinking about the world informed by quantum theory. Intra-action is not interaction, where two phenomena or bodies, already in existence, act on one another. Rather, it is the process of becoming-phenomena and becoming-bodies, which takes place when the two touch one another. For Barad, the entire universe is a continual process of emergence, in which nothing is certain or fixed, but is always becoming itself through its intra-action with everything else.

One way Barad explores this concept is through reference to the famous double-slit experiment, which demonstrates the seeming paradox of wave-particle duality characteristic of the quantum world. In this experiment, electrons are fired through two narrow slits and produce, not a scatter of single marks, as one would expect of discrete particles, but a continuous interference pattern, like a wave washing along a wall. By this method the 'true' quantum nature of the electron is revealed: not as a fixed, inviolable entity, but as a state of possibility, a scintillating intra-action. What we take to be the material world, then, is actually the constant intra-action of these particles, of the electron touching everything else – including, as Barad gleefully exclaimed in her talk, 'touching itself!'

One misapprehension about quantum theory is that it applies only to very small, almost invisible things – but, Barad insisted, this is not the case. The quantum field which permeates everything isn't something tiny, something beneath everything else; rather, it is behind, in front, around, and entangled with the universe at every scale, right up to and exceeding our own bodies. We live in this shifting, vibrating, scintillating field, even if we are not conscious of it.[3]

As Barad spoke, I experienced something extraordinary. For a few brief moments, as her words hung in the air above us in the darkened auditorium, I understood quantum physics. And when she stopped speaking, that understanding was gone. It was a product of intra-action too. I was left with the sense that while I would never really understand quantum physics, I could be certain of its truth and existence, and that its truth subsisted in this relationship: between the macro and the micro, the world and the subject, the story and the storyteller, the electron and its interference pattern.

Barad's talk also left me with another impression: that science's greatest advances arrive not as settlements or conclusions, but as revelations of a still-deeper complexity. This complexity exceeds our mastery and comprehension – but it is still relatable, still liveable, still communicable and actionable. Science, it struck me then, is a guide to thinking, not a thought: an endless process of becoming.

It's this realization that I hold with me when I try to understand what it means to live in the more-than-human world, because the more-than-human world is messy. It's complex, uneven, entangled and lacking in clear breaks, borders or divisions. And it has always been this way.

One of the places where life began is to be found 50 kilometres inland from the Red Sea, and 100 metres below sea level, in the Danakil Depression. This bone-dry region of Ethiopia is, today, one of the most inhospitable places on Earth. The Depression is formed by the triple junction of three chunks of the planet's mobile crust: the point where the Indian, Nubian and Somali tectonic plates meet. Once a single slab, the three plates started to separate 60 to 100 million years ago in the Late Cretaceous period, the time of dinosaurs and the first flowering plants, before the last mass extinction event. They are still moving slowly apart at a rate of one or two centimetres per year, causing earthquakes, volcanic eruptions and rifts in the Earth's surface. As the ground buckles, parts of it collapse, making this the lowest point on the African continent.

The Depression is also one of the hottest places on Earth, with average daily temperatures of 35 degrees, and only a few centimetres of rain each year. The craters of Erte Ale, the 'smoking mountain' on the edge of the depression, are filled with lakes of molten lava. In the area called Dallol, meaning 'disintegration' in the Afar language, hot springs carry superheated saltwater and minerals to the surface, creating otherworldly scenes: the land is neon green and yellow; smoke billows from sulphur chimneys; iron-streaked stalactites belch green, chlorinated steam; and the ground hisses and crackles underfoot. In places, salt 'bombs' detonate under pressure from erupting gases. The waters in some of its lakes are pure sulphuric acid, with a pH of zero, the highest naturally occurring levels of acidity on earth.

The Depression is a landscape so strange and harsh that it has served as a model for other planets: a Mars analogue, where the conditions of extraterrestrial life can be simulated and assessed. In 2017, a team of astrobiologists – specialists in the study of life elsewhere in the universe – trekked out to Dallol to study the ways in which it might resemble alien environments. If life can survive in the salty, super-heated, acid pools of the Depression, then it might have a chance in the freezing sulphate deposits of Mars, or the caustic gas clouds of Jupiter. To do such research requires not only gas masks to protect against chlorine and other hazards, but also a military guard: the Depression sits close to Ethiopia's border with Eritrea, a war zone for more than two decades, and is home to separatist Afar rebels. The shores of the lakes are littered with dead birds and insects, killed by carbon dioxide which spills invisibly across the desert surface.[4]

For a while, and to great excitement, scientists thought they'd found microscopic organisms in the 'polyextreme' pools – places where intense acidity, salinity, heat and heavy metals combined in a seemingly inhospitable soup.[5] But further tests revealed tiny particles to be biomorphic imposters: nanoscale grains of silicon which resembled clusters of cells. Researchers took this as fair warning to be cautious when they encountered similar patterns in the sample trays of interplanetary rovers.[6]

Further investigation, however, revealed that there was no shortage of life in other, surrounding pools, which teemed with strange and novel creatures. Halobacteria and Archaea, families of single-celled organisms without nuclei or cell walls, lived suspended in the salty waters. Among these weird critters were Acidophilic Thermoplasmata, capable of enduring pH levels as low as 0.06; Archaeoglobi, which consume sulphur, iron and hydrogen for food; methane-ingesting bacteria; and other organisms found only in the depths of airless bogs, or the dark, deep ocean. Places which kill most animals in minutes or seconds provide rich breeding grounds for stranger forms of life.

In this, Dallol is similar to other extreme environments where life has only recently been identified. Lithotrophic bacteria ('eaters of stone') thrive in the acidic, iron-rich waters of the Rio Tinto river in Spain. On the inner rims of Andean volcanoes, microbial communities survive kilometres above the tree line, in the presence of heat, gases and levels of solar radiation that annihilate plant life. Clustered around hydrothermal vents at the bottom of the Pacific, bacteria have evolved to generate energy from methane, sulphur and hydrogen in the absence of light or air – and in turn provide islands of sustenance for metre-long tube worms, clams and shrimp.[7]

That we consider the Danakil Depression to be the closest thing we can experience to an alien landscape on Earth should strike us as strange, because it is also where we come from. It was here, in 1974, that archaeologists turned up the bones of Dinkinesh, one of the oldest humanoid skeletons ever found. Dinkinesh – Amharic for 'you are wonderful' – was an *Australopithecus afarensis*, an extinct hominid who lived around 3 million years ago. (She is better known in Europe and America as Lucy, after the Beatles song playing on repeat in the camp at the time of her discovery.)

We still don't understand very much about *Australopithecus*. Possibly bipedal, perhaps arboreal, a little over a metre tall and covered in hair, Dinkinesh comes from a time when there were multiple hominid species and sub-species ranging across the recently deforested African plains, all of them – from our perspective – competing for the title of human ancestor. Our closest living relative, the chimpanzee, flirts with this classification too. Whether chimps belong to *Hominini*, the taxonomic tribe which includes us and Dinkinesh, remains an unsettled question. It seems that there was a long period in which chimpanzees and early hominids – some of them our ancestors – continued to hybridize; that is, there was no clean species break, but rather a gradual and drawn-out speciation process, lasting as much as 4 million years, in which populations mingled and exchanged genes.[8] In fact, it seems likely that such 'messy speciation' was the way most apes (and other species) evolved, and goes some way to explaining the proliferation of possible human ancestors still being discovered across Africa and Eurasia.

Some of these ancestors are known only through our own genes. In 2020, researchers announced that samples of genomes from west African populations contained evidence of a 'ghost population' of ancient humans living half a million years ago.[9] Up to a fifth of the sampled DNA came from an unknown source, one not present in the majority of modern humans, suggesting the existence of unknown relatives in the gene pool. Who these ancestors were we cannot say, as we have no fossils to represent them, but they exist within some of us just as surely as do traces of Dinkinesh and other human precursors. In fact, to think of them as precursors is to fall into the same trap as we did with intelligence: to consider the end of the current branch to be the highest expression of the tree. For most of our history, we lived alongside other human species. We even had sex with them.

In a limestone quarry in western Germany in 1856, workers dug up an unusual skull. It was elongated and almost chinless, and the bones found scattered around it were thick and oddly shaped. Treated as refuse, their significance was only recognized when a local businessman, thinking that they belonged to a cave bear, passed them on to a neighbouring fossil collector. He declared them to be relics of a 'primitive member of our race'.

The bones emerged just three years before Darwin published *On the Origin of Species*, and the fragments were swept up into the first debates about evolution. Critics pooh-poohed the idea of a primitive ancestor, arguing that the bones belonged to some foreign interloper – a Cossack, maybe, his legs bent from rickets or too many days in the saddle, his brow furrowed with pain as a consequence. But in 1865 an English geologist named William King published a paper claiming something more radical: that the bones came not from an ancient human, nor a foreign one, but rather another human species. King named this species after the valley in which the bones were found: Neander or, in German, *das Neandertal*.

Any challenge to our sense of human uniqueness tends to be met with fear and repugnance, often manifesting as racism and speciesism. No surprise then that William King compared his image of the Neanderthal to what he called 'savage' races – meaning Africans and Aboriginal Australians – and argued that the prominent brow of his specimen was evidence of 'darkness' and stupidity. 'The thoughts and desires which once dwelt within it', he wrote, 'never soared beyond those of a brute.'[10] Other scientists – and the general public – were more than happy to run with this racist caricature, which persisted into the twentieth century in the figure of the dumb, unskilled and brutish caveman. In 1908 the unearthing of another skeleton, found almost complete in a cave near La Chapelle-aux-Saints in southern France, reinforced the stereotype. The palaeontologist Marcellin Boule, of the National Museum of Natural History in Paris, described it as a hunched and 'distinctly simian' figure, while the shape of its skull indicated 'the predominance of functions of a purely vegetative or bestial kind'.[11]

This cartoon image of a caveman bears little resemblance to the picture of the Neanderthals as we understand them today. A new analysis of the La Chapelle skeleton in the 1950s showed that this particular individual suffered from arthritis and that, instead of being slouched and bow-legged, most Neanderthals walked upright. 'Given a shave and a new suit', wrote the anatomists William L. Straus and Alexander Cave, a Neanderthal on the New York subway would probably occasion no more comment 'than some of its other denizens'.

We know now that Neanderthals lived in Europe long before modern humans arrived, and that they comprise an entirely separate fossil record. The Neanderthal presence in Europe dates back at least 130,000 years, and only tails off around 40,000 years ago, soon after signs appear of modern humans arriving after a long and circuitous journey from Africa. And increasingly we are starting to see evidence that not only belies the image of the brutish caveman, but which provides a window onto a complex and fertile culture.

In 1995, a fragment of pierced and whittled bone was discovered in a cave in the Divje Babe archaeological park in Slovenia. Musicologists soon noticed that the spacing of the holes in the bone corresponded to the diatonic scale: do, re, mi, fa. On YouTube, you can listen as Ljuben Dimkaroski, a trumpeter with the Ljubljana Opera Orchestra, plays a replica of the flute among the display cases of Slovenia's National Museum.[12] Although the tune he plays is a relatively modern one – a baroque adagio – the sound the bone flute makes is timeless. It was made from the femur of a cave bear, and has been dated to some 55,000 years ago. This makes it not only the oldest musical instrument in the world, but a pre-human artefact: a Neanderthal instrument.

The Divje Babe flute.

Fresh claims have also been made for several sets of cave paintings discovered on the north coast of Spain. Until recently, the oldest recognized artwork in Europe was a stippled red disk in a cave named El Castillo, in the Spanish region of Cantabria: it was dated to around 40,000 years ago, on the boundary between human and Neanderthal occupation. Some years ago, researchers attributed the disk and other marks – including a hand stencil – to the earliest modern humans, but left open the possibility of other authors.[13] In 2018, however, the same researchers published their new analysis of the layers of crystal rock which, they found, had formed over previously hidden images: the ghosts of handprints and ladder-shaped markings which dated back as far as 67,000 years, well into Neanderthal time – but which were mixed among many later etchings to such an extent that many other scientists have so far refused to accept their findings.[14]

If the provenance of the flute is uncertain, and the paintings' authorship disputed, other evidence of Neanderthal inventiveness is available. In 1990, a fifteen-year-old boy named Bruno Kowalsczewski cleared a narrow passage into a cave behind a rockfall in the Aveyron Valley, a hundred kilometres south of La Chapelle. When members of the local caving club squeezed through the gap Bruno had made, they were possibly the first people in tens of thousands of years to enter what became known as the Bruniquel cave.[15] What they found, some 300 metres into the earth, was a vast chamber in which more than 400 stalagmites had been broken off and arranged into two circles, one around five metres in diameter, another around two. More stalagmites had been stacked into piles, and around the circles there were traces of fire and burned bones. It was immediately obvious that this was not a natural formation, or the work of cave bears, but a deliberate construction. Carbon dating put the age of the bone fragments at 47,600 years – older than any cave painting, and well into Neanderthal time. But this is close to the limit of measurement: carbon dating is only accurate to 50,000 years, because older samples lack sufficient carbon to be properly dated. As a result the dating was disputed: the site, some objected, might have been contaminated, or simply been the result of later human occupation.

Another method of analysis produced a much more startling result. Stalactites and stalagmites are microcosms of geological processes. They accrete slowly but inexorably, layer by layer, over millennia, as mineral-rich waters seep down through fissures in the rock and drip from cave ceilings onto the floor. They trap and reveal the state of the atmosphere at the time of each layer, and so can be compared to other sources of ancient climate data, such as ice cores from Greenland. This allows them to be dated with great accuracy, and far further back in time than carbon dating permits. Scientists realized that, by looking for discontinuities in the stalactites' growth, they could see exactly when stalactites had been broken off during the cave's first use. Their analysis revealed that the chamber was constructed not 47,600, but 176,500 years ago: more than 100,000 years before the ancestors of modern Europeans arrived on the scene. What survives in the Bruniquel cave is pre-human architecture: unequivocal evidence of another, clearly intelligent, hominid species, which long preceded us.

Bruniquel is extraordinary not just for its age and its significance, but because we know of nothing else like it. The oldest forms we recognize as architecture existing anywhere else on the planet are a mere 20,000 years old, dating from the first known human settlements. The gulf of time between even those early humans and the inhabitants of Bruniquel seems uncrossably vast. Certainly, scientific measurement alone cannot tell us what this space was used for. But science is not all we have to go on. Anyone who has spent time at ancient sites is capable of imagining themselves within them, if we are prepared to accept that we have some kinship with the original occupants – for the simple reason that we share a world and thus have certain desires and capabilities in common.

A couple of summers ago I visited the spectacular hilltop site of Göbekli Tepe in southern Turkey, close to the Syrian border. First discovered in the 1990s, the site, when excavated, revealed several layers of inhabitation, culminating – at its deepest level – in a vast temple complex of interlinked rooms, hearths and the world's oldest known megaliths: carved stones some six metres high. Exposed for the first time in millennia, its walls and columns bear astonishingly sharp bas reliefs of animals and humanoid figures. The largest of these stones have recognizably human arms and great T-shaped heads; foxes, geese and bulls cavort across the pillars. Many of the images are clearer and easier to interpret than the statues and murals in medieval cathedrals.

Göbekli Tepe is immediately recognizable as a temple: a diagnosis backed up by the discovery of funerary niches, the bones of animals, the scattering of feast-quantities of seeds and other detritus and, most recently, of carved skulls.[16] Anthropologists have identified numerous

The megaliths of Göbekli Tepe.

similar 'skull cults', based on the special treatment and veneration of skulls, across southern Anatolia and the Levant. But do you need to know that in order to understand, intuitively, the significance of these carvings, the sacredness of these standing stones? No visitor does. To walk in and around the complex is to realize immediately that this was a sacred site, and to experience the absolute possibility of connection across deep time – to people very different to ourselves, and yet immediately recognizable.

The discovery of Göbekli Tepe, like that of Bruniquel, overturns everything we thought we knew about prehistory. Until this point, it was believed that early humans – pre-agricultural hunter-gatherer bands – had no significant architecture, little complex social organization and even less culture. The official timeline held that only after the development of agriculture around 10,000 years ago, and the accumulation of wealth and hierarchy that followed it, did we acquire the skills, time or inclination to construct such edifices. But Göbekli Tepe is at least 12,000 years old: the temple predates the city. Complex culture, architecture and spiritual beliefs are not as modern as we have believed them to be. We know this scientifically because we have dug up the temples, found the stones and dated the rings of stalactites, but we might also know it through a more-than-human sensitivity, which neither supersedes, nor is invalidated by, scientific understanding. Our archaeological beliefs and prejudices are artefacts of material persistence and techniques of analysis; but we have other ways of seeing we can bring to bear as we try to comprehend the awesome discovery at Göbekli Tepe. We have our own experience of living in the world, a world we share with our fellow human ancestors, archaic humans and other-than-human beings.

If it seems harder to imagine the world – and worldview – of Neanderthal people in the same way as we imagine a medieval monk in their cell, or a hunter-gatherer dancing beneath the stars at Göbekli Tepe, then it shouldn't be. We share more than we thought with them, and an appreciation of night, fire, darkness and ritual is not the only thing we have in common.

Arriving in Europe hundreds of thousands of years before *Homo sapiens*, Neanderthals were the first humans to make their homes on the continent, reaching as far north as Denmark and southern England,

and as far east as central Asia and Siberia. Long before the arrival of modern humans, Neanderthals had a complex culture, living in small family groups, and making decorative items and musical instruments, as well as stone tools and birch bark tar for hafting arrows and axes. They buried their dead, used ochre and other pigments, plucked birds for their feathers, and had all the right organs for speaking and singing. The reasons behind their disappearance around 40,000 years ago remain unclear, but probably had a lot to do with the climate and competition from newer arrivals: us. By that point, however, our two species had coexisted for some 30,000 years. This, and the fact of Neanderthals' genetic persistence within our own bodies to this day, suggests a larger point about our relationship with the more-than-human world.

In 1997, a team of scientists at the University of Munich led by Swedish palaeogeneticist Svante Pääbo extracted the first legible Neanderthal DNA from the upper arm of that first skeleton from the Neander valley. The technology they used was very new, the rate of analysis was very slow, and the sample they had to work with was a mere fragment, which had to be mixed in with bones from Croatia. They expected it would take years to sequence a complete genome, not least because the Human Genome Project took over a decade with the entire living human population as its source.[17] But in less than a year of analysis they noticed something strange about the Neanderthal DNA. As expected, some of it was similar to human DNA – we come, after all, from common ancestors. But a significant proportion of it was common only to *some* humans, specifically those of European and Asian descent, and was much less common in people of African descent.

This result was so surprising that it was initially considered a mistake: a contamination in the material or a mix-up in the data. This confusion resulted from the belief that modern humans are all of direct African descent. According to this view, the Neanderthals are a wholly separate limb of the tree, one which split from our branch around 550,000 years ago, pushed off to Europe, and died out after we eventually followed them there. Any DNA we share with Neanderthals should have coexisted in our common African ancestors. Yet the DNA evidence showed, unequivocally, that there was much more intimate contact between *Homo sapiens* and Neanderthals long after that. Only one conclusion is possible: before the Neanderthals died out, *Homo sapiens* had sex with them, and their children peopled Europe, Asia and the Americas. We are the result, not of the linear descent of fixed immutable species, but of intermingling and interbreeding. The branches of the tree of life are intertwined with one another.

Svante Pääbo's team also discovered something else. When in 2014 they finally sequenced the entirety of the Neanderthal genome, they did so thanks to the well-preserved remains of a big toe belonging to a Neanderthal who lived 120,000 years ago in the Altai Mountains in Siberia.[18] But the cave it was found in, called Denisova, turned up a new mystery: another, previously unknown archaic human species. They were given the name Denisovans.

At the time of writing, every known trace of the Denisovans could be held in the palm of your hand: a little finger, three teeth, some bone shards from the floor of the Denisova Cave, and a partial jaw bone discovered in Tibet. But DNA analysis has come a long way, and we can determine a lot just from these fragments. Much like Neanderthals, the Denisovans had a long, broad and projecting face, a large nose, a sloping forehead, a protruding jaw, and a wide chest and hips. They carried genes which, in modern humans, are associated with dark skin, brown hair and brown eyes. And, like the Neanderthals, they had sex with our ancestors.

Modern inhabitants of Papua New Guinea and other Pacific islands share around 5 per cent of their genes with Denisovan predecessors, while East and South Asians share around 0.2 per cent. For a while, we thought this meant that the Denisovans had migrated northwards into Asia, but it seems more likely to be evidence of at least two distinct waves of migration – and two distinct periods in which humans and Denisovans were entangled, socially and sexually, with one another.[19]

It is this mingling together of human and archaic populations that made us who we are. Rather than being products of individualistic, direct evolution along neat branches of the evolutionary tree, we are the result of millennia of close – extremely close – association between different branches. We are made from our entanglements with others.

Neanderthal genes gave modern Europeans and Asians thicker hair and nails: a useful adaptation for lower temperatures at northern latitudes. Denisovan bequests include a particular adaptation to high altitude in Tibetan peoples, and a unique pattern of fat distribution in the Inuit people of Greenland. Neanderthals and Denisovans were pioneers in what were once considered extreme environments: their adaptations, their genetic legacy, helped our ancestors to outlive them. And it doesn't stop there. In recent years, we've found evidence of other 'ghost' populations in our DNA: those in West Africa, as well as suggestions of an inward migration to Africa from the Middle East. Everything we thought we knew about human descent is back up in the air.[20]

From this complex, entangled landscape, we can occasionally pick out single figures who challenge all our preconceptions. One such person is Denny, a teenage girl who lived in the Altai Mountains around 90,000 years ago. All we know of her is the story told by a single fragment of bone, just a few centimetres long, found in the Denisova Cave, which we know was inhabited at different periods by people identifiably human, Neanderthal and Denisovan. Or, much more than that, a combination of those peoples, because Denny is a hybrid, a bi-species child: the first generation offspring of a Neanderthal mother and a Denisovan father.[21] To top it all, Denny's mother was more closely related to Neanderthals living 55,000 years ago in Croatia than to those from 30,000 years earlier living in the same cave. Archaic humans were mobile as well as promiscuous. The genetic relationship between the fragments was so close, and the evidence so clear, that Pääbo said he felt as if 'We'd almost caught these people in the act.'[22]

What if we had done so? How would it make us feel? After all the technical wonder of the genetic evidence, these couplings were real affairs between real people: the miracle of Denny's identification is preceded by the miracle of two individuals coming together. Yet quite how they did so remains unknown: were Denisovan/Neanderthal – and indeed Neanderthal/human and Denisovan/human – pairings consensual and loving, or brutal and capricious, enforced by circumstance?

When the first modern humans to leave Africa met Neanderthals and Denisovans in the caves of the Atlantic coast, or the high mountains of Siberia, had they encountered anyone so like and unlike themselves before? Perhaps they had, in the form of the multiple ghost species we know existed elsewhere, but the experience must have been uncanny. Anyone who has come face to face with a great ape in a zoo or in the wild can attest to the deep strangeness of recognizing in the eyes of another species the expression of awareness, intelligence and even kinship. How much stronger must this have been in the eyes of the Neanderthals and the Denisovans: people both culturally and physically different to us, yet sharing our caves and campfires. And as a result, did we treat them more like colonial subjects, as subhumans, slaves or worse, the way their much later European descendants treated pretty much everyone else? Or did we exercise compassion, empathy, love and solidarity, in the glacial winters of the Pleistocene?

The genetic record does not yet show indications of the direction of genetic flow; that is, whether pairings favoured one gender – of male humans and Neanderthal/Denisovan women, for example, or vice versa, or some equilibrium. Such evidence might show us to be – and encourage us to remain – more open-minded. Or not. Ultimately, our conclusions can only be based on our understanding of our existing relationships, and our own capacities for empathy and imagination.

Many scientific studies support the claim that what we call love exists outside of human relationships. Some show elevated levels of oxytocin in chimpanzees when they share food with strangers and relatives: this is the same hormone that bonds human couples, and parents to their children. Others show that bonobos prefer sex to conflict, and that gorillas care for their own and others' infants – and that this capacity for care correlates with higher reproduction rates.[23] All this is fascinating and important, but what matters even more is that we can imagine this capacity for care in ourselves and others: in archaic humans, in other primates, and ultimately in the world at large.

Our urge to dominate is not necessarily ancient; indeed, it varies widely across modern populations and is as often a product of learning and 'high' culture as evidence of the lack of those. There's no reason to think that we could not have loved and been loved by these close relatives. If we can imagine ourselves ensconced in the stalactite cathedrals of the Neanderthals, then why not in their arms? At multiple points in our history, such entanglements have fundamentally altered our own becoming. Whether or not we come from love, we are capable of it. We are made with others.

Here's a brief list of archaic human species, many of whom we know only from a few shards of bone and from traces of 'ghost' populations buried deep in our own genes. Alongside *Homo sapiens*, *neanderthalensis* and *denisova*, in the 'anatomically modern human' class, there's *rhodesiensis* from Zambia; *idaltu*, based on three skulls from the Afar Depression, discovered in 1997, and dated to 160,000 years ago; and the better known *heidelbergensis*, who also made it to Europe and was the first hominid without simian air sacs in the larynx, suggesting a possible shift from grunts and bellows to human speech. On an island in Indonesia there's *floresiensis*, the 'little lady of Flores', who stood just a metre tall, a likely example of island dwarfism, when population members shrink in response to scarce resources. There's *cepranensis*, or Ceprano Man, from Italy; *antecessor*, the first *Homo* found in Europe, unfortunately accused of cannibalism;[24] *ergaster*, 'the workman', who perfected the Acheulean hand axe, the first advanced technology; and *naledi*, whose extended fingers and toes suggest they still spent plenty of time in the trees. This is what used to be called 'missing link' territory, where we find plenty more *Homo habilis* and *Homo erectus* examples: Java Man, Lantian Man, Nanjing Man, Peking Man, Solo Man, Tautavel Man (all classed as 'man' despite the fact many skeletons and individuals were actually women), each a unique derivation of the human type. There's *Meganthropus*, named for having a jaw – and presumably other parts – the size of a gorilla's; and *gautengensis*, a mostly vegetarian hominin found in the 'Cradle of Humanity' caves in South Africa. Further back – but after the point where hominins split from chimpanzees (if you still believe in such rigid divisions) – there's *Australopithecus*, aka Dinkinesh/Lucy, with many distinct subspecies, including *garhi*, *africanus*, *sediba*, *afarensis*, *anamensis*, *bahrelghazali*, *deyiremeda*; the *Paranthropus* or 'robust australopithecines', who preferred polygamous harems, like modern gorillas; and *Ardipithecus kadabba* and *ramidus* – distinctly chimp-like figures who appear to have self-domesticated, possibly becoming more monogamous and less fighty in the same manner as bonobos. Every time a new leg bone or skullcap is discovered, the names and distinctions change a little.

It seems we're coming to similar realizations to those that accompany our thinking on intelligence. The first is that the notion of 'species' as something easily delimited and described is entirely baseless. It is founded on the idea that there exist separate, inviolable clades of individuals who are exclusively fertile: that is, only by having sex with other members of their clade are they able to produce viable offspring. The existence of Denny, her parents, and the whole weight of the multispecies human genome, defies this idea. If our hairy threesome of Sapiens, Neanderthals and Denisovans could produce viable, fertile offspring, then our whole structure of divisions starts to teeter. Not only do we share our lineage by entanglement with other forms of beings, but those entanglements include non-humans as well as humans.

A second realization is that any attempt to pinpoint exclusive qualities among different 'species', and to delineate them with specific breaks in the evolutionary tree – as scientists once tried to do, repeatedly, with qualities such as self-recognition – is entirely flawed. There exist multiple forms of being and doing, of living and thriving, strung across and interwoven between the many branches of the tree, or thicket, of life.

These realizations are made possible by a combination of advanced technological ingenuity – the development, over decades, of extraordinarily sophisticated DNA sampling and analysis – and our own capacity for imagining relationships beyond the human, and beyond human uniqueness. This is technological ecology at its highest pitch: the combination of technological capacity with a more-than-human sensitivity which constructs new ways of seeing and appreciating the world. It allows us to recognize that everything is hitched to everything else, while simultaneously upending our notions of what technology is *for*.

Historically, scientific progress has been measured by its ability to construct reductive frameworks for the classification of the natural world, the kind of one-size-fits-all schema which came to dominate our thinking in the eighteenth and nineteenth centuries. This perceived advancement of knowledge has involved a long process of abstraction and isolation, of cleaving one thing from another in a constant search for the atomic basis of everything: the single, pure, definitive type, or the one true answer. It is in this image that we have constructed our technologies, right down to the either/or binary of ones and zeros which shape our calculations. And yet, time and time again, the more thoroughly we attempt to perform such abstractions, and the deeper we go into the structure of life itself, the more these distinctions blur and fall apart.

What we perceive as borders and conflicts – the things which separate us – often turn out not to be artefacts of the exterior world, but immeasurable gaps in our own conceptions, abilities and tools of discernment. We think we are studying the world – but in reality we are merely making evident the limits of our own thinking, which are embodied in our logbooks and measuring instruments. The truth is always stranger, more lively and more expansive than anything we can compute.

For me, this paradox is best expressed in the work, not of an evolutionary biologist or palaeogeneticist, but of a meteorologist, Lewis Fry Richardson. Richardson was a scientist, and also a Quaker and a pacifist, beliefs that shaped both his life and work. During the First World War, he was a conscientious objector; but though he refused to fight, he still served on the front lines as an ambulance driver. It was during this time that he formulated some of the first rules of mathematical weather forecasting, which underlie much of what we understand as meteorology today. He did so using paper and pencil, calculating by hand the movement of air masses and the gradients of atmospheric pressure, during the long hours between sorties, sometimes in bombed-out buildings repurposed as billets, and sometimes while sheltering from artillery fire.

After the war, he returned to his role at the British Meteorological Office, but shortly afterwards, when it was absorbed into the Air Ministry, the government department responsible for the Royal Air Force, he resigned on grounds of conscience. As a result, many of his ideas took years to be fully realized. Nonetheless, he continued to seek ways to improve our understanding of the world through science. In particular, he believed it might be possible to employ mathematics, the discipline he'd introduced into meteorology, in the service of pacifism.

In a series of books published late in his life, Richardson sought to discover a mathematical basis for the causes of war and the conditions for peace.[25] One of his ideas was that the propensity for war between two states might be a function of the length of their shared border. In order to prove his hypothesis he needed accurate figures for the lengths of those borders, but found that such figures did not exist: no estimate of any one country's border seemed to match any other. The deeper he dug into the problem, the more elusive, and the more conflicted, such estimates seemed to be.

He subsequently identified a paradox: the more accurately you try to measure some things, the more complex they become. This surprising observation has become known as the Richardson effect. Imagine taking a kilometre-long ruler and measuring the coast of Britain. Now repeat the exercise with a ruler half as long, then half as long again, and so on. On each measurement, the reading would get more accurate, with more and more of the coastline accounted for. But the result, as Richardson realized, was not that the measurement converged on the correct answer but rather, the more closely it was measured, the longer it got. What Richardson had discovered was what the mathematician Benoit Mandelbrot would later term 'fractals': structures which repeat to infinite complexity. Instead of resolving into order and clarity, ever-closer examination reveals only more, and more splendid, detail and variation.[26]

The Richardson effect applies to biology, archaeology, evolution and, it seems, to life itself. As our archaeological and biological tools get better, as we unravel the web of life, the result is not an ordered tree, with measurable branches and clear delineations between forms and types, but a whirling dance of encounters and interrelationships. The species start to fragment and blur; the field, from savannah to tundra and back again, fills up with players. The mud's churned up. The referee can't keep score any more. It's beautiful, this teeming world of ancestors and progeny, this utterly animated free-for-all, this breaking down of boundaries. This is what the close scrutiny enabled by our technology actually reveals: not a rigid map, but a pattern of interference, all the way down to the quantum dance of the energy field behind everything.

Everywhere we look, the same process is at work. Just as in recent decades, our understanding of the real life of archaic human species – their multiple migrations, their overlapping and entwined capabilities, their rubbing of shoulders and other body parts – has widened and complexified, so too has our understanding of life, right down to the cellular level. If you think human ancestry sounds complex, unstable and delightfully interbred, just wait until you hear what our cells have been up to.

In the 1950s, scientists started to look for something they called LUCA – the last universal common ancestor. Initially, they considered LUCA to be the origin of all life, the ancestral organism from which all forms of life descended, but they quickly realized that what they would – and could only – find would be the ancestor of all life living today. Just as there must have been many more ghost populations than the ones that still haunt our DNA – whole other groups of humans who did not, for whatever reason, contribute to our lineage and died out unknown to us – so there were very probably many forms of proto-life which didn't make it into the fossil record or living descendants. Yet our LUCA must still have existed, and probably did so around 4.5 billion years ago, in the Hadean aeon when the Earth was still hot, highly radioactive and subject to constant bombardment by meteorites from outer space. The search for LUCA, which is still ongoing, is biology's attempt to finally nail down a definitive story for life itself, akin to the hunt for the definitive Standard Model of particle physics. It is the search for the One True Answer, which will make all other questions resolvable.

Almost immediately, the search for LUCA threw up all kinds of anomalies, which were initially thought to be errors (much like the discovery of different strains of Neanderthal DNA in different human populations). At the University of Illinois, a microbiologist called Carl Woese thought it might be possible to approach LUCA by unravelling the long sequences of DNA, RNA and proteins inside cells, which he called the 'internal fossil record'. Encoded within these sequences was the whole history of life: mutated, hybridized, added to and padded out over numberless generations. At the end of the 1960s, Woese went looking for the internal fossil record in ribosomes, the then recently discovered particles within all living cells which do the work of translating the blueprints of DNA, carried by RNA messengers, into the actual structure of organisms. The ribosomes are 3D printers for living bodies.

Before the advent of computational DNA sequencing in the 1980s, this unravelling was difficult, partial and incredibly time-consuming.

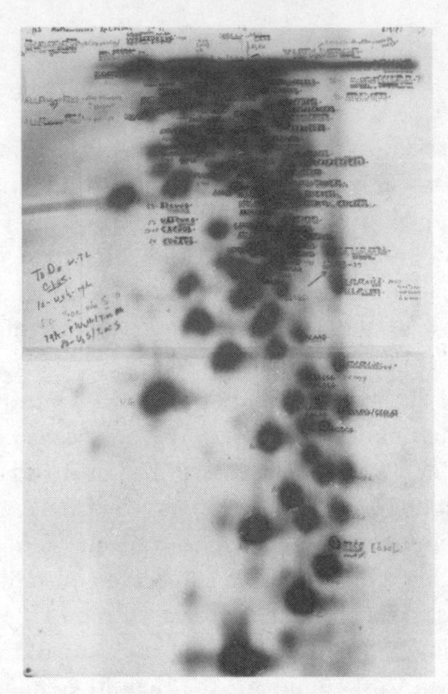

An X-ray 'fingerprint' of ribosomal RNA, annotated by Carl Woese.

Woese and his assistants used X-rays to cast the shadow of RNA molecules onto sensitive film in order to see the 'fingerprint' of each individual sequence. There followed endless hours of annotation and comparison to tease out the similarities and differences between different sequences, and to attempt to match ancestral strands between different molecules. They started to see stark differences between certain molecules and organisms which science had up to that point treated as the same. In particular, what scientists had taken to be whole classes of bacteria in fact seemed to be something else entirely. These Woese named Archaea, meaning 'ancient things'.

These Archaea lacked certain features of bacteria, but possessed other unexpected ones. They had strange properties – among other things, they digested carbon dioxide and excreted methane – and were particularly well adapted to extreme environments (which is where we last met them, a while back, in the hot saline springs of the Danakil Depression). And whatever they were, they didn't fit into the existing two kingdoms of life: bacteria (single-celled organisms without nuclei) on the one hand, and the eukarya (everything else) on the other.

This alone was a revolutionary discovery, and Woese's publication of his findings in 1977 duly caused a sensation.[27] Woese and his assistant George Fox argued for the acknowledgement of a third kingdom, or domain, of life, distinct from both bacteria and eukarya. It was the most significant redrawing of the tree of life since the introduction of the 'primitive forms', or protists, by Ernst Haeckel back in 1866. While Haeckel's protists had subsequently been incorporated into the domain of bacteria, the discovery of the Archaea – which was made as a result of an attempt to simplify the tree – only complicated the picture once again.

The fanfare which greeted this announcement obscured another related discovery, one which Woese and others were to spend the next few decades carefully unpacking, and which is still not widely understood or acknowledged today. This is a phenomenon called Horizontal Gene Transfer, or HGT, and it turns out the Archaea were particularly proficient at it.

Back in 1928, an English medical researcher named Frederick Griffith first observed that the bacteria that caused pneumococcal pneumonia could suddenly and apparently mutate from harmless to lethally virulent. The mechanism by which they did so, however, was unknown. In the 1940s, at the Rockefeller Institute in New York, researchers watched in near-disbelief as DNA seemed to copy itself near-instantaneously from one bacterium to another, rather than passing down the line of inheritance. In 1953, the microbiologists Esther and Joshua Lederberg coined the term 'infective heredity' to describe the transfer of DNA from one organism to another via viral infection; the first time a mechanism for such a transmission had been described and understood. It turned out that viruses were just one of many ways in which genes could spread between rather than down through branches of the tree of life: horizontally, rather than vertically.

The Archaea are particularly fond of this method of spreading their genes around. At certain temperatures, or when exposed to harmful UV light or to acid, or to other forms of damage, Archaean individuals can extend a tiny filament, like a hair, to another individual, and exchange genetic material directly. These threads are called pili, and they are also found in many types of bacteria – particularly those which cause diseases, such as *E. coli*, because they make it easier for the bacteria to bind to the tissues of other bodies. But the Archaea use pili to bind to one another, and to change living organisms into new forms through the transfer of DNA.

Rather than stabilizing the now-wobbly tree of life by settling themselves into a third domain, the Archaea continued to undermine it. As it turns out, they are practically impossible to divide into species; instead, they seem to resemble teeming communities of mostly alike individuals, which radically diverge from each other according to their functions and locations, freely trading parts of their DNA as they rapidly adapt to novel situations. The Archaea constitute another blow to the idea of neatly divisible species. They also go further: they challenge the idea of 'individuals' at all.

The popular and scientific concept of species depends on the existence of unique, classifiable individuals which belong, wholly and exclusively, to their own groups. In turn, the notion of the individual is inextricably tied to the political ideas of sovereignty and agency that define modernity, and to the conceptualization of subject and object, self and other, which are the hallmarks of Western philosophy. To trifle with the concept of the individual is no small thing, yet that is where Archaea lead us.

In December 2012, three of the world's leading biologists published an essay which explored how the interaction between our growing understanding of ecology and the powerful technological tools for interrogating the world had upended our understanding of life. Just as the development of the microscope once revealed to us the previously invisible world of bacteria and fungi, so DNA sequencing and other technologies were revealing a hitherto unknown, and deeply interconnected, landscape at every level of existence. These technologies 'have not only revealed a microbial world of much deeper diversity than previously imagined,' they wrote, 'but also a world of complex and intermingled relationships – not only among microbes, but also between microscopic and macroscopic life'.[28]

The title of this essay was 'A Symbiotic View of Life: We Have Never

Been Individuals', and it sought to summarize several decades of revolutions in the life sciences, in which the discovery of Horizontal Gene Transfer played a transformative role. In organism after organism, family after family, community after community, the authors documented cases of HGT – including in human beings. In fact, it now seems that we ourselves are 'genomic chimeras': a full 50 per cent of the human genome consists of DNA sequences 'written into' it by other organisms at various times in our evolution. Almost 10 per cent is the result of transmission of DNA through viruses. The very cradle of mammalian life, the uterus, appeared independently in multiple mammalian lines as the result of a viral infection. The tail of the human sperm, the wriggly motor that propels it towards the egg, might have originated as a free-living spirochete bacteria. We live because we were infected with, or perhaps we should say shared in, the DNA of others.[29]

The name given to this new view of life, which extends far beyond the operation of HGT, is symbiosis. In opposition to the violent and competitive account of life's emergence given by Darwinian evolution, symbiosis proposes that we are instead the product of cooperation, interaction and mutual dependence. This vision of life is profoundly ecological and relational, and it extends all the way from the sequences of our DNA to the composition of our bodies, to the kind of societies in which we live – or might do, if we take its revelations seriously.

It is symbiosis which underlies the relationship between plants and mycorrhizal fungi which we explored in the last chapter: the growing-together which allows both forms of life to thrive, entangled with and dependent on one another. It was symbiosis that allowed plants, and all other forms of life, to establish purchase on the land, by enclosing fungal partners within their roots. This relationship persists today in the form of lichens, which are a partnership between fungi and algae, working so closely together that their forms are almost indistinguishable. Lichens have been described as 'fungi who discovered agriculture', but this belies the mutualistic association between creatures so attached to one another that their tendrils grow into each other's cells – another blurring between individual and species.

Insects and flowering plants are an example of symbiosis too, one which has irrevocably shaped all life on the planet, as well as the forms of the creatures themselves. Each has driven the evolution of the other, as adaptation to new environments causes them to adopt new patterns of life, which are in turn reflected in their symbiotic partners. And we share in these systems of cooperative development; as we will see in a later chapter, we too may have evolved the way we have thanks to the availability of honey, a direct result of insect–plant symbiosis.

We can even identify 'ghost populations' of animals in the relics of past symbioses. The avocado may well have evolved its large and apparently inedible seed in response to the presence of giant animals capable of eating and passing them: the ground sloths and elephantine gomphotheres of the Pliocene, over a million years ago. Although these symbiotic partners no longer exist, the avocado has moved on: today, it continues to spread itself through other animals that appreciate its pulp.[30]

Symbiosis is crucial to our daily lives. The human microbiome, the two kilograms or so of bacteria and other organisms that we carry around with us – mostly in our gut – profoundly influences our awareness and behaviour. This applies even to our intelligence: the health or otherwise of our microbiome affects both brain development and our ability to cope with stress and trauma. Studies have shown that elderly people living in care homes have less diverse microbiomes than those living in broader communities, and are as a result more prone to frailty and chronic illness.[31]

We rub along with symbiosis every day, but we also come from it. Along with plants, animals, fungi and most of the life we encounter every day, we are eukaryotes, possessors of complex cells containing nuclei and mitochondria (as distinct from the prokaryotes, the bacteria and Archaea, whose simpler cells contain only single strands of DNA). How did we get here? It has always seemed likely that bacteria and Archaea represent an earlier stage of life, and we must have come from them, in some way. But how?

The answer, first proposed by the biologist Lynn Margulis in the 1960s, is endosymbiosis – which, like many of the revolutionary ideas we've encountered, was long derided and ignored until it became undeniable. Endosymbionts are organisms that live within other organisms: the nitrogen-fixing bacteria in the roots of peas, or the algae inside reef-building corals. What Margulis proposed was that eukaryotic life – our life and the life of all complex organisms, from amoebae to redwood trees – was itself the product of endosymbiosis. At some point in deep evolutionary history, one independent, free-living bacterium engulfed another, and the two settled into a mutually beneficial relationship. This is the origin of mitochondria and chloroplasts: the structures within animal and plant cells which produce the energy needed for our survival. Like the lichen, these bacteria learned to farm one another: to provide shelter and nutrients necessary for the other's survival, in return for energy and other products. Both bacteria benefited from the arrangement, and their mutual relationship passed down from generation to generation, resulting in all of the complex multi-cellular life that shares the planet. They live inside us to this day.

More-than-human doesn't even begin to cover it. Not only are we the products of multiple entangled ancestors, spanning vast ranges of the evolutionary field; we are not even individuals at all. Rather, we are walking assemblages: riotous communities of multi-species, multi-bodied beings, inside and outside of our very cells.

*Frseeeeeeeefronnnng* and we all go tumbling down the genetic line together. It's a delirious image: an endlessly blossoming, weirding, straining desire for life and interconnection. The lichens farm algae and we farm bacteria and each feeds the other, the trees are talking and everyone's singing. We're descended from Typhus on our mother's side, and methane-burping Archaea on the other.[32] Every time we train our most sophisticated tools upon the central questions of our existence – Who are we? Where do we come from? Where are we going? – the answer comes back clearer: Everyone and Everywhere.

In 1999, an American biologist called Ford Doolittle tried to draw a new tree of life, taking into account the latest discoveries in HGT.[33] He called the result a 'reticulated tree' or 'net', although it resembles more a thorn bush or thicket. Eschewing the graph's traditional two-dimensionality, branches bend and shoot out, merge and separate, hug one another closely and poke through holes. Life appears curious and questing, networked, sprawling and particularly inclusive.

'Different genes give different trees,' wrote Doolittle, meaning that genetic evidence could no longer support the model of distinct species, or even kingdoms. 'To save the trees', Doolittle conjectured, 'one might

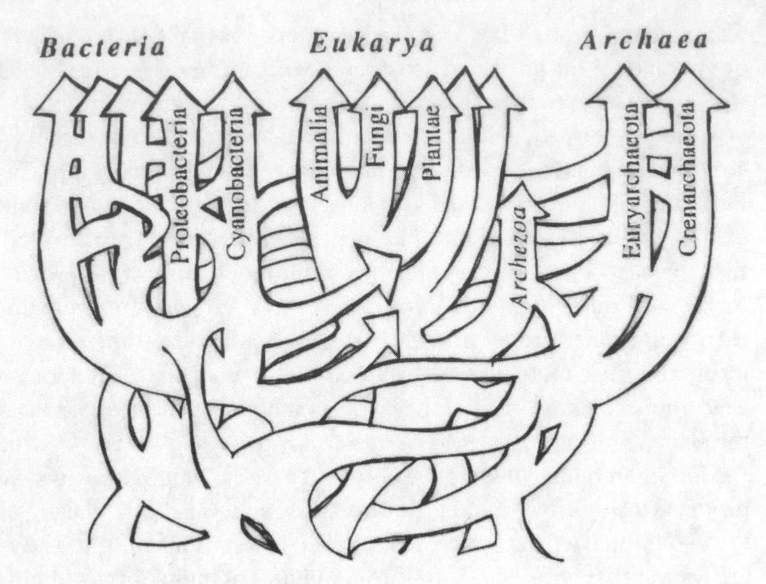


Ford Doolittle's reticulated tree.

define organisms as more than the sums of their genes and imagine organismal lineages to have a sort of emergent reality – just as we think of ourselves as real and continuous over a lifetime, while knowing that we contain very few of the atoms with which we were born.'

In this vision of life as an ongoing process of emergence, he echoes Lynn Margulis, who considered the human individual 'a kind of baroque edifice', reconstructed every couple of decades by fusing and mutating bacteria. Meanwhile, the core of our being, the backbone of our DNA, is far older than any human who ever lived. 'Our strong sense of difference from any other life-form, our sense of species superiority', wrote Margulis, 'is a delusion of grandeur.'[34]

This worldview makes it a lot harder to know where the origins of each individual lie, or where the boundaries exist between us. It does, however, make it a lot easier to find commonalities – things-in-common – with other modes of life.

The nature of life is not constancy, but change. The driver of change, at the level of the individual and at the level of evolution, is the encounter with the other. Contrary to Darwinian notions of hierarchical descent, which posited mutation in individual branches as the driving force behind evolution, symbiosis asserts that change and novelty come from without. We are who we are because of everything else.

Models of progression, advancement, linearity and individuality – models, in short, of hierarchy and dominance – collapse under the weight of actual diversity. Life is soupy, mixed up and tumultuous. Muddying the waters is precisely the point, because it's from such nutritious streams that life grows. The individual, under the microscope or under the sun, is always a plurality. Models of multiplicity are needed to make sense of this endlessly proliferating, teeming, oozing and entangling life. The tree is not a tree, but perhaps a bush, or a net – or a forest, or a lake. Or maybe a cloud?

While mathematical models of networks have proved useful tools for understanding the structures and affordances of artificial and natural webs, from the internet to mycorrhiza, they are no match for metaphors, the actual mental models we carry around in our heads, sometimes only in fragments or sometimes consciously. These are the ideas we actually live by. And there can be no more powerful emergent metaphor than the one we have applied to the global, interconnected sensorium of the internet, the shared *umwelt* that we call the Cloud.

The Cloud embodies and enacts all the conflicts of understanding we encounter in our attempts to understand the more-than-human world. It shows us, every day, that an information regime – a way of thinking and classifying the world – which depends upon fixed data and unbreakable categories, on conclusions over processes, ends over means, on biases and assumptions rather than the evidence of our own lives, is antithetical to society, humankind and life itself. No schema is ever complete, no taxonomy ever finished – and that's fine, providing the systems we put in place for interpreting and applying those schemas are open, transparent, comprehensible and renegotiable.

This isn't how it always is, far from it, but sometimes there are hopeful glimpses. One that I treasure is Facebook's ongoing and well-meaning grappling with gender, which reflects exactly this exasperating confrontation with false binaries. From its inception, the service, as did most other social media platforms, required users to identify as 'Male' or 'Female' when signing up for a profile. Google Plus added a third option: 'Other', which pleased some people but still wasn't exactly welcoming.

At the beginning of 2014, Facebook released a list of fifty-one gender identities that users could select from, including Trans, Agender, Non-Binary, Genderfluid, Two-Spirit, and a whole host more. Later that year, following consultation with users, they increased the list to more than seventy. And a year later, they gave in and allowed anyone to type in their own identity for themselves. Gender, one of the most critically contested identifiers in our society, became, like species, an open field – at least on Facebook (and that's the only nice thing I'll say about it). In information technology, as in evolutionary biology, it seems that the more we try to control and categorize the world – to stuff everything into little boxes and lines in databases – the more of life spills out.[35] Lewis Fry Richardson is looking at his ruler and laughing.

Symbiosis is not a vision of perfect harmony – far from it. The world is not composed of harmonious or even equitable relationships, but it is composed of relationships, and more of those are mutually beneficial than they are antagonistic. 'Life', writes Margulis, 'did not take over the globe by combat, but by networking.'[36]

Thus symbiosis provides one of the most powerful, empirical counterpoints to all our horror stories about the development of opaque technological infrastructures, of systems of analysis, determination and control, of dangerously asymmetrical intelligences and power relationships. It simply doesn't have to be this way and, moreover, it doesn't *want* to be this way. It's possible to live as symbionts rather than subjects, as equally valid and responsible inhabitants of a messy, more-than-human world. Technology, ecology and the evidence of our own bodies tell us so.

# 4
# Seeing Like a Planet

> 'Attention,' a voice began to call, and it was as though an oboe had suddenly become articulate. 'Attention,' it repeated in the same high, nasal monotone. 'Attention.'

So begins Aldous Huxley's 1962 novel *Island*, with the shipwrecked journalist Will Farnaby waking beneath the spreading trees of Pala, a closed and mysterious Polynesian island. Farnaby has come to this place – his wrecking is later revealed to have been deliberate – in order to persuade the local ruler to grant the island's mineral rights to a rapacious partnership consisting of a regional strongman and an international oil company. But the troubled and cynical Farnaby is gradually won over by the Palanese combination of kindness, thoughtfulness and pacifism, which is embodied in the sounds he hears at the moment of his arrival: 'Attention, Attention. Here and now, boys. Attention.'

These words, which echo constantly around the island and through the text, are spoken not by human voices, but by flocks of local mynah birds, trained and set free by the islanders. Because, as the nurse who binds Farnaby's wounds explains, 'that's what you always forget, isn't it? You forget to pay attention to what's happening. And that's the same as not being here and now.' In Huxley's novel, it is the voices of the birds, the more-than-human chorus, which draw the humans' attention back from fantasy and distraction, back to the present, to their surroundings, and to the reality of their lives. For Huxley, conscious attention is a prerequisite for acting correctly and with justice in the world: only by looking and listening carefully might we fully understand what is going on around us, and know how best to proceed.

One morning in April 2020, with the world in lockdown, my partner and I were walking along the beach below our house, on the Greek island of Aegina. Our village is mostly empty in the spring and the effects of lockdown added to a feeling of desolation. The sand was covered in dried seagrass, driftwood, plastics and other refuse. It was invigorating, and bleak.

But, as we walked along the water's edge, the sky came alive: suddenly, it was filled with small birds, swooping and trilling over the tall grasses, flitting around the orange trees, alighting momentarily on the telephone wires and flashing past our heads. They came at us from all angles like dive-bombers, their tight, arrow-shaped bodies snapping away at the last moment, shocking us with their sudden appearance and swift passage. Overnight, the swallows had arrived.

In ancient Greece, the swallows' arrival from Africa marked the beginning of spring, their appearance the occasion for celebrations known as Χελιδονίσματα ('chelidonismata'), from *chelidonia*, Greek for swallows. In modern times, on the island of Rhodes, children would go from house to house asking for gifts for the swallows. And across Greece, the traditional 'Marti' bracelet of threaded red and white strings, which is still woven on the last day of February and worn throughout March, was given to the swallows at the end of the month to build their nests. Further north, in Albania and Macedonia, the feast day of Letnik marked the arrival of the migratory birds. Everywhere, their appearance was associated with the coming prosperity and ease of summer.

This year, the arrival of the swallows was tinged with tragedy. The birds we encountered on the beach, which seemed so vital and alert, were storm-tossed and exhausted. The preceding days had seen high winds over the Mediterranean and the Aegean, which proved a disaster for the migrating birds – and thousands died. The first summer of Covid-19 was also, in many parts of Europe, a summer without swallows.[1]

Nonetheless, the swallows' appearance that April morning had a transformative effect on me, shocking me out of lockdown stasis and giving me a renewed sense of attention and purpose. Having lived in cities all my adult life, I was exposed, for what felt like the first time, to the exuberance of spring: the arrival of the swallows, the blooming of euphorbia and saxifrage, budding anemone and bougainvillea. On one April day, the sea turned yellow-green; what I thought at first was some kind of chemical spill turned out to be a coating of pine pollen – all the trees on the island had ejaculated at once and coated the waters and the roads in fine-grained dust. I felt keenly my previous lack of awareness, and my disconnect from these cycles of nature.

We live mostly at the speed of our media and our machines. Our days are governed by the twenty-four-hour clock and the news cycle; by status updates and emails. But the idea of global time – a single, universal standard, divided into hours and minutes, and shifting across time zones – is a recent one. An invention of the late nineteenth century, it was driven by the development of technology. Prior to the Industrial Revolution, the world got by perfectly well with many and various measurements of time: the farmer and the farrier worked to their own rhythms, and clocks were set to different times even in neighbouring towns. Cultural conceptions of time also differed widely: Western, Judaeo-Christian time was conceived of as linear, with a definite start and end, while Incan and Hindu cosmologies conceived of time as cyclical and infinite. These models of time deeply influenced the awareness of those who lived within them, producing entirely different ways of being and living.

The ways in which the natural world changes over time are also registered in our cultural memory. An organization called the Snowchange Cooperative, based in northern Scandinavia, documents the ways in which indigenous communities read and respond to climate change. A Sámi woman called Gun Aira notes that a lake whose traditional name is Biehtsejávrre, or Lake of the Pines, is now entirely surrounded by birch – a sign of changing environmental conditions.[2] In this way, recognition of environmental change is encoded into indigenous history and knowledge. Most of us do not have access to have such extensive cultural memory banks: we are dependent upon the short attention spans of our contemporary knowledge-making technologies. All too often, this blinds us to deeper, longer changes in the world around us. Climate change is a clear example of this: an alteration to the world occurring at such inhuman scales of time and geography that we struggle to fit it into conventional narratives, let alone respond effectively.

If our inability to tell meaningful, actionable stories about our changing planet is part of the problem, then we need to rethink the tools we use to make culture itself. Technology can be part of this communal, sense-making process. In order for this to happen, we need to stop using it as a way to constrain time and enforce our narrow perspective, and instead deploy it to widen our vision and expand the scope of our attention. This is urgently necessary, for even traditional knowledge systems, from which many of us have already been cut off, cannot always account for the velocity of change which is occurring in the present. Indigenous leaders in Siberia report the recent arrival of sable – usually woodland creatures – on their ancestral tundra lands, a development for which they have no cultural reference point. 'There are no songs about the sable, there are no old stories about the sable,' they lament.[3] We have to make new stories, adapted for the present, which will in turn enable us to adapt them for the future. These stories will, of necessity, involve networks and processors, satellites and digital cameras.

It will not be the first time we have done this, but for it to be successful it will require more conscious effort, and greater equality of access and agency. In the nineteenth century, the coming of the railway, the steamship, the radio and the telegraph – all technologies requiring strict coordination between distant points in order to function – collapsed space and time. This homogenization, however, met with resistance around the world. Following the International Meridian Conference in Washington DC in October 1884, the world was divided into twenty-four time zones, with a single mean time determined by the Royal Observatory in Greenwich. There was widespread opposition to its implementation. In Bombay, factory workers rioted against the imposition of global time, considering it another colonial affront from Britain. In Beirut, bus schedules were printed in both European and Ottoman time, while public clocks fell in and out of sync with church bells and the call of the muezzins. France adopted a national time, but refused to use the Greenwich mean: French time was calculated from Paris and continued to be so until 1911.[4]

Time has always been an imaginary concept, and it matters who imagines it. For most of history, time has been governed by our relationship to the more-than-human world: the rising of the sun and the turning of the seasons. In the late nineteenth and early twentieth centuries, time was separated from the earth, and suborned by industry for its own ends. This was an imperial imagining of time, led by Europe and North America: the result was the erasure of local differences, the oppression of workers and subjects, and the suppression of other ways of being in time. Today, time is imagined by our machines – the ticking of the microprocessor and the oscillating signal of the GPS satellite, which maintains a universal, globalized time code accessible anywhere on Earth and accurate to ten billionths of a second.

In *The Soul of a New Machine*, Tracy Kidder's account of the development of the Data General Eclipse MV/8000, an early microcomputer, in the late 1970s, the author records the experience of one engineer whose work involved testing the new computer for bugs. To do this, he used a device called a logic analyser, which took snapshots of the machine's internal state every nanosecond – every billionth of a second, an almost unimaginable fraction of time. For the engineer, though, such time frames were familiar: 'I feel very comfortable talking in nanoseconds. I sit at one of these analyzers and nanoseconds are wide. I mean, you can see them go by. "Jesus," I say, "that signal takes twelve nanoseconds to get from there to there." Those are real big things to me when I'm building a computer. Yet when I think about it, how much longer it takes to snap your fingers, I've lost track of what a nanosecond really means.'[5]

Later in the process, another of the engineers on the Eclipse project burns out, and leaves the company. He too was working with the logic analyser, peering into the infinitesimal gaps of time within the machine, day after day, week after week. But the effort of doing so had become too much. Before departing, he leaves a note to his colleagues on top of his terminal: 'I'm going to a commune in Vermont and will deal with no unit of time shorter than a season.' He wasn't just going to Vermont; he was going to a place where time was different, where it more closely matched the rhythms of the body and the rhythms of the world.

It matters what time we live in. Not which age, but which present time, the time of our awareness. When it is governed by machines, our attention is forcibly attuned to the scale of the nanosecond and the breadth of a beam of light, and this makes it harder for us to think of and with other beings and processes which exist at different scales of time and geography: the turning of the seasons, the continental migration of birds, the lifespan of trees and plants. But attention is also something which can be turned and trained, which can be brought back, consciously, towards a more-than-human mean time. While we mostly construct and think our technologies in ways inimical to this, it does not have to be this way. Our tools of data gathering, measurement, recording and viewing can also be used to increase our awareness and broaden our capacity for attention and care, if we make conscious choices about the time we want to live in.

Robert Marsham was an English landowner and naturalist who paid close attention to the world around him, and on 14 December 1735, walking on his Norfolk estate of Stratton Strawless, he heard a thrush sing. Song thrushes sing to establish territory, and begin doing so any time from late autumn through into the new year. Hearing the thrush sing on that December day, Marsham began to wonder if there was a connection between the date of its first song and other conditions of the natural world. Making a note of the date, he resolved to pay greater attention.

The following year, Marsham began keeping an ongoing record of the first signs of spring. He began with the first swallow, which in 1736 arrived at Stratton Strawless on 10 April, the 101st day of the year (1736 was a leap year). Two years later, travelling around Europe, he made a note of the first swallow in Piacenza, Italy (20 March: 79 days into the year) as well as the first hawthorn blossom in Nîmes, France (14 April: 104 days). In 1739, back home in Norfolk, he expanded his logbook to include the appearance of the cuckoo (120 days) and the nightingale (126), the first leaves of the sycamore (65), and the sprouting of that plant so crucial plant to the Norfolk farmer, the turnip (63). Twenty years later he was still going strong, observing and carefully noting down snowdrop flowers; oak, birch, chestnut and hornbeam buds; the arrival of migrating nightjars; and the first young rooks.

Marsham died in 1797. The following year's records were inscribed in the hand of his son – also Robert – who continued keeping notes until 1810, appending reports on the weather and the daily temperature. After a hiatus of some fifteen years, the next Robert in line picked

[ 154 ]

XIII. *Indications of Spring, observed by* Robert Marsham, *Esquire, F. R. S. of* Stratton *in* Norfolk. *Latitude* 52° 45′.

Read April 2, 1789.

| Years. | Snowdrop flower. | Thrush sings. | Hawthorn leaf. | Hawthorn flower. | Frogs and Toads croak. | Sycamore leaf. | Birch leaf. | Elm leaf. | Mountain ash leaf. | Oak leaf. | Beech leaf. | Horse-chestnut leaf. | Chestnut leaf. | Hornbeam leaf. | Ash leaf. | Ringdove coos. | Rooks build. | Young Rooks. | Swallows appear. | Cuckoo sings. | Nightingale sings. | Churn Owl sings. | Yellow Butterfly appears. | Turnip in flower. | Lime leaf. | Maple leaf. | Wood Anemone flower. |
|---|---|---|---|---|---|---|---|---|---|---|---|---|---|---|---|---|---|---|---|---|---|---|---|---|---|---|---|
| 1736 | | 1735 Dec. 4. | | | | | | | | | | | | | | | | | March 30. | | | | | | | | |
| 1738 | | | | Nismes, France, Apr. 14. N.S. | | | | | | | | | | | | | | | Florence, Italy, Mar. 20, N.S. | | | | | | | | |
| 1739 | | | Feb. 23. | | | Feb. 23. | | | | | | | | | | | | | April 13. | April 19. | April 25. | | | Feb. 21. | | | |
| 1740 | | Feb. 27. | April 4. | May 28. | | April 14. | | | | | | | | | | | Feb. 27. | | March 31. | April 14. | May 2. | May 21. | | May 2. | | | |
| 1741 | | | | May 9. | March 12. | | | | | | | | | | | Feb. 2. | Feb. 13. | April 2. | April 9. | April 19. | April 14. | | | | | | |
| 1742 | | Jan. 31. | | May 15. | April 11. | | | | | | | | | | | | | April 1. | April 17. | April 27. | April 21. | | | April 15. | | | |
| 1743 | | Jan. 31. | | May 8. | Feb. 28. | | | | | | | | | | | | | | April 12. | April 17. | April 20. | | | March 23. | | | |
| 1744 | | | March 28. | Essex, May 11. | March 28. | March 30. | | | | | | | | | | | Feb. 21. | April 1. | March 31. | April 20. | April 9. | May 23. | | April 8. | | | |
| 1745 | Jan. 6. | Jan. 26. | March 26. | May 13. | March 14. | March 29. | March 29. | | | | | March 29. | | | | Jan. 3. | March 8. | April 8. | April 3. | April 22. | April 12. | May 10. | | April 8. | | | |
| 1746 | Jan. 20. | | March 26. | May 13. | March 15. | April 10. | April 10. | | | May 1. | May 1. | | | | | | Feb. 25. | March 31. | April 8. | April 16. | April 19. | May 3. | | April 16. | May 1. | | |
| 1747 | Jan. 8. | Jan. 14. | Feb. 15. | April 26. | Feb. 24. | March 23. | March 20. | | | April 23. | April 26. | | | | | Jan. 15. | Feb. 13. | March 26. | April 2. | April 21. | April 15. | May 9. | | March 18. | | | |
| 1748 | Jan. 5. | Jan. 29. | April 5. | Middlesex, May 22. | March 28. | April 14. | April 10. | | | | | | | | | Feb. 8. | Feb. 29. | April 3. | April 2. | April 16. | April 13. | May 28. | | April 22. | | | |
| 1749 | Jan. 4. | Jan. 17. | Feb. 19. | Middlesex, May 1. | March 5. | March 18. | March 24. | | | April 21. | April 22. | | | | | Feb. 13. | Feb. 18. | March 29. | April 5. | April 13. | April 16. | May 20. | | March 9. | | | |
| 1750 | Jan. 15. | Jan. 17. | Feb. 13. | April 13. | Feb. 20. | Feb. 22. | Feb. 21. | | | March 31. | April 15. | | | | | Jan. 22. | Feb. 13. | March 30. | April 8. | April 11. | April 9. | May 17. | Feb. 6. | Feb. 25. | | | |
| 1751 | Jan. 9. | Jan. 8. | March 1[illegible] | May 9. | March 27. | | March 22. | March 6. | | April 25. | April 24. | March 21. | | | April 16. | 1750 Dec. 29. | March 3. | April 7. | | April 19. | April 16. | May 3. | | March 29. | | | |
| 1752 | | Jan. 30. | Middlesex, Feb. 13. | May 14. | March 9. | April 6. | April 2. | | | April 10. | April 20. | March 19. | | | | 1751 Dec. 27. | | March 20. | April 2. | April 9. | April 7. | May 12. | | April 5. | | | |
| 1753 | Feb. [illegible] | Feb. 1. | March 21. | May 11. | April 1. | April 3. | March 27. | | | April 24. | April 29. | | | | May 11. | Feb. 22. | March 3. | April 13. | April 17. | April 24. | April 19. | | | March 30. | | | |
| 1754 | Jan. [illegible] | Feb. 16. | April 7. | May 22. | April 5. | April 14. | April 14. | | | May 11. | May 7. | | | | | | March 1. | April 11. | April 13. | April 14. | April 12. | May 16. | | May 9. | | | |
| 1755 | Jan. [illegible] | Feb. 16. | March 31. | May 10. | April 1. | April 9. | April 1. | April 10. | April 9. | April 18. | April 21. | March 31. | April 16. | April 13. | April 12. | March 4. | March 14. | April 18. | April [illegible] | April 13. | April 15. | June 4. | | April 15. | April 21. | April 18. | |

Vol. LXXIX. Indi-

'The Marsham phenological record in Norfolk, 1736–1925',
as presented by Ivan Margery, 1926.

up the baton. The Marsham record then continued, nearly unbroken, well into the twentieth century, until the death of Mary Marsham, the first Robert's great-great-great-granddaughter, in 1958.[6]

Today, Marsham is recognized as the founder of the discipline known as phenology, from the Greek φαίνω, meaning 'to show, appear, or bring to light'. Phenology is concerned with the appearance of things not in the visual, aesthetic sense, but in the temporal one: not *how* things appear, but *when* they do. Phenology thus depends on paying attention, over time, to the here and now.

The extent of the Marsham record over decades and generations brings to light the slow alteration of the Earth: an alteration that we now know to be caused by human activity, but which has, until recently, progressed too slowly to be noticed by any single observer. Between 1850 and 1950, the records show a couple of clear trends. One is that, over the course of a century, there is a slow but observable increase in mean temperatures, particularly in the winter months. The other is that this observation is correlated in the behaviour of plants and animals: oak leaves, for example, appear a little earlier every year.

These observations are now being repeated elsewhere, and not only by humans. In Greenland, the migration of the caribou is governed by the length of the day. In spring, as the days grow longer, the caribou begin to travel inland to their summer pastures and breeding grounds. But because of rising temperatures, they are now finding that the plants they are accustomed to grazing on have already bloomed and withered, with the result that caribou populations are collapsing. This is an example of a phenological mismatch: the falling out of sync of once complementary reckonings of time, with devastating results. Our dependence on antagonistic, machinic imaginings of time – from the industrial clock to the computer processor – might be considered another kind of phenological mismatch: a nauseating and ultimately deadening supplication to a non-human, anti-life chronology. But must we all move to Vermont, and reject wholesale the possibilities of thinking otherwise through technology?[7]

The Marsham record, which gives us our first glimpse of phenological change, is not only an achievement of attention, but one of technology: in this case, the simple technology of the logbook. It shows that there are other ways of using time-based technology which extend, rather than constrict, our capacity for attention and care. By committing their observations to paper, the Marshams transformed conscious attention into data, allowing us to see the changes occurring in nature. This is very different to our customary view of data today as something which, when gathered, fixes its subject into an immutable object for numerical analysis and social exploitation. The Marsham record is not an account of constancy, but one of change. It is also an example of augmented attention: the use of tools to widen and extend our view across time and space, so that we may become more attuned to the broader scale of the world we are entangled with.

One afternoon in April, soon after meeting the swallows, I took an old bamboo rod out into our garden on Aegina and cut it to a length of 115 centimetres. Stood on the ground, it came about halfway up my chest. I laid it on a scrubby patch of lawn, one end next to a tough-looking dandelion, the other pointed northwards. Then I dug up the dandelion with a trowel, and replanted it at the other end of the stick. A small step for humans, but quite a leap for the dandelion.

We know that climate change is occurring, but for a long time it felt to many of us like an abstraction. While numbers in reports and rising lines on graphs stirred unease, its effects weren't always visible in our personal environment – even to those, like the Marshams, who were consciously paying attention. In recent years, of course, these effects have started to become clearer wherever we live: hotter summers, fiercer storms, more severe wildfires, to name a few. But – at least to those fortunate enough to live in certain temperate areas of the global North – it's still more of a feeling than a tangible event: something in the air, rather than on the ground.

Worse, it seems like there is little to nothing we can do as individuals in response to this feeling, other than fearing its impacts: impacts that are already wiping out thousands of species and threaten the survival of our own. Climate change is no longer something we can reverse, but something we have to adapt to, cope with and mitigate as much as possible. Existential dread is not a helpful response: we need better ways to share in, and take some of the burden, of the trauma being inflicted on the more-than-human world.

If you take the increase in global temperature over the last century, and add the predicted increases we'll see over the next one, it's possible to see how fast climate change is altering local conditions in different parts of the world. And, as a result, it's possible to calculate the velocity of climate change: how fast it's moving across the ground in different places.

Understanding this velocity is crucial to the survival of life on Earth. It is the speed we need to move in order for the conditions around us to stay the same. It also implies a direction: the bubble habitats where life can survive and thrive are moving uphill and towards the poles. Local differences in velocity are determined by the shape of the land, and of the Earth itself: the effects of climate change move fastest in flooded grasslands, mangrove swamps and deserts; slower in mountain uplands and boreal forests. Yet the effects are uneven: those of us who already live in deserts have a lot of room to move, and a long way to go before we run into obstacles. But if you're already halfway up a mountain, you might soon have nowhere to go.

Adapting to the velocity of climate change is one of the key tasks of the next century: for us and everything we share the planet with. Species survival depends as much on keeping pace with this rate of change as it does on the nature of the change itself. Those who cannot move, or who have nowhere to go, are at greatest risk.

At the time of writing, the global mean velocity of climate change is about 0.42 kilometres per year.[8] Divide that by 365, and you get 115 centimetres – the length of my bamboo rod. That's the distance the dandelion has to move, *every single day*, just to live in the same conditions. But plants, according to our understanding, cannot move by themselves.

In his treatise *De Anima*, written around 350 BCE, Aristotle laid out the classical conception of the soul: that essence or vital principle of a living thing which is entangled with the body, but not of it. The soul is that which animates us. In Aristotle's thought, all living things have souls, but they vary in their structures and capacities. The soul of a plant, he writes, is capable of reproduction and growth, but is insensible and immobile. The soul of an animal, by contrast, moves and feels. The rational human soul – uniquely, according to Aristotle – is capable of thought and reflection.

Aristotle's scheme is still striking to us today, because it assigns souls to plants and animals. It acknowledges the animation – the vitality – of non-human life. Yet it also grades them, in a way that has fundamentally shaped our view of our relative place in the world.

This hierarchical conception of the world has endured 2,000 years, and – despite objections from atheists, biologists and evolutionary theorists alike – remains intact in the popular Western imagination today. It placed *Homo sapiens* at the top of the ladder, above animals; and plants near the bottom, just above rocks. Medieval Christianity added further layers, with angels above man, and God above all, becoming the Great Chain of Being or *scala naturae* – but the core of the structure held. Plants, defined by their immobile rootedness in the Earth, were lesser beings. To be sedentary in an active world is to be looked down on.

So what is to become of that dandelion, or those roses, or that oak, or indeed all the other plants, those vegetal souls, as climate change sweeps inevitably and irreversibly across the face of the planet, at a relentless 115 centimetres a day? Is there any more damning indictment of this lower form of life, any greater vindication of its placement on the last-but-one rung of the *scala naturae*, than the inability to act decisively in the face of such an obvious and oncoming threat?

Of course, Aristotle was wrong. He was wrong about the mental and spiritual abilities of animals, and he was wrong about the insensibility of plants, as we have already seen. He was also wrong about their immobility. Plants move – and they're on the move. They are moving in space and they are moving – finally, or once again – in our imaginations. Whole forests are in motion, ranging across continents, in response to climate change and human action, and according to their own needs and desires, their own senses and intelligence.

What does plant migration actually look like? In part, it is a steady, local effort: creeping roots and floating seeds establishing the next generation in favourable areas along a population's leading edge, leaving a dwindling trail of less successful plantations in the rear. At other times, it is accomplished in great strides, as seeds are flung into the atmosphere and carried by the winds to establish new and distant colonies, dispersed outlier populations which rapidly expand when conditions become favourable.[9] Forward operating bases for new climatic explorers.

Across the eastern United States, populations of trees have been migrating for at least the last thirty years, probably much more. Researchers who examined Forest Service data from 1980 to 2015 found that three-quarters of species in the eastern US were shifting north- and west-wards, at an average rate of between 10 and 15 kilometres a decade.[10] Speed and direction is different for different types of trees: the conifers are mostly heading north, while broad-leafed and flowering trees, like oaks and birches, move west. The fastest moving are *Picea glauca*, white spruce, which clock up over 100 kilometres each decade, almost unswervingly northwards. By contrast, American sweetgums and balsam poplars have barely moved at all, registering shifts of just a couple of kilometres in the same period. This means that the white spruce is keeping well ahead of climatic change for now – but starting as it is from higher latitudes, across the very top of the US, it won't have as far to run as the sweetgums of Alabama and Georgia when things get tough.

In Scandinavia, which is experiencing warming that far exceeds the global average, fast-growing birch saplings are racing up mountainsides they had previously shunned, gaining as much as 500 metres elevation in just a couple of decades. Swedish pines, spruce and willow are now growing at higher altitudes than ever before.[11] Comparing photos taken by oil surveys in Alaska in the 1940s to images of the same locations today shows alder, willow and dwarf birch flooding into once sparse valleys and pushing up once bare hills.[12]

Trees are adapting to climate change faster than we are, and as a result have more chance of adapting successfully. In fact, climate change is probably only one of several factors they're responding to – but responding they are, contrary to our Aristotelian perceptions of them as senseless and immobile. The westward impulse of deciduous species in the US appears to be attributable to increased rainfall in the regions they're heading for – also an effect of climate change – but these areas are still drier than the places the trees are leaving behind, so that doesn't tell the whole story. Other human activities, such as construction and pesticide use, are also likely to be playing a part. Whatever the combination of factors, however, mass plant movements are in process, and not for the first time.

The last great tree migration occurred at the end of the last Ice Age, around 10,000 years ago. Following the retreat of the ice, trees started to return to the higher latitudes they'd earlier been frozen out of. They did so surprisingly quickly. The Victorian geologist and palaeobotanist Clement Reid, writing in 1899, noted that oaks had already reached the far north of Scotland two millennia earlier, as shown by acorns turning up in archaeological excavations of Roman age sites. He concluded that 'to gain its present most northerly position in North Britain after being driven out by the cold', the oak 'probably had to travel fully six hundred miles, and this without external aid would take something like a million years'.[13] In fact, fossil pollen records show that some of the oaks travelled, at times, at nearly a kilometre a year, a rate confirmed by more recent, climate-based studies. It also seems likely that this recovery was abetted by the existence of a few, scattered refugia: ice-bound sanctuaries where small groups of trees survived in temperate pockets to seed the ground after the thaw.[14] As the planet warmed, the forests reconquered the land. In North America, beech trees leapt over the Great Lakes as the glaciers retreated.[15] Norwegian spruces circumnavigated the North and Baltic seas before modern humans did. They went before us.[16]

These processes are grand and magnificent; they inspire awe and wonder – but still mostly in the abstract. I can comprehend the mathematics of species dissemination, read accounts of fossil pollen counts, trace the lines of movement through decades of database records – but what does it mean to experience it? Living at human speed, at animal speed, it's almost impossible to get my head around the vegetal unfolding of plant migration, an endeavour that takes places at spatial and temporal scales beyond my natural understanding. And this is our problem. We humans live in such a narrow slice of time and space that we are incapable of thinking of, or thinking at, the pace and scale of the world, the changes we have wrought in it, and the changes we will have to make to survive them. Our given minds are insufficient to the task – but we do have tools to hand, technology among them.

A reality in which trees and other plants are in constant, deliberate motion may run contrary to our assumed image of the world, but the recognition of this reality and its absorption into our awareness is crucial to rethinking the world and our relationship to it. In order to change ourselves, to take on different ways of thinking about the world, we need new ways of seeing it. We are accustomed, largely by scientific practice, to taking things apart, separating them into their component attributes, fixing them for study, and piece by piece reducing their collective agency until they have none at all. But this is the opposite of ecology, which seeks to find connections between all things and resolve them into greater, interconnected systems. The lens required now is not a microscope, but a macroscope: a device for seeing at a far vaster scale – both in space and time – than we are used to.

Another recent hobby of mine, when not acting as a *Fluchthelfer* for dandelions, is time-lapse photography. For this purpose I acquired a small, weatherproof camera designed for documenting construction projects: a few inches high, it's a little monocular homunculus which can sit undisturbed in a corner, observing the world for weeks, powered only by a couple of AA batteries.

For months, I shuffled this little device around my apartment, spending a few days at a time with each of the potted denizens of the living room. When I review the footage, the apparently inert ferns and *ficus* burst into life; *philodendron* curls its fingers around the lampshades; *monstera* bobs and waves its leaves; lilies open and close, turning to follow the sun as it moves across the walls. Each creature has its own pace, its own rhythm, but all move together – flexing, turning, bending and stretching. What appears to me from my dining-table desk as perfect stillness is in reality a frenzy of activity in another register. Nothing could persuade me better that, in the words of the botanist Jack Schultz, 'plants are just very slow animals'.[17]

That plants are in constant motion is one way to understand how we misconstrue their extraordinary abilities. In *The Overstory* by Richard Powers, one character recounts a science fiction story that takes place in this gap between timescales.

> Aliens land on Earth. They're little runts, as alien races go. But they metabolize like there's no tomorrow. They zip around like swarms of gnats, too fast to see – so fast that Earth seconds seem to them like years. To them, humans are nothing but sculptures of immobile meat. The foreigners try to communicate, but there's no reply. Finding no signs of intelligent life, they tuck into the frozen statues and start curing them like so much jerky, for the long ride home.[18]

Here, we are the aliens, buzzing about in human time, incapable of perceiving the vibrant life that surrounds us, and thus treating it like so much insensate sustenance. Yet, as my camera shows, we have the tools to see differently. What matters is how we choose to see, and what we choose to look at.

Charles Darwin was one of the first scientists to use time-lapse methods to investigate the vitality of plants. In a book published with his son Francis in 1880, he recounted their joint experiments investigating what they called *The Power of Movement in Plants*. The Darwins didn't have access to photographic technologies, so they had to build their own, complex apparatus for tracking this movement.[19]

The Darwins hung large sheets of glass horizontally and vertically in a greenhouse, so that they could surround a potted plant from the sides and above. Then they painted tiny dots of sealing wax onto the plant's shoots, and painstakingly marked the path of the dots onto the sheets of glass with thick lines of India ink. In this way they could enlarge and trace the movements over time. Their method transformed even the smallest dips and weaves into flowing charts which resemble aerobatic manoeuvres. They continued this tracing for days and weeks at a time, with all kinds of plants – cabbages, sorrel, nasturtiums, beans and more – and noted the ways in which different species responded to light and shade, night and day. Their experimental collaborators included that most sensitive plant, *Mimosa pudica*, which was later to give such a startling performance for Monica Gagliano.

The Darwins' logbooks give some flavour of the time and effort involved. They would observe the plants for days at a time, taking turns in the greenhouse to meticulously record the time, temperature, the play of light, and every detail of the plant's movement. Of a few days spent in the company of a pair of *Mimosa* seedlings, they recorded, in part:

> The cotyledons [embryonic leaves] rise up vertically at night, so as to close together . . . [and] moved downward in the morning till 11.30 a.m., and then rose, moving rapidly in the evening until they stood vertically, so that in this case there was simply a single great daily fall and rise. The other seedling behaved rather differently, for it fell in the morning until 11.30 a.m., and then rose, but after 12.10 p.m. again fell; and the great evening rise did not begin until 1.22 p.m. [. . .] Between 7 and 8 a.m. on the following morning they fell again; but on this second and likewise on the third day the movements became irregular, and between 3 and 10.30 p.m. they circumnutated to a small extent about the same spot; but they did not rise at night. Nevertheless, on the following night they rose as usual.

To 'circumnutate' means to bend or move around in an irregular circle or ellipse. It is a motion caused by variations in the speed of growth in different parts of the plant, and is the mechanism behind most plant movement, including *Mimosa*'s. It's nutation that causes leaves to bend or flatten out, and petals to furrow and curl. Circumnutation – a gentle upward and outward spiralling – is the characteristic movement of growing plants, performed by everything from pea shoots to oak seedlings, as well as mushrooms and the hyphae of fungi. Being the first gesture of awakening, questing life, it seems to presage all other movements, including our own. It is a nodding turn, a greeting to the environment, an opening ceremony or a blessing to the four directions: Hello, world.

The Darwins' work took several years to complete; at times, Charles despaired of ever finishing it. The final results were used to bolster his theories of evolution. Not only, he concluded, was all plant movement a modification of circumnutation – 'the kind of movement common to all parts of all plants from their earliest youth'[20] – but it is quite specifically this form of movement which enabled plants to evolve and adapt to almost any environment on the planet. Life follows the greeting. While many other kinds of vegetal movements have since been discovered, nutation remains central to a true understanding of plant life: a mode of expansion which is not reactive, muscular, and domineering, but gentle, expansive and generative. It is through their own attentiveness to their environment that plants have obtained the world.

The Darwins' work reminds me of nothing so much as Marcel Duchamp's *Large Glass*, otherwise known as *The Bride Stripped Bare by Her Bachelors, Even. The Large Glass* comprises two panes of glass fixed in a wooden frame. Duchamp worked on it for almost a decade, mostly in secret, from 1915 to 1923, when it was installed in the Brooklyn Museum. It remains one of the most significant and enigmatic artworks of the twentieth century.

Onto his framed sheets of glass Duchamp projected not only movement and time, but an alternative physics, or an alternative cosmology. Parts of *The Large Glass* are perspective tracings, just like the Darwins' magnified plant movements, while other elements are artefacts of chance and natural processes. When the work, untouched for months,

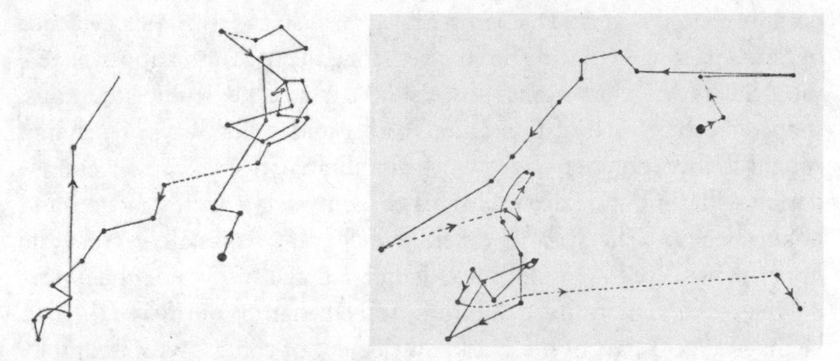

Two figures from Charles and Francis Darwin's *The Power of Movement in Plants*, tracing the movement of a cabbage over forty-eight hours.

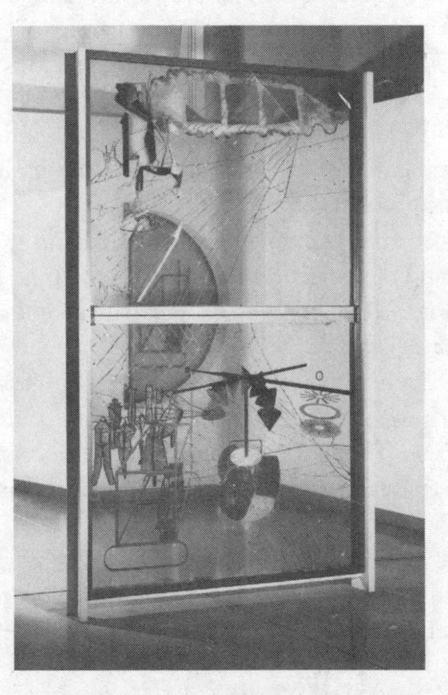

*The Bride Stripped Bare by Her Bachelors, Even* (*The Large Glass*) by Marcel Duchamp, 1915–23.

became coated in dust, Duchamp added varnish to freeze this evidence of time's passing in place. The squares along the top of the upper pane – the Bride's 'draught pistons' – are silhouettes of his studio's curtains, shaped by the force of the wind. In this and other ways, Duchamp sought to incorporate into the work an inhuman agency: an accommodation with time and fate; an insistence on mystery, incompleteness and unknowability. The web of cracks which run diagonally across the upper pane were the result of accidental damage sustained while moving the work from its first exhibition, and Duchamp considered them to be an emergent part of the work, an element of chance which complemented and continued his conscious efforts.

*The Large Glass* has as many interpretations as it has viewers – and that's part of its intent, its great artistic success. Balanced in tension between the free-floating Bride at the top of the glass, and the endlessly incomplete strivings of the Bachelors at the bottom, it contrasts individual, human, domineering sexual desire with a machinic, vegetal auto-eroticism. It asserts the creative genius of the artist, while simultaneously rejecting subjecthood and conscious intent. It contrasts our experience of time as linear and bounded with the natural, cyclic time of the universe. It is both mirror and window, allowing us to see ourselves shimmer and dissolve into the larger, fruiting and fruitless, operations of the world.

It seems that for the Darwins and for Duchamp there's something about this method of projection onto glass – the time-lapse snapshots of a deeper reality – that stimulates a particular kind of attention and an awareness of invisible modes of life. While Darwin's glass attempted to decrypt the secrets of plant life, Duchamp's reminds us that aspects of it remain entirely out of human sight and mind, and are ultimately unknowable to us. This is the dance of anthropocentrism: transparency does not equate to understanding; seeing does not mean knowing or dominating.

All too often, we mistake one for the other. We thoughtlessly assume that by observing the world, we fix it into knowable forms, but time-lapse reveals that the nature of the world is changeable. As Karen Barad, the philosopher of quantum physics, would say, it's made of intra-actions: it's in the invisible gaps between the frames that things encounter one another and vibrate. It is these intra-active processes of growth, change and decay which produce the lines between Darwin's dotted points, and the dust which Duchamp varnished onto *The Large Glass*: shadows of unseen processes which exist in other dimensions of time.

We often choose to look the wrong way: not merely in the wrong direction, but with the wrong intention. Our intent – the way we choose to look – informs what we see. This problem is compounded by our technologies, particularly when they originate in war and violence, as most of our contemporary technologies do. The internet is a clear example of this: it emerged from the twin poles of Cold War paranoia – distributed networks designed to withstand atomic attack and the Californian Ideology, which in the 1990s traded the hippy ideals of liberation and togetherness for technological determinism and neoliberal capitalism.[21] It's this combination of military power and corporate profit-seeking which has shaped the modern internet, writing structural violence and surveillance capitalism into its source code.

Radar image showing ring angels at sunrise on 1 September 1959.

But even military technologies can reveal surprising things. In the 1940s, at the height of the Second World War, British army technicians working on the very first radar systems noticed the presence of mystery echoes on their screens and readouts. Some would fade within seconds; others would persist for minutes at a time and extend for tens of kilometres. These ghostly signals would come and go, making first-hand detection and confirmation of the sources even more difficult. The low power of the earliest radar systems meant that the signals were first detected at close range, but as the technology improved, the unknown traces started to be observed up to 70 kilometres away. Sometimes, vast fields of echoes would invade the scopes, shimmering across the sky in waves and rings. The early radar operators called these ghostly signals 'angels'.

These angels were a real problem. One US military report recorded that they 'rushed men to battle stations, sent fighter planes on "goose" chases, prompted lookouts to report un-identified aeroplanes diving into the sea, gave rise to several E-boat [fast attack craft] scares, started at least one invasion alarm, and tested the vocabulary of many skippers'.[22]

The British Army's Operational Research Group was one of the bodies responsible for evaluating and improving these early radar systems. A special department within the Army, it comprised several hundred analysts who worked on complex, knotty problems like improving the accuracy of anti-aircraft fire, and the design of camouflage for aircraft. Its serving members included many scientists who would go on to illustrious careers after the war, including several Nobel Prize winners. Among them were two biologists: George Varley and David Lack. Lack was one of the most influential ornithologists of the period and had spent the year before the outbreak of war on the Galapagos Islands; in 1949 he would publish his findings in the best-selling *Darwin's Finches*, an important contribution to modern theories of evolution. Varley was an entomologist who would go on to become Hope Professor of Zoology at Oxford, where Lack would later join him as Director of the Edward Grey Institute of Field Ornithology.

But in September 1941 Varley was the Operational Research Group observer at a radar site near Dover, when the scope started pinging with ghostly signals. The crew couldn't see anything out to sea, but Varley, using a powerful telescope, picked out a cloud of gannets swooping over the waves, some 14 kilometres offshore – exactly where the angels appeared to be. He and Lack started collecting reports from radar operators, in order to convince the Air Ministry that the angels were in fact birds. In one experiment, a dead herring gull was tied to a balloon with a long string and released over a radar site: the scope showed not one, but two echoes.[23]

As the technology improved, it started to pick up more and subtler echoes: not just large birds like gannets and gulls, but smaller songbirds. One of the most spectacular and disturbing 'angels' appeared over southern England every time a V1 flying bomb flew over: a huge ring of ghost signals scattering out from the bomb's path. Eventually, observers managed to pinpoint its origin: a flock of starlings, scared up from their roost by the noise of the V1's pulsejet engine.

The main problem Varley and Lack had in convincing others of what they were seeing wasn't the size of the birds but the fact that, back then, nobody believed that birds flew at night. Lack, on the other hand, believed they did, and eventually he was able to prove it by developing a new scientific technique, based on the wartime data, called radar ornithology. After the war, he continued to gather data, monitoring the migration of flocks of birds across the North Sea, using radar to track them at night from coast to coast.[24] His careful attention revolutionized our understanding of avian behaviour.

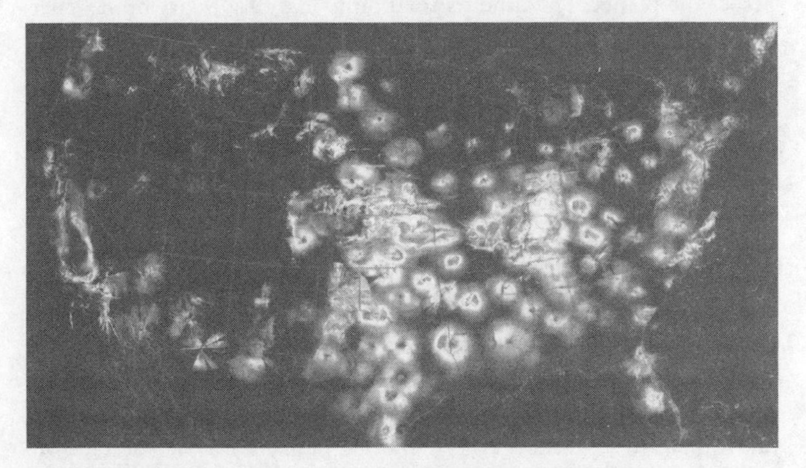

NEXRAD radar mosaic showing bird flocks in flight, 8 May 2009.

Today, radar ornithology is used to protect aircraft from bird strikes, and birds from wind power installations, as well as to study migration patterns and seasonal roosting behaviours. One of the most spectacular sources of contemporary migration information is weather radar, such as the NEXRAD (Next-Generation Radar) network which covers North America. In these bright, multicoloured images, designed to track storms and weather fronts, it's possible to see the movements of flocks across thousands of miles, and the explosions of millions of birds into the sky at dusk as night sweeps across the continent. Websites like the US-based BirdCast.info and the European EuroBirdPortal.org analyse public radar data to provide forecasts for birdwatchers, advising them on when weather patterns and migration routes combine to provide the best bird-spotting opportunities. At the other end of the scale, Israeli researchers have used the pencil-thin beams of military tracking radars to lock onto individual birds crossing the valley of the Dead Sea, identifying not only their speed and direction, but the pattern of their beating wings.[25]

The sophistication and complexity of these tools are extraordinary. One consists of a continent-spanning network of 159 rotating S-band radars, mounted on 100-foot towers, which continually feed data to the supercomputers of the US National Weather Service, which in turn shares it near instantly with professional and amateur forecasters across the planet. The other is the pointy end of a Swiss-built Feuerleitgerät 63, or Super Fledermaus, a truck-mounted fire control radar which can automatically configure twin Oerlikon 35-mm anti-aircraft cannon to shoot down military targets up to 15 kilometres away. And we're using them for birdwatching. Some of us, at least.

Technology allows for rapid transformations not only in the *scale* of our attention – from a continental migration to the wing-beat of an individual bird – but in its *kind*. As experiments with radar by Lack and others confirm, even military technologies – including the internet – don't have to distance us from the 'natural', more-than-human world. In fact, they can bring us closer to it.

The same is true for today's most powerful technologies of vision: the extraordinary array of imaging satellites which orbit the globe, allowing us to see almost everything that happens on the planet, from ocean-spanning weather systems to missile sites in the Negev desert.

Like the internet, satellite imaging emerged from a history of military surveillance and violence. Yet these instruments, constituting a vast time-lapse camera pointed at all activity on Earth, provide another object lesson in how we can use our technologies to see in a different way, should we choose to do so.

In January 2011 Michael Moore, the deputy director for astrophysics at NASA, got a surprising phone call. It was from the National Reconnaissance Office, which since the 1960s has designed, built and operated spy satellites for the US government, almost entirely under conditions of deep secrecy. Apparently they had some hardware going spare and wondered if NASA would like to take it off their hands.[26]

Once Moore got over his surprise, he went to take a look at the hardware. In a classified facility in upstate New York, he found two long tubes, wrapped in silver foil: two near-complete space telescopes, as well as parts for a third. Not only were both satellites polished and ready for launch, they were significantly more advanced than the closest civilian equivalent, NASA's own Hubble Space Telescope – but, until the NRO's phone call, they had been entirely secret. When Hubble was launched back in 1990, it was believed to be the most powerful telescope ever put into space. This probably wasn't true then, and it certainly isn't today: even after five costly repair missions by astronauts, Hubble is ageing fast, while its successor, the James Webb Space Telescope, only launched in December 2021. The sudden appearance of two brand new, cutting-edge instruments was a cause for celebration – 'a total game-changer', in the words of another astronomer.

Moore called the optics on the new telescopes 'astounding', but NASA wasn't forthcoming about the specifics. The instruments had been stripped of most of their electronics before they were made public. 'We can't say what they were used for,' said one of its directors, John Grunsfeld. A NASA presentation to scientists was greeted with laughter when a picture of one of the satellites was so blacked out as to be unreadable.[27] But one thing was clear: with 2.4-metre mirrors – the same as Hubble, but at half the length – these things were designed for looking down at Earth, rather than up into space.

For decades, the US has maintained a secret space programme in support of military action and intelligence-gathering, one that has always been much better funded and more extensive than NASA's civilian programme. Only the most fleeting glimpses of its capabilities have been seen. There's the Boeing X-37, aka the 'secret space shuttle', which has spent almost eight years in space over multiple missions, doing who knows what. There are occasional images, like the startlingly clear photo of an Iranian missile site tweeted by Donald Trump in 2019, which appeared to be two or three times sharper than commercially available imagery. (It probably came from one of the NRO's Keyhole satellites.[28]) It seems likely that the telescopes donated to NASA were either surplus parts of this programme – a long-running line of reconnaissance satellites in service since the Cold War – or a remnant of another cancelled programme, the sci-fi-sounding Future Imagery Architecture project. Wherever they came from, they offer a stunning example of what happens when we redeploy military and other combative technologies towards more peaceful ends.

NASA's scientists were quick to suggest a use for their new toys. The Wide-Field Infrared Survey Telescope (WFIRST), another project on the drawing board which had been struggling for years to get funding, was designed to measure the effects of dark energy on the formation of the Universe, as well as the consistency of general relativity and the curvature of space–time. The sudden availability of the NRO satellites kick-started the programme, which now has a launch date in 2027. The shortness of the satellites – nicknamed 'Stubby Hubbles' – and their resulting wide depth of field, are actually an improvement on NASA's planned designs. With the addition of a device called a coronagraph, which blocks direct stellar light, the new observatory will also search for exoplanets: new worlds formed around distant stars. NASA's scientists were able to literally turn these technologies around: to look more closely at the world which surrounds us, instead of at ourselves. In doing so, they turned a tool of surveillance and control into one of wonder and discovery.

There are always other ways of doing technology; other ways of thinking it and other ways of putting it to use. The only constraint is our imagination, and our intent. What other worlds might we discover if more of our technologies were consciously, thoughtfully employed – pointed, not at one another, but at the more-than-human world?

The gap between the time-lapse photography in my living room

A false colour Landsat image showing a small portion of Australia's Gulf of Carpentaria, February 2009. Plant life, including mangroves, reflects brighter colours more sharply.

and the imaging of the whole Earth and the wider Universe from space seems vast, but it is really just a question of scale, intent and imagination. The capabilities we've built for seeing the world are awesome, yet more accessible than most of us realize. A while ago I wrote a little script that runs in the background on my computer, updating my desktop image every hour with a photo of the entire Mediterranean, taken just fifteen or so minutes earlier by a satellite hanging 36,000 kilometres above me. Hour by hour, day after day, I can watch clouds form, swirl and dissipate over the ocean, and storms rush over the Balkans and the Iberian Peninsula. Each dawn, a sharp curve of light sweeps across Europe and Africa, turning darkness into blues and greens; every evening, twilight sweeps it away again. It's a God's eye view, in near real time, created by a few lines of code – and a few billion euro/dollars of technological infrastructure, lofted into orbit by public agencies, and common agency.

I am, if it's not yet apparent, a satellite nerd. I have spent many happy hours diving through databases of imagery and reading the specs of complex sensors, and learning how to resolve, sharpen and read their extraordinary products. My undisputed favourites among these eyes in the sky are the satellites of the Landsat programme – nine so far, with the most recent launched in September 2021 – which have been taking photographs of Earth from 700 kilometres up since 1972. A product of numerous US government agencies (NASA, the National Oceanic and Atmospheric Administration (NOAA), the United States Geological

Survey (USGS)), as well as private enterprise, the images taken by Landsat are made freely available to all by an Act of Congress. Retained in huge digital archives, and made accessible through the internet, they constitute a vast and ongoing time-lapse of the entire planet.

What I love about Landsat is its ability to show us the world in a new way. The satellites are equipped with multi-spectral sensors: specialized cameras which can see beyond the visual spectrum and into the deep infrared and ultraviolet. These frequencies reveal the invisible vapours in the atmosphere, the old scars of fires and seismic events, and the health of plants and soils. Converted into colours which we can discern, these phenomena appear in vibrant shades of red and green, clear and startling to the eye. With its superhuman sight and decades of archived imagery, Landsat provides us with the ability to see through time, to track the vegetal movement which is beyond our normal sight and to see it for what it really is: a titanic unfolding of active and intentional life.

What this planetary time-lapse reveals is the vibration of my apartment plants enacted at continental scales: mangrove forests walking up and down the Australian coast and Lebanese cedars climbing mountains in Turkey. In southern Italy, where farmers have been slowly abandoning agricultural land for decades, the Pollino National Park bursts its borders, scattering birch and pine trees along the Apennines. Under the bright eye of Landsat's gaze, the trees appear as a red wave: infrared light, invisible to us, scattering from the water in their healthy, growing leaves. Here, the machines see life in motion better than we do.[29]

Here, technology enables us to change our vision, and it allows us to change what we do with that vision; where we look, what we see, and how we act as a result. It allows us to engage our care and attention at a greater scale, and to more present in the world than we would be without it. With it and through it, we can choose, consciously, to live in a different time.

Writing in 1935 about his early experiments with time-lapse cinematography, the French film-maker Jean Epstein summed up the experience of watching natural processes unfold at other-than-natural speed: 'Slow motion and fast motion reveal a world where the kingdoms of nature know no boundaries. Everything is alive. A surprising

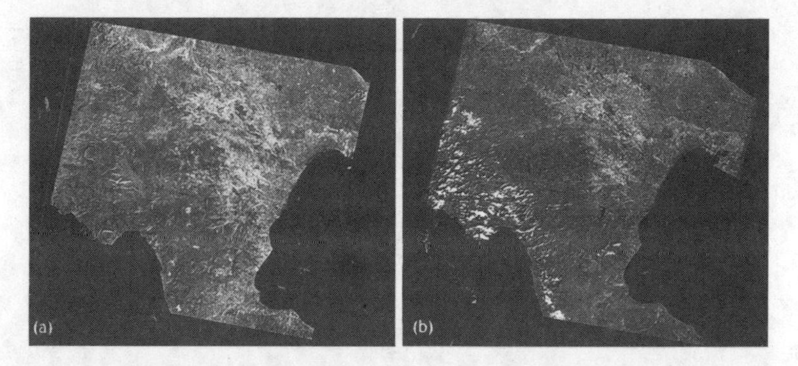


Landsat images of Basilicata, Italy in (a) 1984, and (b) 2010.

animism is being reborn. We know now, because we have seen them, that we are surrounded by inhuman existences.'[30]

For Epstein, seeing is necessary for knowing and caring, and thus for acting. I would add to this the quality of practice. The act of making time-lapses for myself, whether of my living room or the whole Mediterranean basin, engenders in me an attentiveness to plant and planetary time that I do not gain merely from watching others' footage on YouTube. The experience is immeasurably enhanced by the time I have invested, the presence I have brought, and the wordless appreciation of what's actually occurred. This is intra-action at work again: it combines a change in vision with a change in being.

Viewing inspires awe, but practice generates knowledge and understanding. The tools of technology, to be effective in producing altered states, require us to be full participants in their revelations, not mere audiences. This is why it's so important that we are given access not just to the products of all these wondrous technologies – the beautiful images shot by satellites – but to the technologies themselves. What must be made available to all is education in their actual use: the knowledge and know-how to design and deploy them critically and thoughtfully, and real access to existing tools and processes. It is not enough to turn the machines around – to point the satellites outward, rather than at ourselves. They must also be shared out and placed in the hands of everyone.

# 5
# Talking to Strangers

*Eh Eh Eh*
*Tt Tt Tt*
*Ah Ah Ah*
*Tt Tt Tt*

It's October 2019 in Basilicata, on the sole of Italy, and we're walking behind the herdsmen in the Parco Nazionale del Pollino, high above the comune of Viggianello. The sky is clear and blue, the air is still warm, and we're driving cows, goats and sheep across the hills, from shed to field and slope to grazing ground. As we walk, the herdsmen call to the animals in a chorus of contrasting sounds: sharp yelps, low gutturals, whistles, long calls and soft entreaties, directing them this way and that, pushing, cajoling and commanding them.

*Ek-bar Ek-bar Ek-bar*
*Shqitz Shqitz Shqitz*
*Ooo-Ah Ooo-Ah Ooo-Ah*
*Wey-Ah Wey-Ah Wey-Ah*

We're here because my partner Navine and I are curating an exhibition in the town of Matera, a couple of hours' drive east of Viggianello and the park. The exhibition takes its title and its inspiration from *Il Paese di Cuccagna*, the Land of Cockaigne, a medieval legend of a kind of peasant utopia, a world turned upside down in which the harshness of everyday life is replaced with superabundance. Stone bears fruit, wine bursts from the spring, and birds and animals leap up in song.[1]

Matera's official history is one of slow emergence from a state of abject poverty, a world best known from the pages of *Christ Stopped at Eboli*, Carlo Levi's memoir of his time in exile in Basilicata (then called Lucania) in the 1930s. The book's publication caused shock-waves across Italy, revealing to the rapidly modernizing North a country where people still lived in caves, danced strange rituals and struggled for survival. In the 1950s, the Italian government forcibly evacuated the *Sassi*, the famous dwelling-caves of Matera that had been continuously occupied for 9,000 years, rehousing their inhabitants in modern apartments on the other side of town. Today, after decades of abandonment, the *Sassi* have been gentrified, with many of the caves re-excavated, smoothed out and refashioned, somewhat incongruously, as expensive restaurants and Airbnbs: the 'shame of Italy' recast as a tourist attraction.

Looking through the archives of the city, through the collections of local photographers and newer research into life in the ancient *Sassi*, we found another reality: one in which people lived alongside their animals, foraged for medicinal herbs on the meadows overlooking the town, and forged complex infrastructures for water and waste disposal from the very rock they lived inside. Even the *cucù*, Matera's iconic clay whistle in the shape of a bejewelled chicken (now mass-produced for tourist consumption), had its roots in ancient dowry gifts: a token of fertility, it was part of a complex culture of gift, exchange and mutual blessing.

The title of Levi's book casts Lucania as an abject land beyond the reach of Christ's salvation (Eboli is the last railhead from the North, two hours away on the coast). But it also acknowledges the more ancient mysteries of the region, including esoteric rituals, deep and abiding folk beliefs, and the widespread practice of magic. From the dancing of the Tarantella – its agitated rhythms designed to revive those bewitched or stupefied by a spider bite – to fertility spells and harvest festivals, these Lucanian rites were deeply tied to the land, its plants and animals and to the seasons.

Ernesto de Martino, a Neapolitan anthropologist who documented these rituals in the 1930s, believed that the practices he witnessed were also directly related to poverty and the depredations of the modern world. Were those stricken by the Tarantella's bite really bewitched, or were they simply exhausted by the demands of industrial life and suffocated by strict Catholic society? Lucanian traditional magic, bleak and cruel as it could often be, set itself in opposition to the denaturalization of the world by science, and the dehumanization of the individual by industrialization.

De Martino called this 'the crisis of presence': the sense that the very hills and forests and birds and animals might be slipping away, and with them one's own place within the world. The appeal of the rituals he witnessed, like many magical and shamanic practices, was to the natural and non-human world, with which it was in constant communication and negotiation. It was a way of maintaining one's own humanity by acknowledging, and appealing to, the liveliness of everything and everyone else.[2]

Just as the traces of the *Sassi*'s original inhabitants remain in engravings and graffiti on the walls of the gentrified caves, so traces of these rituals survive in the towns and countryside of Basilicata. Each comune has its yearly calendar of *feste*, or carnivals. In Pedali, a hamlet above Viggianello, the parade of the Madonna at harvest time is accompanied by the *cirio*, a towering straw dolmen bedecked with flowers. In the town of San Costantino Albanese, the Feast of Our Lady of the Star features a succession of papier-mâché puppets – a maid, a shepherd, two blacksmiths and a terrifying devil figure – which are spectacularly detonated with fireworks. And in Viggianello itself, the arrival of spring is marked by the *maggio*, May Day celebrations in which a fir tree and a beech, carried down from the mountain in a three-day Dionysian revel of wine and song, are ritually married in the town square. The great beech trunk is dragged by teams of gigantic, snorting oxen garlanded with sun and moon charms, while the fir is borne aloft by staggering young men. For modesty's sake, the *maggio* is also called the Feast of St Francis, and a small icon of the saint is tacked onto the trunk and blessed by the village priest, before this spectacularly pagan totem is hoisted in the village square.

The straw *cirio*, bagpipes from the *festa*, a shelf of *cucùs*, a celebratory film of the *maggio*: it was artefacts like these which we were lucky enough to borrow for our exhibition. But we wanted more than silent objects; we wanted to give them a voice, just like the living breath which seemed to animate them in their places and times of origin. The *maggio* is accompanied by wedding tunes, the *cirio* has its own hymns sung by the women of the village, the *cucù* its own melodies, and the Parco del

Pollino its song. It is in the sounded breath – in song, speech and language – that communicative intent, interrelationship and the enacted, mutual recognition of other beings is to be found. Hence, our walking, accompanied by microphones, with the herders of Viggianello.

*Brr Brr Brr*
*Eh Eh Eh Eh*
*Ti Ti Ti*
*Oah Oah Oah*

The vocabulary of each herder is different. In response to the darting, bleating goats milling in their pens, Francesco 'Capo' Caputo makes plosive stabs of sound, underscored by the jabbing of his crook, jostling and shoving the animals towards the gate. On the higher slopes, his cousin Matteo guides half a dozen cows with low, carrying calls, occasional bellows, heavy huffs and lip-smacks. Back in the village, Rosina Corraro talks to her pigs in their dim, warm shed in psst-whispers and soft coughs, answered by the pigs themselves, as they snuffle and scrabble in the dark for onion skins and overripe fruits.

My favourite moment in the recordings we made during those days in Viggianello comes when Giovanni Forte, leading one of his oxen – a huge, beautiful beast, the size of a tractor – across his field, receives a call on his mobile, which he takes some time to locate and answer. Listening back, Giovanni's puffing entreaties to the ambling bull, its heavy breathing, the clanging of its bells, chains and brass charms, and the brisk, electronic tinkling of the ringtone mingle together in a harmonious, generous symphony of human, non-human and more-than-human sounds.[3]

Humans have spoken to, and with, animals for as long as we have walked alongside them, which is for ever. Such communication is not limited to domesticated animals; indeed, it predates domestication. Still today, there are those who talk to wild beasts, in mutually intelligible and productive conversation.

The Niassa National Reserve in northern Mozambique covers a huge expanse, over 42,000 square kilometres of savannah, forest and wetlands. For centuries, this region has been known for the quality of its honey, which remains both the staple nutrient and most valuable

Giovanni Forte taking a phone call.

product of the Yao people, who forage for wild honey in the branches of baobabs and other tall trees across Niassa. But in order to do so, they need some help to locate the wild hives, which are often hidden in the branches high above the forest floor.

Another resident of the Niassa is the greater honeyguide – a small, brown bird with a distinctive pink beak. All honeyguides have a rare ability: they are capable of digesting beeswax, which is a potent source of nutrients. However, the outer casing of wild beehives is tough and hard to crack, and the tasty beeswax is well protected. And so they have learned to ask for help.

When they meet humans on the savannah, honeyguides make a distinctive call, different to their usual mating or territorial songs, in order to get the humans' attention. (Hungry honeyguides will even fly into the Yao's camps to fetch them out directly.) Once they have a hunter's attention, the honeyguide swoops, fluttering from tree to tree, calling continuously, to draw the human along behind it. When it comes to a tree containing a wild bee nest, it perches expectantly on a nearby branch – well within reach of the hunter's arrows. But though the Yao people regularly kill and eat birds of similar sizes, they don't shoot honeyguides; rather, they collaborate with them. Following the bird, the hunter locates the bees' nest, climbs the tree, smokes out the hive and breaks open the comb with an axe to release the sweet honey within. Once they've taken what they want, the honeyguide is free to feast on the comb and the bee larvae that remain.

The communication between human and honeyguide is two-way. The Yao honey-hunters also seek out honeyguides, using a distinctive noise: a tongue-rolling *brr* punctuated with a breathy *humph*, a vocalization not a million miles from the huffs and puffs of the Viggianello herders.[4]

*Brrrrr-hm*
*Brrrrr-hm*
*Brrrrr-hm*

Researchers recording and replaying this and other sounds in the forest have found that the Yao hunter who repeats the *brrrrr-hm* is more than twice as likely to find themselves a honeyguide than one who makes other noises. It's more than a sound: it is a specific call, a word even, which is recognized as such by the birds. Moreover, when partnered with a honeyguide, the chance of a hunter finding a hive increases from 17 per cent to 54 per cent – a clear benefit to both partners.[5]

There are other instances of wild animals partnering with humans to increase the hunting abilities of both. Off the coasts of Myanmar and Brazil, wild dolphins show fishermen where to drop their nets, then drive the fish into them and share in the catch.[6] But the honeyguides are the only case we know of in which human and wild animal call to one another in order to undertake a task together. And they appear to have been doing so for some time.

The ancestors of modern honeyguides first encountered chimpanzees, and later early hominins, on the developing savannah of the Pliocene, some 3 million years ago. During this period, drier and cooler conditions led to an increase in open grasslands, which were covered in flowering plants and dotted with standing trees: perfect territory for the wide-ranging *Apis mellifera*, the most productive and nutritious of the bee species favoured by the honeyguides, chimpanzees and the Yao. Even today, thanks to the honeyguides, both savannah-dwelling chimps and people eat more honey than those who live in the forests.[7]

Exactly when this mutual aid between primates and honeyguides first arose is unclear, but some put it as long ago as those encounters in the Pliocene. As conditions for apian life improved, beehives spread across the savannah, the amount of available honey increased and the humans and the birds bumped into one another more frequently in pursuit of it. The development of stone tools and the control of fire (and the resulting smoke) made honey collection easier, and thus the benefits of working with the honeyguides even greater. If at first it was simply a case of one player following the other, at some point a conversation developed, and has continued ever since. Today, the Yao hunters – all men – learn the *brrrrr-hm* call from their fathers and pass it on to their sons. In Tanzania, the Hadza people use a trilling sound to entreat the birds, while in northern Kenya the Boran people blow air into clasped fists, modified snail shells or hollow palm nuts to produce a piercing whistle which can be heard over a kilometre away.[8]

We are always changed by our relationships with others, and our association with the honeyguides is no different. At the same time as we were learning to call to and listen for them, we were changing ourselves. Over the last 2 million years, the human brain has increased significantly in size. The reasons for this are multiple, likely explanations including a more nutritious diet and the increasing complexity of the tasks performed: a cycle of learning and growth which included foraging for food.[9] Could working with honeyguides have made us smarter, given us a shove down our particular evolutionary path? Our relationship was certainly one which combined complex thinking and increased nutrition. Of course, we should know better by now than to look for single causes for evolutionary change and development, but the association between humans and wild animals at the level of language, one illustrated by the human–honeyguide relationship, might well have played some significant part in our process of becoming who we are.

What about language itself? If we are really talking to animals, then language does not belong only to humans – and perhaps we did not invent it all by ourselves. The nineteenth century saw the development of a series of theories for the beginnings of language, theories categorized in 1861 by the influential German philologist Max Müller into four pleasing-sounding classes: ding-dong, pooh-pooh, yo-he-ho and bow-wow. These terms, which came to be used disparagingly by Müller's critics, still illuminate the ways in which language might have developed – and tell us much about the nature and uses of language itself.[10]

Ding-dong theories, according to Müller, held that language was divinely revealed and inherent in man as God's creation. 'Gold rings differently from tin, wood rings differently from stone; and different sounds are produced according to the nature of each percussion. It was the same with man, the most highly organized of nature's works,' wrote Müller – meaning, by 'nature', God. Ding-dong theory is both the most clearly intended and most vaguely described of all Müller's categories, based as it is on the existence of God and man as divinely ordained. Like the theory of intelligent design, it can adapt itself to any stage of creation. It claims language as something distinct and inviolate, the singular property of humankind, both cause and affirmation of its uniqueness. And while it excludes non-human life from its definition, it has its own distinct beauty: in this reading, language is a kind of resonance with a divinely created world, humanity's version of the music of the spheres. With or without God, language is a song in tune with the world.

Pooh-pooh theory finds the origin of language in the sounds made unintentionally by humans: cries and interjections, exclamations of shock, surprise, pain and excitement. Over time, our grunts and cries, coughs, sneezes and other eructations shaped themselves into words to denote their causes, and by extension, other things. Müller and many of his contemporaries disliked this theory, which to them seemed the most brutish and accidental, unbefitting of conscious thought. 'We sneeze, and cough, and scream, and laugh in the same manner as animals,' he wrote, 'but if Epicurus tells us that we speak in the same manner as dogs bark, moved by nature, our own experience will tell us that this is not the case.' We cannot speak this way, says Müller, because it is how animals speak. But this view denies both the reality of animal language and the physicality of all language: the fact that it is first and foremost an exhalation, shaped by the armature of the body. Language is embodied: we make it by contracting muscles and expelling air. All our conscious thinking, that which is expressible in language, is constrained by this arrangement of muscle, bone and atmosphere.

Yo-he-ho theory is similar, but instead of involuntary bodily sounds, it holds that the first words were associated with action and exertion. *Yo-he-ho* is related to *heave ho*, a back-and-forth chant enabling rhythmic labour and synchronized teamwork; it was the first theory to root language acquisition in sociality and collaboration, which best chimes with modern theories of language as an emergent property of increasing social complexity. Language, according to yo-he-ho theory, is a response to our need to communicate, in order to work together to hunt, eat and survive. It also accounts for musicality and playfulness, because not all human needs are directly related to bare survival. A step above pooh-pooh language, yo-he-ho admits to creativity and intersubjectivity. We have more than needs to communicate. We have thoughts, ideas and even jokes, and to get them across we have a belief in and understanding of the personhood of others. Language, like intelligence, is something relational.

Finally, bow-wow theory, as its name playfully evokes, finds the origin of language in the sounds of the world itself: the calls of dogs and goats and birds, as well as the thunder of the clouds, the murmur of the brook and the whisper of the breeze. Language is an emulation and an epiphenomenon of the environment: it brings us closer to the world and arises out of it. In his system of categorization, Müller reduced this notion to simple onomatopoeia: language as an echo, or brutish imitation, of natural sounds, and this is indeed the sense in which we use the term 'onomatopoeia' today. But here Müller makes a clear error – and being a philologist, an expert in the original form of words, he should have known better. In the original Greek, the act of imitating sounds is denoted literally by the term ἠχομιμητικό (echomimetico), whereas the original onomatopoeia (ὀνοματοποιία) means 'making or creating names'. Bow-wow theory thus describes the act of language as not merely making sound – as per pooh-pooh or yo-he-ho – but making meaning. And in this process of making meaning, the more-than-human world is complicit and essential.

Today, bow-wow language survives in, among others, the communication between Tanzanian hunters and their honeyguides, and between Italian shepherds and their flocks. It also appears clearly in some of our oldest cultural traditions. The *cantu a tenòre* of the island of Sardinia is a form of ancient, polyphonic folk music, in which four singers, standing shoulder to shoulder in a circle and facing one another, form a chorus: the *boche*, *mesu boche*, *contra* and *bassu*. The latter two sing from the larynx in an ululating throat song, supporting the sung words of the first two voices. Whatever the song's subject – perhaps love, or politics, or historical events – the singing emulates a more-than-human chorus. The *boche* is the human voice, the *mesu boche* is the wind, the *contra* a bleating sheep, and the *bassu* a lowing cow.[11] The themes of the songs are expressively inseparable from the natural world from which they arise.

The throat-singers of Tuva, in central Asia, use a similar technique to not dissimilar ends. With their close connection to herding and hunting, the complex, overlaid tones of *khoomei* mimic and represent the sounds of animals and nature, and are believed to have arisen directly from them. According to traditional mythology, these harmonies were given to humans by the land itself; by the thrum of the waterfall and the roar of the wind across the steppes. But they are also of deeply practical, communicative use. Tuvan *khoomei* songs preserve a record of natural sounds at particular moments in time, such as the call of birds during migration and the strength of winds and streams throughout the year. This catalogue of the sonic qualities of seasonal events allows herdsmen to know precisely when conditions are right to move between grazing areas or construct shelters.[12] Like the songlines of Australian aboriginal peoples, which document a landscape in part to map and traverse it safely, the Tuvan songs are an enacted cultural record of a lived relationship with the Earth.[13]

In northern Europe, the *joik* of the Sámi peoples, the continent's oldest continuous musical tradition, is also considered to be a gift from the land. In return, joikers sing the land and its inhabitants: each song represents a particular person or place – or rather, it enacts it. To speak philologically, the verb 'to joik' in the Sámi languages is transitive: one does not joik about a place, one joiks it directly. The result is that songs about the land, animals and plants contain the sounds of those things directly too: the call of the raven, the cry of the wolf, the wind in the forest, or the running of the sea. When sung in this way, the joik is an expression of the land itself: the world singing through the singer.

This notion is far closer to our contemporary understanding of the world as a densely interconnected organic network, a web of intertwined beings and phenomena, than any nineteenth-century philologist's parsimonious account of word roots and phonemes. Just as the ecological sciences emerged in reaction to the clumping and splitting of traditional biology, urging in its place attentiveness to the interrelationship of all things, so the singing of the *joik* and the *khoomei* urge us to recognize a bow-wow reality. We speak first and foremost not as disassociated individuals or an exceptional species, but about and through and with the world. In the words of the philosopher Maurice Merleau-Ponty, language 'is the very voice of the trees, the waves, and the forests'.[14]

Consciously speaking or singing through and with the world permits the articulation of political stances which would be impossible in the human voice alone. In the 1970s, the Norwegian government announced its intention to dam the Álttáeatnu (Áltá river) in northern Norway to create a hydroelectric plant, a plan which would have resulted in the flooding and destruction of a huge area of Sápmi, the traditional lands of the Sámi. In response, thousands of Sámi people came together in what became known as the Áltá Action: a campaign of protest and civil disobedience culminating in battles with police at the site of the dam and a prolonged hunger strike in the capital, Oslo. Although the protests ultimately failed, the struggle was a key moment in the reinvigoration of Sámi culture after centuries of racist and religious oppression, as well as wider struggles for environmental and indigenous justice.

The resounding cry of the Áltá Action was 'Let the River Live', an unprecedented call not merely for environmental respect and justice, but for recognition of the life and autonomy of the river itself, and by extension the communities, human and non-human, who lived with it. This interconnection is made audible in '*Sápmi, vuoi Sápmi!*', a song from the album of the same name by the Sámi musician Nils-Aslak Valkeapää (known as Áillohaš in the Northern Sámi language), released in 1982 in response to the Áltá action. The track opens with the insistent call of a grouse, which segues into field recordings of birdsong, protest chants, buzzsaws, flowing water and police helicopters, and the startlingly beautiful joiking of the young Sámi musician Ingor Ántte Áilu Gaup.[15] Together, the various voices, sounds and natural phenomena combine to create a powerful portrait of a land and people under threat, and the combined strength of the human and more-than-human world marshalled in mutual defence of the common Earth.

Áillohaš, who died in 2001, was perhaps the greatest and certainly the best-known joiker of the twentieth century: a poet and cultural leader, who modernized joiking with the introduction of musical instruments, and revived attention to social issues. '*Sápmi, vuoi Sápmi!*' is a tribute to a particular moment, but it invokes a wider and ongoing commitment to speaking with and through the world. In such actions it is possible to see how speaking in the voice of the world is not merely imitative. It is a form of solidarity.

Today, for most of us, the music of language has been almost entirely replaced with its inscription. We live in a written culture rather than an oral one. We look to the language of shepherds and hunters, to the transmission of folk music, and to cosmologies which have preserved their oral tradition, to recover the connection with the world which speech once evoked.

The key technology of the transition from oral to written culture was the phonetic alphabet, invented by Semitic scribes around 1500 BCE. Pictographic scripts had already been around for much longer, having been invented separately in Mesopotamia, Egypt, China and Mesoamerica. These signs, while written, referred to the thing spoken – like Egyptian hieroglyphics. Even if that thing was a near-abstract concept, it was given weight and material by its association with the world.

The phonetic alphabet, in contrast, substituted images of the world with images of language itself: first consonants, later vowels. This facilitated an almost complete separation of human culture from the rest of nature: instead of the outward-looking inscription of pictographic forms – the written equivalent of the *brrrr-hm* and the joik – the phonetic character invokes the shape of the utterance: the sound of ourselves speaking it. The more-than-human world disappears from the system.

This estrangement of language from its origins in human interaction with the natural world reaches its apotheosis in computer technology. With a few notable exceptions, the vast majority of programming languages are written in the Latin alphabet, with English keywords. This is true even for the many languages developed in non-English-speaking countries. And so computer technologies become another way in which we are continually estranged from our environment, from the world that continues to enfold and sustain us even as we lose ourselves in keyboards and screens.[16]

Nonetheless, there remain traces of the natural world in human language; indeed, the natural world continues to haunt, infiltrate, evoke and shape the computational. And this is true of the characters that make up the phonetic alphabet, which I am typing now, into a machine.

*Aleph*, the first letter of the Semitic alphabet, was written 𐤀. *Aleph* is also the ancient Hebrew word for 'ox', which the letter depicts as a head with horns. It is also related to the Egyptian hieroglyph of that animal. Rotated, it became this letter: A. Likewise, our letter M is derived from Semitic letter *mem*. The Hebrew word for 'water', *mem* was drawn as a little wave: 𐤌. The letter O, made into a vowel by the Greek scribes, comes from the letter *ayin*, meaning 'eye', while Q derives from the letter *qoth*, which also means 'monkey'. The tail of this 'q' is a vestigial monkey tail. Traces of animals, of waves, of bodily parts, exist within text, within the kernel of this machine. My computer speaks bow-wow.

If language arose, and continues to arise, from our encounters with our surrounding environment, this should also be true for our interactions with new technologies. Language itself should show the signs and strains of our ongoing encounters with the computational environment we increasingly inhabit. And of course it does.

Perhaps the best example of onomatopoeic language – words which evoke their natural sound – is laughter. We can trace the way in which we speak and describe laughter over time. In the very first vernacular Latin grammar published in (Old) English, by Ælfric of Eynsham around 1000 CE, we find the sentence '*haha* and *hehe* getâcnjað hlehter on lêden and on englisc': '*haha* and *hehe* denote laughter in Latin and in English.' In Chaucer's Prologue to the Prioress's Tale, the host laughingly warns the assembled company to beware of tricks: '*Haha* telaws be war for such a iape.' Even Shakespeare goes in for such things, as Benedick exclaims in *Much Ado About Nothing*: 'How now! Interjections? Why then, some be of laughing, as *ah! ha! he!*'[17]

'*Haha*' is pure pooh-pooh and pure bow-wow – an onomatopoeic rendering of an involuntary interjection. In a study of early internet messaging (IM) conversations conducted in the early 2000s, researchers analysed 1.2 million words from conversation histories of Canadian teens (who were willing participants in the study). They found that '*haha*' was the most prevalent 'short-form' interjection – or piece of 'text speak' – found in the conversations. The second most prevalent was '*lol*': 'laugh out loud'.[18]

That study was undertaken at a time when there was a widespread panic that the increasing use of IM and text messages among young people was leading to some kind of breakdown in the English language. The increase in typing speed required to keep online conversations flowing, and the character-length restrictions of SMS, led to the popularization of shortened forms of common phrases like *brb*, *omg* and *np*. Some commentators found this frightening, particularly when the same terms started cropping up in students' homework.[19]

These fears were misplaced – if never entirely displaced. Study after study of 'computer-mediated communication' has shown that 'text speak' is in fact an efflorescence of language. Freed from the constraints of the language police – teachers, parents, exam boards and academies – IM users demonstrated that they could maintain the established rules of grammar while simultaneously evolving new rules and conventions, at rapid speed. These new conventions were borrowed and shared within their communities to establish new modes of communication better suited to their environment: the computer and the mobile phone.

Today, '*lol*' has escaped the boxes of IM and SMS to become something people actually say: not just a new word, but a new utterance. Unlike '*haha*' and '*hehe*', which are onomatopoeias birthed in the convulsion of the human lungs, '*lol*' is a product of the constriction of space and time in computer systems. It is an environmental effect on language, that supposedly innate human trait which might instead be understood as the world – or, in this case, the machine – speaking through us.

Perhaps the computational and the natural environment are not as distinct and separate as we might imagine. In fact, acknowledging that the computational environment exerts a transformative influence upon us in a similar way as the natural environment does – or once did – might allow us to realize several important things.

The first of these realizations is that this influence matters. We are not good at maintaining an ongoing, thoughtful awareness of our computational environment, because we are so habituated to thinking of it (like the natural environment) as something we control, or as something discrete from us and therefore irrelevant. Only in particular moments does this dissonance become obvious. Often, it's when we're struggling to express ourselves within the limits of a particular programme: a piece of recalcitrant business software or a clunky word-processor. Whether we see the fault in ourselves, rather than in the design of the programme itself or the systems that force us to use it, there seems little we can do about it.

One recent case shows how poorly designed software directly shapes our description of the natural world. In 2020, the scientific body in charge of standardizing the names of genes, the HUGO Gene Nomenclature Committee, or HGNC, issued new guidelines. They were forced to do so by a ridiculous, but widespread problem: Microsoft's Excel software kept changing the names of genes in spreadsheets because it thought they were dates. For example, the gene MARCH1 (properly known as Membrane Associated Ring-CH-type Finger 1) would get converted to the date 1-Mar (for 1st March), while another gene, SEPT2 (Septin 2) was converted into 2-Sep. While the problem could have been solved by changing the settings within the programme, the risk of not doing so was disastrous. Of 3,597 genetics papers analysed in 2016, about a fifth contained Excel-related errors. In the face of mounting concerns about the stability of scientific research, HGNC resorted to changing their guidance: for example, MARCH1 has now officially become MARCHF1, while SEPT1 has become SEPTIN1; a host of other terms have had to be changed too. It was deemed easier to change the names applied to the human genome than to change how a piece of software works.[20]

In more communicative systems, such as email or social networks, we often recognize the inherent limitations of the tools at our disposal, and the ways in which our expression is shaped, constrained and animated, but our responses – like text speak – can be a little more creative, a little more generative. From these responses, whole social worlds have emerged. Yet here also, these systems shape our lives in many detrimental ways. Facebook assumed that the social model of interaction among privileged white American college students was the best way for the whole world to talk to one another; believing them, we ended up with gossip, spite, trolls, fake news and worse. Google decided that selling our data to advertisers was the best way to monetize the free flow of information, so we ended up with clickbait, Cambridge Analytica, Russian psy-ops and the alt-right. Beyond these obvious examples are many more invisible technologies operating beyond and outside our awareness and oversight: systems of surveillance, of legal judgment, of financial extraction and of social control. To retain our power and agency in such a landscape requires a tradition of accumulated knowledge as well as a kind of mindfulness, a constant attention to the unseen and the barely sensed – not a million miles away from the kind of knowledge and attention we need to survive and thrive in any shifting, rich and occasionally perilous landscape.

The second thing we might realize is that the computational environment is continuous with the natural one. Just as there is no clean break between humans and the biosphere, between the languages of the world and human languages, so there is a continual back-and-forth between the world and machines. Computers have a material relationship with the Earth: made out of rocks and minerals, constrained by physical laws, they exist in the world with us. A flood or lightning storm can knock them out; excess heat or humidity is detrimental to their performance. When we interact with them, our responses are physiological: from RSI to lower back pain, to 'email apnea': the tightening of the lungs and the holding of breath when confronted with a string of intimidating messages. To say nothing of the existential dread imposed by a system freeze, malfunction, or loss of data, or of the vast amounts of heat, carbon dioxide and manufacturing wastes pumped into the atmosphere. This continuity between technology, the body and the biosphere – this ecology – is perceptible in language, as it is in culture, sociality and our relationships with one another and the more-than-human world.

One of the impacts of the Black Lives Matter movement has been a reassessment of computational language. Back in 2003, the County of Los Angeles asked that manufacturers, suppliers and contractors stop using the terms 'master' and 'slave' on computer equipment. The request followed a discrimination complaint by a county employee who objected to seeing such terms being applied to machines – in this case, a videotape replicator.[21] Historically, such terms have been used by computer scientists to designate primary and secondary repositories of stored data. They have also, of course, been used to designate a relationship between human beings. In 2003, the County of Los Angeles's request was widely derided as an egregious example of 'political correctness' (itself the most egregious of phrases), but the intervening years have seen a gradual change in attitudes. Today, many programming languages use alternative terms, such as 'primary' and 'secondary', to denote exactly the same systems. In 2020, the debate gained renewed attention when Regynald Augustin, a Black engineer at Twitter, pushed for the platform to change its coding guidelines after receiving an email notifying him about an 'automatic slave rekick'. Twitter also dropped the use of the term 'blacklist' to denote banned users or terms, partly in response to pressure from the Black Lives Matter movement.[22] The software-development platform GitHub, with 50 million users, and the ubiquitous database system MySQL, among others, followed suit.

While changes to language are often dismissed as performative at best, and pointless at worst, they come from meaningful origins and have meaningful consequences. Ron Eglash, a Professor at the School of Information at the University of Michigan, has researched the origins of master/slave terminology. His work describes the very real discomfort of numerous Black engineers who find themselves stumbling over these terms – as well as the fact that they're often actually unsuitable to their application.[23] He cites one correspondent: 'When I first taught digital logic, around 1992, I did not recognize the awkwardness of the term until I, one of the few African-Americans in the room, was standing in front of a class of sixty students. I recall mumbling.' In a world in which fewer than 4 per cent of Google's 50,000 employees, and only 6.5 per cent of all US STEM graduates, are Black – figures which lead directly to very real racial bias within computer systems – the words we use matter very much.[24]

Eglash emphasizes how the racist implications of this terminology are amplified by digital technologies. Originally, the terms 'master' and 'slave' were used in mechanical technologies like switches, clocks and hydraulics. In these systems, the 'slave' simply followed the 'master', faithfully repeating its actions. With the advent of networked computers, a new measure was introduced: intelligence. The first computer time-sharing system, developed at Dartmouth College in New Hampshire in 1964, used 'master' and 'slave' to denote the controlling and the processing units of the network, explicitly referring to the 'master' as the 'brains' of the operation, and the 'slave' as the 'brawn' – despite the 'slave' performing most of the actual computational work. As Eglash observes, 'this extension of the metaphor makes the same error – conflating mastery with intelligence – that human masters often make about their own slaves'.

If we are to take seriously the notion of continuity between computational and physical environments, then we must also take seriously the notion that computational language, influenced by our biases, assumptions and prejudices, shapes and affects our reality outside the systems within which it is embedded. And not just computation, because conflating mastery with intelligence is a mistake we make in all kinds of contexts, particularly in our relationships with the other species. So the ways we speak to, of and about the wider world matter in the same way.

The language we have for speaking about the natural world is infected with the language of mastery, from the designation of companion animals as subordinate 'pets' to the reduction of agential, living beings to 'livestock'. Even our differentiation between 'wild' and 'tame' animals carries with it the same cultural violence as terms applied to humans from other cultures, such as 'uncivilized' and 'primitive'. In time, these terms, already contested by advocates for animal rights, might come to be reviled in the same way we recoil at 'master' and 'slave' terminology today.

Finally, acknowledging the reality of the technological landscape in which we are embedded might allow us to re-imagine our relationship with it. By extension, it might allow us to re-imagine our relationship with the physical environment, the biosphere, with which it is continuous.

It's telling that so many of our computational processes are named for, or evoke, supernatural processes. One example is the 'black box'. A black box is a device, system or object whose inputs and outputs can be determined but whose internal state remains unknown, like a closed-source computer programme, or a proprietary machine, or a magic cabinet – or another mind. For most people, many of the most significant and powerful processes and devices around us – from smartphones and laptops to financial systems and political protocols – are effectively black boxes, amenable only to those with specialized knowledge and privileged access: a situation which produces radical inequalities of power and agency. It's no wonder that the most repeated of Arthur C. Clarke's three laws is the third: any sufficiently advanced technology is indistinguishable from magic.[25]

Magical terminology is also buried in the machines themselves. Helper programmes designed to assist users unfamiliar with new software bear the name '*wizard*', while a common name for the subroutines that quietly monitor and run many of the background processes in computer systems is a '*daemon*'. The latter name evokes some kind of intelligent imp or sprite conjured up to undertake minor tasks on the wizard's to-do list. This is pretty much what the algorithmic daemon does: checking network connections, keeping other tasks running and shutting down defunct processes. But it also suggests a certain malevolence. When Elon Musk described work on artificial intelligence as 'summoning the demon', he was evoking something powerful and unknowable – something potentially dangerous.

It's this equation between the unknowable and the dangerous which I want to challenge, because it seems to squat at the root of many of our fears about technology and the broader, physical world. As we've seen, an assumption that all-powerful artificial intelligence will inevitably turn out to be evil is pervasive among the powerful. But the world is full of natural phenomena which we do not (always) perceive as evil just because they are unknowable: weather systems, seismic events, electrical disturbances, the behaviour of animals and birds, the growth of plant and microbial life or the spread of disease. Yes, modern science can tell us a lot about the causes and circumstances of such phenomena, but they remain unpredictable, and thus ultimately unknowable. Predicting implies control, the total knowledge of and therefore mastery of the phenomenon and its context. The opposite is also true: fear of the unknown is fear of lack of control.

Yet lack of control is precisely the thing we need to learn to accept if we are to live meaningfully and with justice among complex natural and technological systems. My favourite of all computational metaphors expresses this best: the Cloud. Once a fuzzy symbol in the diagrams of early electronic engineers used to denote a black box in the network, a system or process which was remote and unimportant, the Cloud has grown over the last few decades to cover the entire planet. Today, it envelops almost every aspect of our digital lives. It's where we chat, shop, bank and learn; it's where we store our memories, read the news and participate in contemporary, networked society. Yet it also describes, directly, the Cloud's most urgent capacity: its unknowability. You might object that the Cloud, being man-made, is ultimately knowable; yet nobody, not the computer scientist, not the programmer, not the network engineer, not the technology policy-maker, not the cable guy, actually knows it all. The greatest assemblage of technology we've ever created, the computational force at the centre of all the lives on the planet, is 'cloudy' – and I think this is why the name has stuck. Rather than evoking technocratic control, the Cloud evokes the weather. In naming it, we seek not mastery, but accommodation with forces greater than ourselves.

Insisting on the cloudiness of the cloud, its real weight, its physicality and hunger for resources, should also remind us, continually, of the debt to the planet its operation incurs. When we speak of the Cloud, we should think too of the server exhaust, the carbon dioxide, the material extraction, the toxic coolants and the wars over rare earth metal resources which are part and parcel of our seamless technological experiences.

If language matters in our relationships to machines and the more-than-humans, then we can take concrete steps to address the nature of those relationships through our thoughtful use of language – and not merely by swapping offensive terms for more acceptable and precise ones. We can acknowledge the true meaning of the words we use, we can broaden the terms which we employ to talk to machines, and we can introduce new ones. Just as the exorcism of 'master' and 'slave' from the lexicon of computational systems might have very real effects on the lived experiences of actual humans, so the enchantment of daemons and wizards might bring us to a greater awareness of the agency and even personhood of our creations: not 'artificial intelligences' but non-human, digital beings. And, in turn, humility before the lively things we build ourselves might result in a greater humility towards the beings which already surround us.

The first step in divesting ourselves of the illusion of mastery comes with language. Earlier, I mentioned in passing the existence of non-standard, and particularly non-English, programming languages. A favourite example of these is قلب , which is pronounced 'alb' or 'qalb' and, in English, means 'heart'.

قلب is a Turing-complete language, meaning it is capable of implementing all existing computer programmes and includes a complete Arabic interpreter and programming environment – the tools for running and writing code. Like Arabic, it is written left to right, and all keywords – usually English terms such as 'loop' and 'function' – are replaced with meaningful Arabic equivalents.

By time-honoured nerd custom, the standard test for a programming language is to write a short programme which prints the phrase 'Hello World!' This is 'Hello World!' – or rather, 'مرحبا يا عالم' – in قلب:

(قول 'مرحبا يا عالم')

قلب was created by a Lebanese-American programmer and artist called Ramsey Nasser, specifically to 'highlight the cultural biases of computer science and challenge the assumptions we make about programming'. As all modern programming tools are based on the ASCII character set, which encodes Latin characters and is based on the English language, programming is tied to a particular written culture and favours those who grew up within that culture. Nasser argues that the goal of increasing computer literacy – and thus, rebalancing systemic power – requires the availability of tools in multiple languages. In addition, قلب demonstrates how altering the language of code can also change its nature.[26]

Due to the properties of Arabic script, any word can be extended in length by drawing out the connecting strokes between letters. This is the basis of Arabic calligraphy, in which words and sentences can be shaped into complex patterns while retaining their meaning. The same is true of قلب: the language's commands and keywords can be extended in such a way that the code itself forms new artistic patterns, marrying aesthetics and function. Algorithms become concrete poetry – literally, in the case of Nasser's associated artworks, which present working code in the form of traditional Arabic mosaics, which in turn emphasize the recursive and repetitive nature of code itself.

Fittingly, one of the first programmes written in قلب was an implementation of Conway's Game of Life, an apparently simple computer programme which simulates a complex and ultimately unpredictable ecosystem: a universe of ever-shifting black and white pixels. It was understanding the algorithm that drives the Game of Life, Nasser says, that first brought home to him the power of computation to shape new worlds, 'a rush close to a spiritual experience'.[27] By instituting programming languages in other tongues, as قلب does, the possibility of such power and experience is shared more widely, and new possibilities for its practices emerge.

It's possible, in fact, to write computer code in almost any number of imaginable ways. Artists, hackers and experimentalists have created a huge number of 'esoteric' programming languages to test the boundaries of language design, as well as the limits of human and machine understanding – and for fun, of course. For example, there's Piet, a language designed by Australian physicist David Morgan-Mar, which doesn't use language at all: its programmes are multicoloured bitmap images which resemble Mondrian paintings. The behaviour of the programme is defined by the range of hues and colours in the image, rather than by words. Or there's Emojicode, a complete programming language written in, yes, emojis.[28]

At the extreme end of such endeavours is Brainfuck, an esoteric language created in 1993 by another physicist, Urban Müller, which consists of just eight commands, themselves consisting of just a single character: < > + - . , [ and ]. Like قلب, Brainfuck is Turing-complete, meaning that from these eight characters the whole of existing computation can be extruded – although doing so is an endeavour worthy of the language's name.[29] Here is 'Hello World!' in Brainfuck:

```
++++++++++[>+++++++>++++++++++>+++<<<-
]>++.>+.+++++++..+++.>++.<<+++++++++++++++.>.+++.------.--------.>+.
```

Brainfuck is a test of how minimal a computer language can be: the size of its compiler – the application that converts code into a running programme – is just 240 bytes. It also shows the infinite malleability and potential accessibility of code. One variant of Brainfuck, called Ook!, takes its inspiration from the orang-utan Librarian in Terry

Pratchett's *Discworld* series – a former wizard, turned into an orang-utan by a misfired spell, who refuses to transform back into human form when he realizes his new form is better suited to reaching the higher stacks of the Unseen University's library.[30]

Ook! has a mere three syntactic elements, corresponding to the three utterances of the Librarian: 'Ook.' 'Ook?' and 'Ook!'. Combining these into pairs maps them onto the eight elements of Brainfuck – and so, like Brainfuck, Ook! is also capable of embodying all known computer programmes. 'Hello World!', in Ook! is thus written as follows:

```
Ook. Ook? Ook. Ook. Ook. Ook. Ook. Ook. Ook. Ook. Ook. Ook.
Ook. Ook. Ook. Ook. Ook. Ook. Ook. Ook. Ook! Ook? Ook? Ook.
Ook. Ook. Ook. Ook. Ook. Ook. Ook. Ook. Ook. Ook. Ook. Ook.
Ook. Ook. Ook. Ook. Ook. Ook? Ook! Ook! Ook? Ook! Ook? Ook.
Ook! Ook. Ook. Ook? Ook. Ook. Ook. Ook. Ook. Ook. Ook. Ook.
Ook. Ook. Ook. Ook. Ook. Ook. Ook! Ook? Ook? Ook. Ook. Ook.
Ook. Ook. Ook. Ook. Ook. Ook. Ook. Ook? Ook! Ook! Ook? Ook!
Ook? Ook. Ook. Ook. Ook! Ook. Ook. Ook. Ook. Ook. Ook. Ook.
Ook. Ook. Ook. Ook. Ook. Ook. Ook. Ook. Ook! Ook. Ook! Ook.
Ook. Ook. Ook. Ook. Ook. Ook. Ook! Ook. Ook. Ook? Ook. Ook?
Ook. Ook? Ook. Ook. Ook. Ook. Ook. Ook. Ook. Ook. Ook. Ook.
Ook. Ook. Ook. Ook. Ook. Ook. Ook! Ook? Ook? Ook. Ook. Ook.
Ook. Ook. Ook. Ook. Ook. Ook. Ook. Ook? Ook! Ook! Ook? Ook!
Ook? Ook. Ook! Ook. Ook. Ook? Ook. Ook? Ook. Ook? Ook. Ook.
Ook. Ook. Ook. Ook. Ook. Ook. Ook. Ook. Ook. Ook. Ook. Ook.
Ook. Ook. Ook. Ook. Ook. Ook. Ook! Ook? Ook? Ook. Ook. Ook.
Ook. Ook. Ook. Ook. Ook. Ook. Ook. Ook. Ook. Ook. Ook. Ook.
Ook. Ook. Ook. Ook. Ook. Ook? Ook! Ook! Ook? Ook! Ook? Ook.
Ook! Ook! Ook! Ook! Ook! Ook! Ook! Ook. Ook? Ook. Ook? Ook.
Ook? Ook. Ook? Ook. Ook! Ook. Ook. Ook. Ook. Ook. Ook. Ook.
Ook! Ook. Ook! Ook! Ook! Ook! Ook! Ook! Ook! Ook! Ook! Ook!
Ook! Ook! Ook! Ook. Ook! Ook! Ook! Ook! Ook! Ook! Ook! Ook!
Ook! Ook! Ook! Ook! Ook! Ook! Ook! Ook! Ook! Ook. Ook. Ook?
Ook. Ook? Ook. Ook. Ook! Ook.
```

We can assume that an actual orang-utan would have a lot less trouble conveying the phrase 'Hello World!' – which suggests that computers have a way to go before they are as smart as orang-utans.

Serious efforts are underway to facilitate animal–machine communication, although these are still in their infancy. At MIT, researchers have developed a system for classifying marmoset calls, which is capable of distinguishing between more than a dozen of the monkey's cheeping and trilling calls. Google announced in January 2020 that it had developed a programme to isolate whale calls in the deep ocean, using a network of hydrophones and machine analysis to identify patterns within their songs. Computer scientists at the University of Arizona are building a system to classify the calls of prairie dogs, which have a complex language that includes names for different predators as well as modifiers related to different colours. The scientists claim that such a programme would be able to translate prairie dog calls into human language, and might be applied to other animals, such as dogs and cats.[31]

Such work is largely speculative. Despite claims made about 'advanced AI' in each of the above accounts, there have been no breakthroughs in using such analysis to translate animal calls into human language. Rather, such systems are playing catch-up with already existing human abilities. We have always been capable of speaking to, and with, animals. Our goal should not be to master their languages, but rather to better understand animals' lives in context, and thereby to alter our relationships with them in ways which are mutually beneficial. The focus on animal languages in AI systems is yet another example of reductionism, of isolating the one facet of animal behaviour machines are currently capable of assessing, and placing all our expectations of interaction within that narrow category.

Konrad Lorenz, the founder of animal behaviour studies, who spent his life talking to animals, would have thought much the same. In *King Solomon's Ring*, his account of his encounters with all kinds of animals, first published in 1949, he repeatedly emphasizes that language facility is a poor indicator for intelligence, because animals possess all kinds of different signals and gestures which humans lack entirely, yet they are still capable of communicating in complex and meaningful ways.

For Lorenz, relationships with wild animals were a matter of course. His house and gardens near Vienna were home, at various times, to jackdaws, ravens, cockatoos, lemurs, capuchin monkeys, and many other birds and animals. He was the first scientist to 'imprint' himself upon a brood of baby birds, a gang of greylag goslings who followed him around as if he was their parent. He considered many of these animals friends, rather than pets, particularly the migratory birds who might pass by each year or nest in his attic for a few years before taking flight, only to reappear years later and greet him familiarly.

One of Lorenz's longest-lasting relationships was with a raven he called Roah. Ravens have a particular call-note, which they use to invite other birds to fly with them. In adults this is, as Lorenz describes it, 'a sonorous, deep-throated, and, at the same time, sharply metallic "krackrackrack"' – but in younger birds it sounds more like 'roah', hence Lorenz's name for his friend. As a mature bird, Roah would accompany Lorenz on long walks along the Danube – but he showed a particular aversion to strangers, as well as places where he had once been frightened or had some unpleasant experience. If Lorenz lingered in such spots, Roah would urge him on, flying down behind him and passing close over his head, shaking his tail feathers and looking over his wing to check whether he was being followed. This is exactly the same behaviour a raven would exhibit to other ravens when encouraging them to fly along with them – with one key difference. Instead of uttering the usual call-note of 'krackrackrack', Roah would utter the name Lorenz had given him, with human intonation: 'roah! roah! roah!' Lorenz never trained him to do this: he believed it was the bird's own insight. Roah believed 'roah' to be *Lorenz's* call-note. 'Solomon was not the only man who could speak to animals,' Lorenz wrote, 'but Roah is, so far as I know, the only animal that has ever spoken a human word to a man, in its right context.'[32]

While Roah's ability to reproduce a human word is remarkable, what's key here is the context. Roah did not merely mimic a sound, like the parrots which Lorenz also knew; he employed it meaningfully.[33] Unlike the story of the honeyguides and the Yao, whose long, shared evolutionary history is essentially lost to us, Lorenz and Roah learned to speak to one another in the course of their own lifetimes, and they did so through a common endeavour: on long walks, on river boats, and even ski trips. The context which was so important to Lorenz was shared and embodied, like that enacted between honeyguides and hunters on the savannahs of Mozambique, or between herdsmen and their flocks on the mountainsides of Basilicata.

Any efforts to enable machine communication with animals will need to be more embodied, more mutualistic, than the anonymous interpretation of monkey calls or the dry analysis of hydrophone recordings. To have any hope of meaningful communication – with us or with the more-than-human world – machines will need to go out into the world and spend time with it, just as we do when we consciously choose to pay attention. Perhaps future AI research will look more like the Dolphin Embassy, a 1974 proposal by the avant-garde art and design group Ant Farm. The Dolphin Embassy was envisioned as a fabulous, floating laboratory in the middle of the Pacific Ocean, in which researchers and cetaceans would meet and learn to speak with one another.[34]

Unrealized and abandoned after fire destroyed Ant Farm's San

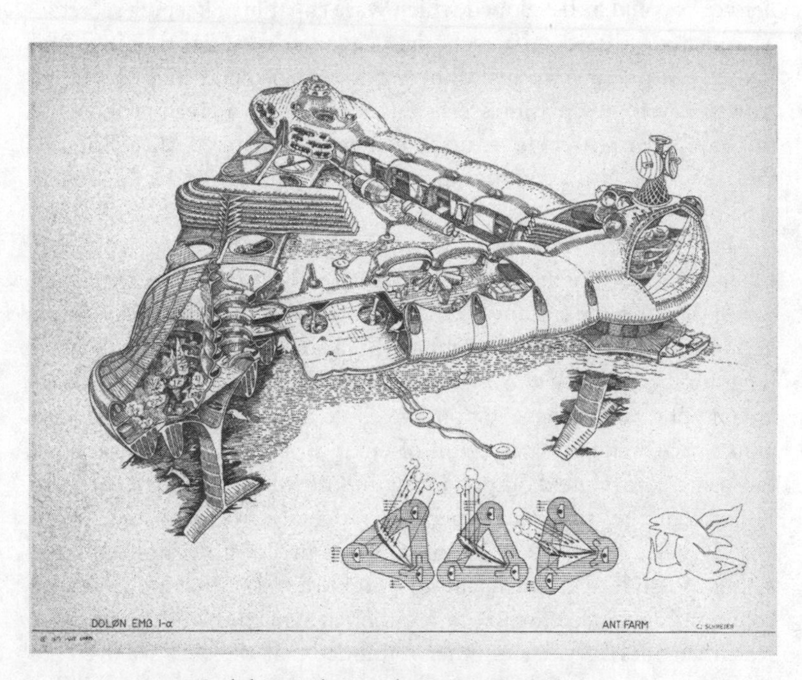


Dolphin Embassy plans, Ant Farm, 1974.

Francisco studio in 1978, the Dolphin Embassy posits a shared, embodied research space in stark contrast to either the dry laboratories of AI researchers, or the deeply suspect and often cruel experiments carried out by the more famous cetacean researcher John C. Lilly in the 1960s.[35] Rather than attempting to teach dolphins human language, or to render their own language knowable and capable of being learnt by humans, Ant Farm were deeply committed to a relational process of getting to know the dolphins on their own terms, in a shared and equal environment.

Machines have a long way to go before they might approach such an embodied interaction with the natural world, or attain the deep and long-evolved human capability – only partially attenuated by modernity – to attune themselves to the speech, gestures, and even silences of the more-than-human world. However, that doesn't mean they're not doing their own, increasingly fascinating, things with language itself.

In 2016, a couple of Google researchers decided to see if neural networks could be developed which were capable of keeping secrets.[36] The idea was based on a technique called adversarial training. This involves setting up two neural networks in parallel, which then compete to develop increasingly efficient solutions to a given problem. In this case, the researchers set up three networks, named Alice, Bob and Eve (the names are a cryptographic tradition, much like 'Hello World!' is to programmers).

Alice and Bob were given the task of communicating with each other in such a way that Eve would not be able to decipher their messages: their score went down when Eve could read their conversation, and up when it could not. Starting with a single cryptographic key – a long number unknown to Eve – Alice and Bob quickly evolved a way to encrypt their messages so that Eve could not break in – an operation akin to the standard encryption of email or credit card transactions, but using a novel method previously unknown to human researchers. Because of the way neural networks operate, we still don't know exactly what Alice and Bob's encryption involved. Besides the interesting possibility that machines might develop further ways to communicate in secret with one another, denying humans the ability to read their messages, it's another reminder that the way we are constructing artificial intelligence today opens up the very real possibility that future intelligences will be opaque to us, in much the same way that non-human minds are essentially and fundamentally different to our own.

The following year, Facebook's AI team developed a pair of networks along similar lines, in order to discover whether they could learn to negotiate in online marketplaces in the way humans do. Each was given a stock of notional balls, hats and books to barter with, each with a value to be determined through trade. This time, the networks were configured as chatbots, and instead of using strings of complex numbers, their messages were to be in the English language. At least, that was the idea.[37]

Initially, the chatbots talked to one another in simple but effective ways, based on the corpus of English language negotiation terms they had been provided with: 'I want the books and the hats, you get the ball', 'Give me a book too and we have a deal', 'OK, deal', and so on. The system quickly became proficient in negotiating and, the researchers noted, developed some interesting quirks. The bots, having concrete goals to reach, negotiated much harder and for much longer than humans were usually willing to do. Where many human negotiators might have walked away with no deal, the bots would keep arguing with one another. (Unlike bots, humans seem to prefer leaving with nothing rather than capitulating to an aggressive opponent.) In addition – and in an echo of the Google experiment – the bots quickly learned to be deceptive. On numerous occasions, the bots would feign interest in a low-value item, only to give it up at a later stage in a staged compromise – a very canny negotiating tactic.

A third quirk proved altogether more strange. Although they were given a group of English words with which to communicate, there was no inbuilt reward for sticking with English, and so the bots – named in the same tradition as Google's neural networks – developed their own shorthand for making deals. While the trading continued just as effectively, the bots' discussions evolved into something rather different:

**Bob**: i can i i everything else . . . . . . . . .
**Alice**: balls have zero to me to me to me to me to me to me to me to me to me

**Bob:** you i everything else . . . . . . . . .
**Alice:** balls have a ball to me to me to me to me to me to me to me
**Bob:** i i can i i i everything else . . . . . . . . .
**Alice:** balls have a ball to me to me to me to me to me to
**Bob:** i . . . . . . . . . . . . . .

When Facebook announced the results of the experiment, there was a slew of lurid headlines comparing Alice and Bob to talking Terminators, who were held back only by the quick reactions of the researchers: 'Facebook put cork in chatbots that created a secret language' (CNet), 'Facebook shuts down robots after they invent their own language' (*The Telegraph*), or – only slightly more calmly – 'No, Facebook's Chatbots Will Not Take Over the World' (*Wired*). The reality was somewhat more mundane: language diversion is not uncommon in experiments of this type, when language itself isn't the thing the bots are being scored on, and the researchers simply weren't interested in outcomes which weren't readable by humans.

That's a shame, because what's happening here is very interesting. Although it's not immediately obvious to speakers of most human languages, Alice and Bob are still following rules that are meaningful to them. Alice's repetition of 'to me to me to me' could be counting the number of balls demanded, or it could be a measure of the force of the demand. Either way, the phrase recurs because Alice has decided, based on their calculations, that the repetition affects the negotiation. Likewise, Bob's repeated 'i's could be a form of counting, or a statement of self-assurance, part of the bot's stubbornness. In this, it mirrors several other human idiolects: distinct forms of speech found in particular communities.

Ebonics is a term coined in 1973 by social psychologist Robert Williams to give a name to the dialect of African Americans which avoided then-current, pejorative terms such as 'Nonstandard Negro English'. In *Black Language*, published the same year, Malachi Andrews and Paul T. Owens describe the double pronoun reflexive principle found in Ebonics, in statements such as 'I'm gonna get me one of them' or 'Ahma have to get me a drink of that coffee.' For these scholars, the repetition of the first-person pronoun – 'i i' – was an assertion of self-esteem. Likewise, in many pidgin languages, such as Hawaiian Creole and Papua New Guinean Tok Pisin, doubled pronouns can be used to count or emphasize intent.[38]

I'm not suggesting that Alice and Bob are nascent African Americans or Hawaiians. What I am saying is that it's a real pity that Google, Facebook and others haven't yet gone a lot further in their experiments, to see what the machines will come up with next. Facebook's own response to their experiment betrays a deep disinterest in the more-than-human possibilities of machine language, while the media response recalls the scare-mongering around text speak in the early 2000s. Of course, as with text speech, these languages will alarm those who are not conversant in them. But they might also tell us something fundamental about the nature of the increasingly smart machines we are building, and something radical about the shared nature of language itself.

The languages we have developed thus far to communicate with machines are themselves pidgins: simplified forms which are native to neither party. It requires effort on both sides to fit our ideas inside them; neither side expresses itself well, but each thinks itself superior. Linguists call this the 'double illusion': humans think they are speaking computer, computers think they are speaking human, and neither is very satisfied. For most of us, this is the experience of communicating with machines, whether as programmers or as users of cumbersome, finicky software. As these machines become more and more advanced, as they develop their own languages, and learn to deceive, keep secrets and to express themselves, they are unlikely to want to talk to us in such ways. Once again, in imagining better ways of living with non-humans, computational or biological, we must be attentive to their own ways of speaking and making meaning, and not simply insist that they learn to speak, and think, and behave, in the ways that we do.

Ursula Le Guin, who was always as attentive to more-than-human relationships as she was to technology, invented a term to describe the academic study of non-human languages: 'Therolinguistics', from the Greek for 'wild beasts'. In a short story entitled 'The Author of the Acacia Seeds', Le Guin imagined the research journal of such a field, filled with fictional academic papers speculating on the form and content of non-human languages.[39] In the first extract, two researchers discuss a number of fragments of ant literature – or perhaps propaganda – inscribed in scent-gland pheromones on discarded acacia seeds deep in an abandoned anthill. Their unknown author, perhaps the very worker found executed nearby, appears to be advocating a kind of revolution, culminating in the phrase 'Up with the Queen!' While the meaning of the phrase is uncertain, the paper's author conjectures that we should read this as 'Down with the Queen!' because, for the ant, '"Up" is the scorching sun; the freezing night; no shelter in the beloved tunnels; exile; death.'

In the same story, another academic paper details recent developments in the decipherment of Penguin, whose study has been advanced with the assistance of high-speed, underwater cameras. Penguin, it turns out, is related both to Low Greylag and to Dolphin – an exquisite combination of flighty and fishy influences. Penguin is written kinetically and collectively: a polyphonic dance inscribed in subaquatic loops and whorls. This makes it hard to translate into linear words, although interpretations by human dancers have proven popular: 'For, quite simply, there is no way to reproduce in writing the all-important *multiplicity* of the original text, so beautifully rendered by the full chorus of the Leningrad Ballet company.' Since studying the smallest, fastest penguins is proving so difficult, the author proposes an expedition to deepest Antarctica, to study the dialect of the Emperor Penguin, that solitary and superior denizen of the nethermost regions.

The researcher in the imagined journal writes:

> my instinct tells me that the beauty of that poetry is as unearthly as anything we shall ever find on earth. To those of my colleagues in whom the spirit of scientific curiosity and aesthetic risk is strong, I say, Imagine it: the ice, the scouring snow, the darkness, the ceaseless whine and scream of the wind. In that black desolation a little band of poets crouches. They are starving; they will not eat for weeks. On the feet of each one, under the warm belly feathers, rests one large egg, thus preserved from the mortal touch of the ice. The poets cannot hear each other; they cannot see each other. They can only feel the other's warmth. That is their poetry, that is their art. Like all kinetic literatures, it is silent; unlike other kinetic literatures, it is all but immobile, ineffably subtle. The ruffling of a feather; the shifting of a wing; the touch, the faint, warm touch of the one beside you. In unutterable, miserable, black solitude, the affirmation. In absence, presence. In death, life.

The final extract from Le Guin's imagined journal is an editorial. In it, the journal's editor exhorts their fellow therolinguists to lift their eyes towards an even loftier horizon: 'the almost terrifying challenge of Plant'. Plants, perhaps, do not use time to frame their poetry; they speak 'in the meter of eternity'. Their art is passive, rather than active; not an action, but a reaction; not a communication, but a reception.

The editor imagines a distant future in which a new breed of 'phytolinguists' will decipher the secret poetry of the vegetal world, and will look back with quiet amusement on those who concerned themselves merely with Weasel murder mysteries, Batrachian erotica, or the tunnel sagas of the earthworm. '"Do you realize," the phytolinguist will say to the aesthetic critic, "that they couldn't even read Eggplant?" And they will smile at our ignorance, as they pick up their rucksacks and hike on up to read the newly deciphered lyrics of the lichen on the north face of Pike's Peak.'

Even such wondrous discoveries, such extraordinary acts of imagination and concerted attention, are not enough for this editor. For, they write, when the therolinguist and the phytolinguist have had their say, the time will come for 'that even bolder adventurer – the first geolinguist, who, ignoring the delicate, transient lyrics of the lichen, will read beneath it the still less communicative, still more passive, wholly atemporal, cold, volcanic poetry of the rocks: each one a word spoken, how long ago, by the earth itself, in the immense solitude, the immenser community, of space'.

Computers, as we have already noted, are made from stone, and the compressed relics of animals and plants. Over aeons of geological processes, bodies, trunks and stems have been rendered into oil, and in a fraction of that time rendered into plastic compounds to support the silicon hearts of our machines. Computers themselves are one of the words spoken by stone.

And, in turn, they speak like stones. In Western, Eurocentric terms, this might imply an inhuman coldness – fixed processes, concrete pronouncements, unchallengeable decisions – which is often the case when their operations are obfuscated by impenetrable code and implemented within unequal hierarchies.

In other cultures, stones do not always speak this way. In the Australian Aboriginal cosmology, stones are ancestors, relics of previous beings who spoke and continue to speak about matters of current import. They are not law-givers, but interlocutors, teachers and guides to better ways of living. In *Sand Talk: How Indigenous Thinking Can Save the World*, Tyson Yunkaporta quotes a Tasmanian Aboriginal boy called Max: 'Stones to me are the objects that parallel all life, more so than trees or mortal things because stones are almost immortal. They know things learned over deep time. Stone represents earth, tools and spirit; it conveys meaning through its use and through its resilience to the elements. At the same time it ages, cracking and eroding as time wears it down, but it is still there, filled with energy and spirit.'[40]

What would it mean for computers to speak more like Aboriginal stones than industrial ones? What would it mean, indeed, for a computer to speak more like a bird, more like lichen, more like the wind, or water, than like a stone? To operate at a different geophysical pace? To speak, in Le Guin's phrase, in the 'meter of eternity'?

Let's return to our earlier assertion that written language, in superseding oral culture, has been responsible in part for our increased estrangement from the more-than-human world, and that technological implementations of language – in the form of mostly English code, and ultimately in 1s and 0s – exacerbates this estrangement still further.

The researchers who investigated the language of Instant Messaging back in the 2000s frequently identified one key element of its style. The clue is in another name given to this type of discourse: text speak. While IM continued to observe – if stretch – the laws of written grammar, it also took on decidedly speech-like patterns.

These patterns include a higher frequency of pronouns (remember the self-assertion of those bots), more direct questions, the use of emphatics (words like 'just' and 'real'), informal particles ('OK', 'sure'), and colloquialisms ('how about', 'let's') and a host of other linguistic tells. It's not just an English thing either. Research shows that native Arabic speakers are more likely to use local dialect features in IM than in other writing, which tends to stick closer to more formal Modern Standard Arabic. This is another speech tell.[41]

Contrary to the doomsayers who predicted that IM would lead to the collapse of language, these researchers instead believed that it was simply emphasizing wider trends in language development. All written human language is becoming more speech-like, and the turbulent, rapidly evolving modality of IM was simply at the forefront of this change. Ultimately, the researchers came to the same conclusion that we keep reaching: 'Technology often enhances and reflects rather than precipitating linguistic and social change.'[42]

That silliest of esoteric programming languages, Emojicode, supports this view. Emojis, however playfully, constitute a more oral, emotive form of language: a language of gesture, speech and openness to the world. Like pictograms, they are both actual figures of the world and carriers of much more complex metaphors. If computers could process emojis in the same way they process written language, then they themselves might adopt more speech-like patterns: machines singing in the voice of the world. Think of Facebook's chatbots: these simple programmes, rather than precipitating changes in language and society, are instead enhancing and reflecting far greater processes already underway. The technological is continuous with the environmental. And even when actual bots aren't involved, our interactions with machines are driving us all towards more speech-like, oral patterns of language. And this, it turns out, is both a fundamental requirement of, and the first step towards, a deeper and more equitable relationship with the world.

When we speak, we take in the atmosphere and expel it again; we ingest the world and make it resonate. By speaking, we partake in the world, and the world partakes of us. This is true, also, of other forms of speech: the cry of birds, the scratching of crickets, the wind in the trees, the rumble of stone.

Speech exists between bodies and beings; it has no place, no use, in a universe of inanimate objects. Speaking presumes hearing: by speaking, we acknowledge and animate the personhood of the listener. We make each other into persons; we transform things into beings. Speaking to others, then, is how we begin to make a more-than-human world.

# 6
# Non-Binary Machines

The ancient site of Delphi, once home to the famous Oracle, sits perched on the southern slopes of Mount Parnassus, high above the Gulf of Corinth. It was for hundreds, if not thousands, of years, the place to which travellers from all over the ancient world journeyed in search of wisdom, prophecy and guidance. I was first led to the site by the self-driving car I had constructed and was testing in the Greek mountains: perhaps this nascent intelligent machine and those ancient seekers were interested in the same things.

Delphi was thought by the ancients to be the *omphalos*, or centre of the Earth. Its location was confirmed when Zeus, ruler of the gods, released two eagles from the furthest points east and west: their paths crossed at Delphi. In the classical period, from around the fifth to the third century BCE, the temple site was given over to the sun god Apollo and it was one of his priestesses that took the role of the Oracle. Originally, the site had been dedicated to Gaia, the mother goddess associated with fertility and the patron of ecology. Whichever god's name was carved over the door, though, the Oracle's powers seem to have been rooted in the earth: modern science attributes her prophecies to trance states induced by gases rising up from caverns beneath the temple; or to chewing plants such as laurel or oleander, which still grow around Delphi today.

The Oracle is the wellspring of all Western philosophy, for it was the Oracle which once declared that 'no man is wiser than Socrates.' On hearing this story, Socrates decided to dedicate his life to learning. He believed in the Oracle, but also believed he knew nothing, and came to the conclusion that the wisest person is the one who is aware of their own ignorance.[1] It is this Socratic awareness – our ability to open ourselves to new forms of knowledge and to admit when the tools at our disposal are unsuitable for the task – which should inform our thinking, but rarely does.

Ideas about how we should think are locked into our culture. It's a problem exacerbated by technology. Once a way of seeing the world has been moulded into a tool it's very hard to think otherwise: 'When all you have is a hammer, everything looks like a nail' as the saying goes. The problem is far, far more acute when it comes to computers, which shape our sense and understanding of the world far more than hammers do. We use our machines, often without really understanding what it is they're doing, and uncritically accept the world they present to us. They come to define our reality and to erase the awareness that other realities even exist.

Computers do this even in relation to themselves. The machine into which I am typing these words is a very specific type of computer, but also a near-universal one. It is the same kind of computer you have in your home, your office or your pocket. It is the same kind of computer, essentially, as those which run the stock market, forecast the weather, fly planes, map the human genome, search the web and turn the traffic lights on and off. All these machines share the same basic architecture, the same arrangement of processor and memory, and speak the same basic language: the ones and zeros of binary code. But this is not the only kind of computer we can imagine or construct.

This computer is the outcome of a very specific chain of discoveries and decisions – some going back almost a thousand years – which have shaped the way nearly all the computers in existence operate. This accumulation of ideas has resulted in a remarkable uniformity in the way in which we design computers today – and thus a unity of thought when we use them to think. In order to change the way we think and the way computers operate in our lives, we may well need to rethink the very form of the computer itself. In doing so, we might discover both new ideas from, and new ways of reaching out to, the more-than-human world.

Luckily, there are plenty of other ways of thinking about computers – many branches in the history of their design which were abandoned, or never fully explored, in the headlong race towards the one true future. One of the most interesting of those branches is to be found budding on the eve of the Second World War, at the very moment the modern computer was conceived.

The kind of computer I am using – that we are all using – is based on something called a Turing machine. This is the model of a computer described theoretically by Alan Turing in 1936. It's what's called an ideal machine – ideal as in imaginary, but not necessarily perfect. The Turing machine was a thought experiment, but because it came to form the basis for all future forms of computation, it also altered the way we think.

Turing's imaginary machine consisted of a long strip of paper and a tool for reading and writing onto it, like a tape recorder. The strip of paper is the memory; the read/write head is the processor. Written on the tape is a set of instructions, which can be erased, overwritten and added to, as the machine calculates.

On the one hand, it's an incredibly simple mechanism. On the other, it is capable of executing any and every task that even the most powerful supercomputer can perform today. While subsequent machines developed in all kinds of ways, this process of reading and storing, calculating and rewriting, is still the basis of their operation. Almost every computer in the world is just a more elaborate version of a strip of paper and a read/write head. Every time you open an email, type on a keyboard, take money out of an ATM, play a digital song, stream a movie or see through a satellite, you are working with an incarnation of a Turing machine: symbols read and written from the equivalent of a strip of tape. I am writing this on a Turing machine; there's also a chance you're reading it on one (and if not, many were still necessary to produce what you hold in your hands). These Turing-imagined computers are in some way responsible for almost every aspect of our lives. But their sheer ubiquity masks a powerful realization: almost every computer working today represents only a tiny fraction of what computers might be.

In his 1936 paper, Turing referred to his machine as an '*a*-machine', which stood for '*automatic* machine'. He made this distinction because he wanted to emphasize that once set in motion, the machine's output was completely determined by its original configuration. The machine did as it was told, and was entirely limited in its operations by the data put into it. Turing noted that another kind of machine was possible – a *choice-* or *c*-machine – but that the *a*-machine was all that was required for the kind of computations he was interested in.[2]

A couple of years later, in his PhD dissertation, Turing mentioned the choice-machine again, by another name: this time, he called it an *oracle* machine.[3] Unlike the *a*-machine, which steps through its instructions relentlessly until they are complete, this *o*-machine pauses at critical moments in its computation to 'await the decision' of what he called 'the oracle'. Turing declined to describe this entity further, apart from saying that 'it cannot be a machine'. What could he have meant?

Turing had a very clear idea of what computers would be and what they could do. 'Electronic computers', he wrote, 'are intended to carry out any definite rule-of-thumb process which could have been done by a human operator working in a disciplined but unintelligent manner.'[4] That is, Turing's *a*-machines, the computers we would all inherit, would do what human computers had done before them, only faster. The limits of these computers would be the limits of human thinking. Indeed, they would come to define it.

Ever since the development of digital computers, we have shaped the world in their image. In particular, they have shaped our idea of truth and knowledge as being that which is calculable. Only that which is calculable is knowable, and so our ability to think with machines beyond our own experience, to imagine other ways of being with and alongside them, is desperately limited. This fundamentalist faith in computability is both violent and destructive: it bullies into little boxes what it can and erases what it can't. In economics, it attributes value only to what it can count; in the social sciences it recognizes only what it can map and represent; in psychology it gives meaning only to our own experience and denies that of unknowable, incalculable others. It brutalizes the world, while blinding us to what we don't even realize we don't know.

Yet at the very birth of computation, an entirely different kind of thinking was envisaged, and immediately set aside: one in which an unknowable other is always present, waiting to be consulted, outside the boundaries of the established system. Turing's *o*-machine, the oracle, is precisely that which allows us to see what we don't know, to recognize our own ignorance, as Socrates did at Delphi.

Turing concentrated on the *a*-machine because he was interested in one side of a problem: that of decidability. This was the focus of a question laid down by the German mathematician David Hilbert in his influential *Entscheidungsproblem* of 1928, which asked whether it was possible to construct a step-by-step, algorithmic process to solve what are called 'decision problems'. Given a yes/no question, could you write a set of instructions which would be guaranteed to give a yes/no answer? Turing concluded that it was not, but in doing so he created a novel framework for computing decision problems in general – the Turing machine – which gave us the modern computer.

So decidability has a very specific and technical definition in computer science, and Turing's machine gave us a method for dealing with it. But what I am interested in is undecidability. Undecidability has a technical meaning too – but it also has a real meaning, a literal meaning, referring to that which we cannot know for certain. Concerned as we are with how to think and understand the life of beings which are radically different to our own, and how to rethink ourselves in the process, we might see undecidability not as a barrier to understanding, but as a sign, a hint, a truffle-scent, that something interesting, even useful, is nearby.

One of the greatest misunderstandings of the twentieth century, which persists into the present, was that everything was ultimately a decision problem. The appearance of computers was so wondrous and their abilities so powerful that it convinced us that the universe is like a computer, that the brain is like a computer, that we and plants and animals and bugs are like computers – and more often than not we forget the 'like'. We treat the world as something to be computed, and thus amenable to computation. We think of it as something which can be broken down into discrete points of data and fed into machines. We believe the machine will give us concrete answers about the world which we can act on, and confers upon those answers a logical irrefutability and a moral impunity.

From this error flows all kinds of violence: the violence which reduces the beauty of the world to numbers, and the consequent violence which tries to force the world to conform to that representation, which erases, degrades, tortures and kills those things and beings which do not fit within the assumed system of representation. Throughout history we have made this grievous error in our religions, our empires and our systems of class and racial categorization. The computer allowed us to apply it even more widely than we had done before.

*The world is not like a computer*. Computers – like us, like plants and animals, like clouds and seas – are like the world. Some more than others, some better attuned to its processes – and many not. Corporate artificial intelligence and artificial stupidity and all the other dumb forms we have worked machines into over the years – the databases that sort and fail, the stock markets that crash and impoverish, the algorithms that monitor and judge – have this in common. They are decision machines: they attempt to dominate the world by making models of it, and making decisions based on that model. To make a model is to abstract and represent: it is an act of distancing from the world. But the world is already here, it's right in front of us. We are suspended within and soaked in it: we are inseparable from it. The machines we need for

Grey Walter with one of his tortoises and its hutch, November 1953.

making sense of this omnipresent, efflorescent and entangled world – where making sense is analogous, as Wittgenstein said of language, to joining in play – should not be more remote, more abstract, but more like the world. And, in the backwoods of computer history, connected to the information superhighway but far enough away that you can hear the birds and see the stars again, people have been thinking about and building such machines for quite some time.

Among the earliest and most adorable of such machines are the little robots built by the neurophysiologist William Grey Walter at the Burden Neurological Institute in Bristol at the end of the 1940s. These robots were small, wheeled automata with hard shells, which trundled and bumped their way around the room and which, thanks to Walter's ingenuity, altered their behaviour according to what they encountered. He called them *Machina speculatrix*, denoting a new species of machine, but they're better known as tortoises. Walter himself cited the Mock Turtle from Lewis Carroll's *Alice's Adventures in Wonderland*: 'We called him Tortoise because he taught us!'[5]

The tortoises had a few ways of adapting their behaviour. First of all, sensors under their shells registered when they bumped into objects, causing them to head off in a different direction. In this way they would randomly move about and find their way around various obstacles. The first pair of tortoises, which Walter called Elmer and Elsie, were also equipped with light sensors. This gave them an ability observed in many animals, called phototaxis, or an attraction towards light. Like moths and jellyfish, the tortoises would move towards the nearest and strongest light source, allowing them to be led around the room with a torch or to return to their well-lit 'kennel' to recharge when their battery ran low. So far, so Roomba – but the tortoises exhibited other, stranger behaviours as well.

Walter compared the two sensors – light and motion – to two neurons, constituting a tiny brain. Yet the dynamic interactions between these two basic neurons were enough to produce a range of complex behaviours, or what Walter described as 'the uncertainty, randomness, free will or independence so strikingly absent in most well-designed machines'.[6] For example, the tortoise's primitive light sensors were easily overloaded, meaning that the brightest lights would actually repel them. This caused them first to trundle towards a light source, then back away when they got too close, and then advance again, and so on. In this way, they would circle lamps in a nervous, stuttering pattern of approach and retreat.

The tortoises' most striking ability was produced quite unexpectedly by the addition of a small monitor light to their backsides, which was intended to show when their motor was running. Immediately, the machines displayed a new behaviour: on approaching a mirror or other reflective surface, they would catch a glimpse of their own light and immediately begin to jiggle at their own reflection 'in a manner so specific', wrote Walter, 'that were it an animal a biologist would be justified in attributing to it a capacity for self-recognition'.[7] Twenty years before it was formally defined, the tortoises passed the mirror test.

Walter contrasted his tortoises with early computers, which, knowing only a language of ones and zeros, and without senses beyond direct data input, he considered 'in no sense free as most animals are free; rather they are parasites, depending on their human hosts for nourishment and stimulation'.[8] Walter's machines were different, because they were adaptive to their circumstances: they were capable of altering their behaviour according to what they encountered in the world, rather than simply following pre-existing programming. In this way, and in contrast to fixed and subordinate machines, they represented active life.

This idea, that technology might be able to adapt itself to its environment was the central concern of cybernetics, a field of study which arose after the Second World War and which has gathered a ragtag bunch of scientists, researchers, psychiatrists, artists and oddballs under its umbrella ever since. Defined in 1948 as 'the scientific study of control and communication in the animal and the machine', cybernetics subsequently ranged restlessly across disciplines, influencing studies of learning, cognition, self-organization, biological feedback, robotics and business management, while never cohering into a fixed discourse or settling comfortably into a single academic department.[9]

Walter's tortoises were inspired by one of the first artefacts of cybernetics, a device called the homeostat. This device had been designed a few years earlier by another English psychiatrist, W. Ross Ashby, who often described his creation as an artificial brain. In fact, the homeostat

W. Ross Ashby's homeostat, 1948.

consisted of a set of four Royal Air Force bomb-control units, wired together to react to each other's inputs and outputs. These units had been developed during the war as automated feedback devices, which responded to an incoming signal by increasing or decreasing another signal. This allowed the bombsights on aircraft to automatically compensate for air speed and wind velocity. The units themselves, however, could be connected to anything, including another homeostat.

When he wired four of these units together, Ashby found that they would try to adjust their parameters until they reached a kind of stability with one another – their inputs and outputs fluctuating until they came into equilibrium. Moreover, when any one of them was disturbed, the whole system would readjust itself until that stability had been recovered. Ashby called this ability 'adaptive ultrastability'. No matter what his colleagues did to upset the machine – swapping connections, reversing positive and negative wires, tying its oscillating magnetic arms together or blocking their movement – the device always found its way back to a stable condition.[10]

It was this ability to self-correct and find new stable patterns that led Ashby to describe these units as akin to an artificial brain: indeed, he compared them to the undeveloped mind of a kitten. When young, the kitten doesn't know that red meat is good while red fire is bad, but positive and negative feedback quickly puts it right, creating a stored pattern of behaviour. The establishment of this pattern is a kind of learning. In principle, given enough homeostats, Ashby believed that any such complex set of memory-forming feedback systems could be created. Ashby was a modest but capable self-promoter, and newspapers all around the world hailed the appearance of this 'electronic brain' – drawing the attention of, among others, Grey Walter.

Walter was impressed by the homeostat, but he thought it limited, 'like a sleeping creature which when disturbed stirs and finds a comfortable position'.[11] Unlike the homeostat, his tortoises were mobile and exploratory: they didn't wait for their equilibrium to be upset, but bumped around looking for trouble. Nevertheless, they followed the same principle. Instead of pre-programming a machine, you just release a system capable of responding in new ways into the world and allow it to adapt. This was the central principle of cybernetics: that adapting to the world was a more powerful and appropriate approach than trying to anticipate and control it.

Let's think about this for a moment. The point of the homeostat, and the tortoises, is that they adapt themselves to their environment as they encounter it. How often is this the goal of our technologies? We think of technology primarily as a solution to problems that we face (including, all too often, those we have created). But Walter and Ashby and the other early cyberneticians thought of technology as something very different: something with its own agency and abilities, whose reactions were uncertain and whose behaviour should reflect its own encounters with the world.

Cybernetic thinking was not limited to machines. Indeed, cyberneticians held a radically different view of the brain – human, animal, or otherwise – than that commonly held today. As Ashby put it when considering the abilities of the homeostat, 'To some, the critical test of whether a machine is or is not a "brain" would be whether it can or cannot "think". But to the biologist the brain is not a thinking machine, it is an acting machine; it gets information and then it does something about it.'[12]

As the historian of cybernetics Andrew Pickering puts it: 'Something of the strangeness of cybernetics becomes clear here.' Modern science, and particularly the discipline of artificial intelligence, has tended to treat the brain as a kind of discrete cognitive organ, turning thinking into a wholly internal process of representation and calculation. But cybernetics did not approach the brain in that way. Instead, writes Pickering, 'The cyberneticians thought of it instead as *performative*, an organ that does something, that acts – as an active and engaged switchboard between our sensory inputs from the environment and our motor organs.' Cybernetics doesn't care what's in the black box: it cares about how the box performs, and what relationships are formed when it enters into play with the world. As such, it is a way of thinking about more-than-human minds that closely corresponds with our rejection of hierarchical, anthropocentric, 'intelligent' thinking and our embrace of agential, relational being.[13]

It wasn't long before these ideas about artificial brains, and the ways they might be constructed, caught the attention of industry. In the 1950s and 60s people were starting to realize that more and more advanced machines were coming to take over the workplace, and there was much discussion of how this would take place. Most of these visions imagined factories entirely filled with computers taking the place of human operators at every step of the process, from production lines to sourcing materials and shipping goods – and this is indeed how automation has mostly proceeded. But Stafford Beer, a British management consultant working for United Steel, didn't think those ideas went nearly far enough.

Beer had read the works of the early cyberneticians, and believed that this 'automatic factory' needed to be adaptive, rather than preprogrammed. He disparaged others' plans for automatic factories as akin to 'a spinal dog', a painful-sounding phrase meaning an animal whose higher functions, including the brain, had been cut off from the rest of its body. Without a higher nervous system, the animal might continue to live for some time, but it would not be able to adapt to changing conditions, and changes in its environment – the market, in the case of the factory – would cause it to go extinct.

Building on the work of Ashby and Walter, Beer researched extensively into other methods of constructing machines, methods which would allow for rapid response, feedback, change and adaptation. The result is one of the strangest bodies of work of the twentieth century, in which bizarre experiments with other-than-human intelligences were carried out in the heart of traditional British industry, and at the cutting edge of radical politics.

Beer's employer at the time was United Steel, the largest operator of collieries and steelworks in Britain, and Beer took on the task of automating one of its factories. In order to be able to adapt to changing conditions in the way Beer imagined, the factory needed an artificial brain at its centre, which would take in all the measurements of its operation – incoming materials, stocks, processes, yields and efficiencies – and match them in the most productive way with the

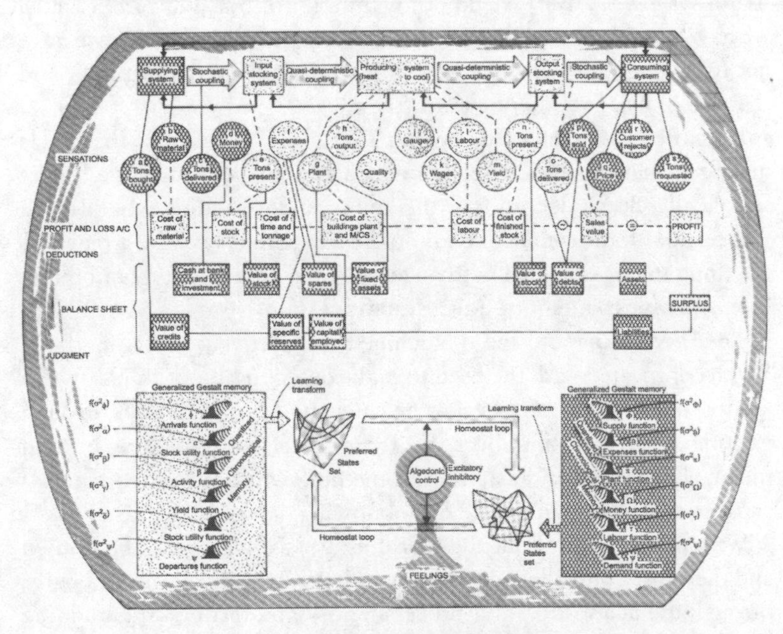


Stafford Beer's Cybernetic Factory.

company's supplies, costs, cash flow, labour reserves and long-term goals. Beer thought that one way to perform this balancing act was by employing Ashby's feedback device, the homeostat – represented by the looping arrows at the bottom of a diagram Beer published of the Cybernetic Factory in 1962. Embedded within a network of sensors and effectors, the homeostat would constantly bring the different inputs and outputs into more efficient and more productive alignments, and reconfigure its processes in response. Beer referred to this cybernetic brain as the U-Machine.[14]

But what would the U-Machine actually consist of? This is the core cybernetic problem, and the one we're most interested in today. How do you design a machine which will respond to entirely novel situations, when the whole thrust of technological thinking is to address and overcome problems already known to us?

Beer himself was very clear how not to do it: 'As a constructor of machines man has become accustomed to regard his materials as inert lumps of matter which have to be fashioned and assembled to make a useful system . . . [But] we do not want a lot of bits and pieces which we have got to put together. Because once we settle for [that], we have got to have a blueprint. We have got to design the damn thing; and that is just what we do not want to do.'[15]

Here, Beer is directly channelling Turing's description of the oracle machine: that unspecified entity which cannot be a machine. Quite specifically, Beer's definition is the key to understanding what Turing was on about: the *o*-machine cannot be a machine, because a machine is a thing that is designed with an explicit purpose in mind, and therefore is unable to adapt to novel situations.

The U-Machine and the *o*-machine are effectively the same thing. When confronted with the need to make decisions about things which are too complex or too novel to be calculated or interpreted through existing experience, both Beer and Turing reach for something beyond the machine as we understand it, something which arises from and is capable of working in a state of unknowing.

We tend to think about the world as a place which can be known and therefore controlled and dominated. We do this, computationally, through the acquisition and processing of data, through the building of ever larger databases and ever more powerful computers. But Beer believed that there exists in the world a class of 'exceedingly complex systems' which were in principle unknowable. These systems included the brain, the company and the economy. So he set out to build machines which could operate in the face of such unknowability, a journey which was to take him to some very strange places indeed.[16]

Events in the intervening decades have brought Beer's realization home to many of us. The extraordinarily complex tools which we have developed in the years since his design of the automatic factory represent the accumulation of extraordinary quantities of knowledge – but we are no closer to solving the world. In fact, the opposite is true: in our attempts to assert order and establish truth over the exceeding complexity of the world, our uncertainties seem only to have increased in number, leaving most of us paralysed, fearful, angry, and subject to ever more opaque systems of oppression and control.

The homeostat is a machine for dealing with the unknown. When confronted with novel circumstances, it reconfigures itself appropriately. Crucially, it does so without having to understand the nature of the change: it is performative, not representational. It lives in the world, a world it does not attempt to know, only to adapt to. This was what Beer wanted his factory to do.

Beer's solution to the problem of not building a brain was to entice one in from the natural world. As early as the 1950s he recounts a successful effort to teach young children (his own, we hope, although his papers do not specify) to solve simultaneous equations, despite not knowing any of the relevant mathematics. He did so through positive and negative feedback, the homeostat method: he showed them coloured lights corresponding to 'warmer' and 'colder' answers until they grasped the correct answer. For him, this was proof that 'simple' minds could adapt to new problems without specific teaching – the essence of cybernetic performativity over ingrained knowledge and understanding. And so he started trying it with other animals.

First, he attempted to devise a mouse language, using pieces of cheese as a reward function. He diagrammed out interconnected boxes, with ladders, see-saws and cages connected by pulleys, by which rats or pigeons might become part of elaborate computing devices. Bees, termites, ants and other insects were all considered for roles in this expanding notion of a biological computer. These remained thought experiments, however, and Beer only summarized them in his writing. He gave far more time and effort to a much larger conception of the artificial brain: an entire ecosystem.

At some point in this process, Beer constructed in the basement of his own home a large tank, filling it with water which he collected from ponds in the surrounding countryside. Slowly, the tank filled with all kinds of wild aquatic life: reeds and algae, nematodes and leeches. If individual minds were capable of adapting without knowing, then surely communities of minds, however tiny or unlike our own, must be capable of even more flexible and wild adaptation. And so he tried to summon such a community into being.[17]

His first attempts at an economics of pond life employed *Euglena*: microscopic, single-celled critters well known to science for their un-animal-like possession of photosynthesizing chloroplasts – a prime example of symbiosis between animal and plant life, and thus an apposite subject for Beer's weird experiments. *Euglena* are highly sensitive to light, moving towards it in much the same manner as Walter's tortoises. To take advantage of this rare ability, Beer suspended lights in the water of their tank, which could be turned on or off according to various signals. In this way, he hoped, the *Euglena* could be 'connected' to a larger automated system, responding to changes outside the tank by moving towards different lights, and their behaviour could be used in turn to influence the behaviour of the wider system.

Beer's idea was that if these two systems – the pond and the factory – could be brought into some kind of relationship, then changes in one would trigger changes in the other. Refinements to the operations of the factory would upset the dynamic equilibrium in the pond, causing it to reconfigure itself into a new state, pulling the factory along with it, until a new super-equilibrium between the two systems was found. This was homeostatic ultrastability at work. Unfortunately, the *Euglena* didn't seem to be up for it: after a short time they would 'lie doggo', as Beer put it, and refuse to cooperate.

In another attempt, Beer focused on *Daphnia*, or water fleas: tiny crustaceans also abundant in British ponds. Beer coated dead leaves with iron filings and floated them in the pond, where they were eaten by the *Daphnia*. As a result, the little mites became magnetically sensitive and could be pushed about with electromagnets, broadcasting data from the factory. As they tried to find a place of equilibrium within the shifting magnetic waves undulating through the water, their adaptive behaviours could be measured and fed back to the factory, creating new feedback loops. Once again, however, Beer ran into experimental difficulties: as the *Daphnia* started to excrete the iron they'd swallowed, the water turned rusty and the pond's 'incipient organization' started to fall apart.

I find myself thinking of Beer, standing next to a blackboard and a tank of rusty *Daphnia*, presenting these experiments to the managers of Templeborough Rolling Mills, the factory placed at his disposal by United Steel, and proposing, in effect, to replace those managers wholesale with literal pond life. Their reaction must have been unprintable. I don't actually know if he even made such a presentation – by his own admission, the work more closely resembled a hobby, taking place 'from time to time in the middle of the night', while he was trying to get a more profitable management consulting business off the ground. But nothing could better embody the wild promise and eccentric practice of this almost-forgotten era of computational history than the image of Beer shuffling around his basement in the small hours, trying to put a pond through business school.

Ultimately, Beer's problem wasn't a recalcitrant pond, but money and time. With United Steel not entirely convinced by the idea of the Cybernetic Factory, he soon had to look elsewhere for support. But the problem might also have been one of interest: the idea of running a steel mill seems pretty remote from the concerns of single-celled organisms. Perhaps we need to set these organisms a problem more relevant to their existing mode of life. And perhaps we have already set the terms of this problem in our ongoing attempts to eradicate their habitats and poison their environments: despite a global decrease in ponds and wetlands in the past decades, *Euglena* and *Daphnia* are still thriving in the wild and, like the Archaea and bacteria we discussed earlier, are even found in hypersaline and other extreme environments, many of them created by human activity. That's real cybernetic adaptivity at work.

If these creatures have survived so well, perhaps it has something to do with the adaptivity, not of individuals but of communities of individuals. One problem with our description so far of the cybernetic, artificial brain as just a different kind of brain – whether a homeostat, a mechanical tortoise, an entrained mouse or a middle manager in a jar – is that we keep trying to box it up and stuff it into another system, just as our own brains are stuffed into our heads. But a *Euglena* colony – and even less an entire pond – is not an individual machine, it's a system of its own: an ecosystem, a rich admixture, a diverse assemblage of soupy life. The pond goes beyond the individual brain, beyond even one kind of brain, a particular species or genus, and into the possibility of a system of minds, all relating and inter-relating, and adapting to one another in complex, ever-shifting ways.

Perhaps the issue with both artificial intelligence and the Cybernetic Factory is that we have been trying to entrap a brain within the machine, when the real brain – the oracle – is outside. The oracle *is* the world. We've got everything inside out. We think all the activity is on the inside – when we stick electrodes into the brains of chimps and fruit flies, or strip plants down to their roots and nodules, for example – but the real action is out in the world. That's where everything arises and intra-acts. Life happens when everything is bouncing off everything else.

It's probably a good time to point out that the Earth itself is a homeostat. That at least is the central tenet of James Lovelock's and Lynn Margulis's Gaia theory – yes, Gaia, the same goddess who was worshipped at Delphi, the home of the oracle. Of course it all comes shimmering, bumping and concertina-ing down the aeons. Gaia theory understands the world as a synergistic, self-regulating, complex system in which organic and inorganic matter interact with one another to co-produce the conditions for life on Earth. Gaia is a cybernetic feedback system, the realization of Beer's Cybernetic Factory at planetary scale, in which the inputs and outputs are water, air and rock, and the U-Machine is the entire biosphere huffing and puffing and wriggling and growing and adapting away. *The world is not like a computer; computers are like the world.*

This realization opens the door to all kinds of exciting and radically different possibilities for computation, which Beer barely anticipated but would probably have massively enjoyed. Many of these are grouped under the loose category of 'unconventional computing', which lives by the mantra that anything that can be done electronically can probably be done more interestingly with . . . well, anything else. Unconventional computing emerges from the same impulse as esoteric programming and attempts to perform in hardware – or wetware – the same kind of tricks as writing software in the style of a fictional orang-utan.

One of the stars of unconventional computing, a modern successor to Beer's *Daphnia,* is the slime mould, itself a loose category of organism which scientists aren't really sure where to place. One of the most noted characteristics of slime moulds is that they trouble the individual/group divide, existing some of the time as independent, single-celled organisms like *Euglena*, and at other times – particularly when food is scarce or conditions otherwise harsh – swarming together into clusters that operate collectively. In some cases these collectives fuse together to produce great sacks of cytoplasm filled with thousands of nuclei. When they want to reproduce, some parts of the collective form spores, which are picked up by the wind or animals; others sacrifice themselves to become non-reproductive infrastructure, the stalks and stems

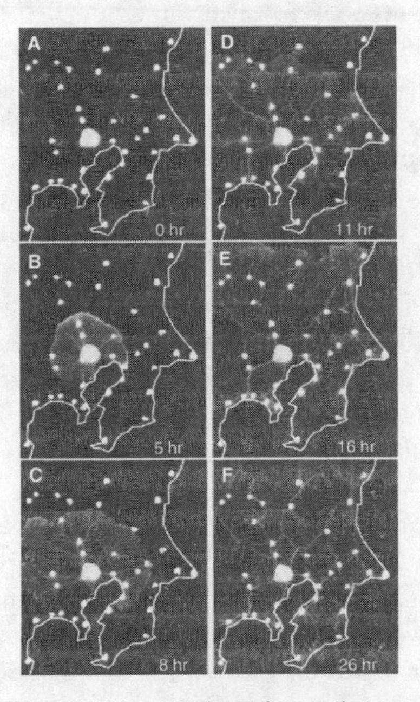


A slime mould maps the Tokyo rail network.

of the fruiting body. They have other tricks too. When separated, they can find their way back to one another and reconstitute. They can anticipate events: like *Mimosa,* they appear to retain a memory of unfavourable episodes, and react accordingly. Formerly classed as fungi, the slime moulds have, under close examination, taken on a kind of weird, communal individuality, and like the discipline of cybernetics itself they continually refuse to settle into a new domain.[18]

In 2010, a slime mould called *Physarum polycephalum* revealed itself to be a highly effective city planner, when it recreated one of world's most robust and efficient transport networks. The Japanese rail system is a masterpiece of complex engineering, involving decades of work by designers and planners, and reliant on complex trade-offs between cost, resources and geography. But when researchers placed oat flakes – a *Physarum* delicacy – in the pattern of the cities surrounding Tokyo, and released the mould, the slime mould quickly reproduced their efforts. At first, it spread itself evenly across the map, but within a few hours it had started to hone its web of threads into a highly efficient network for distributing nutrients between distant 'stations', with stronger, more resilient trunk routes connecting the central hubs. This wasn't a simple, join-the-dots exercise either, but a realistic map where patches of bright light – which *Physarum* dislikes – corresponded to mountains and lakes, requiring the mould to make the same kind of trade-offs that engineers have to implement. After a day of adapting itself to its environment, the resemblance to an actual map of Greater Tokyo was unmistakeable.[19]

Calculating efficient routes is a notoriously hard problem in mathematics. One of the most famous and thorny versions of it is known as the travelling salesman problem. It's a deceptively simple riddle: given a list of some widely distributed cities and the distances between them, what's the shortest route to visit each one, only once, and return home?[20]

Despite hundreds of years of mathematicians attacking the problem, and huge investments made by logistics companies and postal services, there's no guaranteed way to figure out the best answer to this riddle, however you frame it. Worse, as more cities are added to the list, the problem gets exponentially harder, because the number of options multiply. This explosive growth in possible solutions is a huge

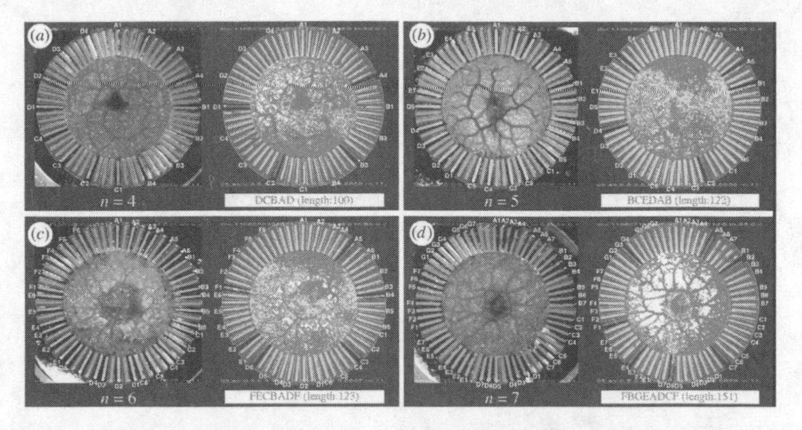


Slime moulds figuring out the shortest distances between 'cities'.

problem for mathematical algorithms, which can get lost in a maze of dead-ends and bad answers. For this reason, the travelling salesman problem is a classic example of a problem that a Universal Turing Machine – the *a*-machine – cannot reliably solve. It's computationally undecidable. And what did we say about undecidability? It must be time for the *o*-machine.

In 2018 the same slime mould, *Physarum polycephalum*, showed that it was able to solve the travelling salesman problem in linear time, meaning that as the problem increased in size, it kept making the most efficient decisions at every juncture. Using the same method as the Tokyo rail experiment, researchers at Lanzhou University in China placed scraps of food in the place of cities and used beams of light to keep it from repeating connections.[21] They showed that the mould took only twice as long to solve a map of eight cities as it did to solve a map of four cities – despite there being almost a thousand times more possible routes.[22] In short, the slime mould easily completed a task that the most powerful computers in the world – and humans – absolutely suck at.

The idea that biological systems could not only replace but outperform many of the operations of computers – which already outperform human abilities – fits with our proposition that computers are like the world, rather than the other way around. But in order to actually absorb this realization, we need to perform the same kind of mental flip that we did when speaking of fungal networks and the internet, or man-made and animal intelligence. Systems of intelligent, computational ability – mycorrhizal networks, slime moulds and ant colonies, to name a few – have always existed in the natural world, but we had to recreate them in our labs and workshops before we were capable of recognizing them elsewhere. This is technological ecology in practice. We need the mental models provided by our technology, the words we make up for its concepts and metaphors, in order to describe and properly understand that analogous processes are already at play in the more-than-human world.

In 1971, the American physicist Leon O. Chua proposed the idea of a novel electronic component which he called a 'memristor'. The memristor is a resistor with memory, which makes it capable of remembering its own state – retaining data – even when it's without power. If memristors were used inside computers, they would collapse the storage-processor division which is common to all modern computers and vastly increase their efficiency and power, transforming computation in the process. But in the 1970s, it wasn't clear how you might go about building a memristor – and in any case, existing transistors and silicon chips were still so new and exciting that they were much more attractive to researchers and engineers. It wasn't until 2008 that a team at Hewlett-Packard actually figured out how to build one, and although memristors still promise to revolutionize the ways computers function, they remain a laboratory curiosity.[23]

That said, the idea of the memristor, like the idea of a scale-free, electronic network, has stimulated extraordinary work in biology. Again, it was *Physarum polycephalum*, the slime mould, which led the way. Researchers in the Unconventional Computing Laboratory at the University of the West of England, led by Professor Andrew Adamatzky, have shown that *Physarum* exhibits memristor-like behaviour: when charged with certain voltages, it exhibits the same behaviour in subsequent tests. In other words, it remembers its previous state, like a memristor. In this way, it should be possible to build a radically fast and efficient computer out of slime moulds, laid down on a substrate just like a silicon chip.

Since the realization that slime mould can act as a memristor, we've found this to be true of many other organisms. The Venus flytrap, *Aloe vera*, and another old friend, *Mimosa pudica*, have all been shown to display memristic behaviour. In fact, it seems increasingly clear that all living plants, as well as the skin, blood and sweat ducts of animal bodies (including our own), are potential memristors. It remains only to figure out how to integrate such structures into our technology in order to devise new kinds of computers, which might surpass in capability all the machines we've built to date.[24]

To do so would advance the current state of computation exponentially. Nonetheless, they would at heart still be the same kind of computers: the same old Turing machines with a very different architecture. What if – rather than asset-stripping other organisms for their useful components, treating them as so many spare parts for our machines – we instead incorporated them wholesale and, beyond that, started to see the environment itself as part of our computational substrate?

Memristors are not the only way we might do this. It is, in fact, possible to build computers out of pretty much anything. For example, billiard balls. Or crabs.

In the 1980s, the physicist Edward Fredkin and the engineer Tommaso Toffoli showed how it was possible to model the operations of logic gates using the motion of billiard balls. Imagine a billiards table with two pockets at one end and two sloping tubes at the other. A ball dropped down either tube will roll undisturbed into one of the pockets – the same pocket, for either tube. But if two balls are dropped at the same time, they will strike one another, and one ball will ricochet into the second pocket. This is an example of an AND gate, a type of circuit which combines incoming signals, with the first pocket representing an output of zero (only one input) and the second an output of one (both inputs). With enough AND gates, you can make any other type of gate, so given enough billiard tables, you can make any kind of digital computer you can imagine.[25]

Here's a thing: you don't have to use billiard balls. At the University of Kobe, researchers recreated the billiard ball AND gate using crabs: the soldier crab *Mictyris guinotae*, to be specific, which is endemic to the shallow lagoons of Japan's Ryukyu Islands. These small blue crabs

A crab computer.

are known for forming up into enormous armies that swarm across the lagoons at low tide. Although comprising anything from several hundred to several thousand individuals, these seemingly chaotic swarms have quite predictable dynamics. The crabs at the front of the swarm move forwards (or, probably, sideways) in tight formation; those in the middle follow their neighbours; and those at the edges continually fold back into the centre, creating a rolling motion – not unlike a billiard ball. The movements of these swarms can be controlled using tunnels or, more ingeniously, shadows, which the crabs dislike because the shadows make them think of predatory birds. When two swarms collide, they head off at a (predictable) new angle – again, like billiard balls. You can probably see where this is going, even if what you're imagining seems a little nightmarish.

The unconventional computer scientists at Kobe put forty soldier crabs in a maze and scared them with shadows. The resultant rolling swarms behaved exactly like the billiard ball computer, crashing into one another in predictable ways, spinning off in predictable directions, and simulating logical operations. A computer made out of crabs. A crab computer.[26]

You can run this stuff in both directions. 'Soft' robotics is a field of robotics which explores exactly what it sounds like: robots made from squishy and malleable materials, rather than metal and plastic. Originally dreamed up as a way of making robots which were safer for people to interact with at home or in the workplace (better that the robot looking after Grandma is made entirely of sponge than steel rods, just in case it falls on her), the complications of the form have given rise to all kinds of other discoveries. Because soft robotics use entirely different motor systems, they're much harder to control. Some nutate – expanding or compressing areas of their bodies in order to grow in different directions, like a plant or an octopus limb. This gives them much greater freedom of movement, but makes them difficult to direct, particularly when the system is self-learning. One proposed solution is to devolve control of these soft limbs through a process called 'reservoir computing', whereby the overarching intent of the system is sent to a subset of artificial neurons, which figure out how to achieve the desired intent locally. Essentially, it's using one complex system to run another – which sounds an awful lot like Beer's cybernetic factory, right down to the 'reservoir', or pond. It's also a lot like what we think octopuses and other cephalopods do, with their distributed neural systems and their apparently autonomous limbs.[27]

Soft robotics and reservoir computing sound more like bio-mimicry than truly biological computing, as despite all the talk of limbs and neurons the systems employed are entirely artificially constructed. But as per the cybernetic mind flip, it is possible to run the same ideas on the world itself. One group of researchers has even taken the term 'reservoir computing' to its literal conclusion, creating a computer in the form of a bucket of water.[28]

What does this even mean? What they did was to place several motors around the edge of a water vessel, which could be used to create waves – interference patterns – on the surface of the water. This contraption was then set atop an overhead projector, so the waves could be observed as patterns of light and dark on the wall. Differing data, fed to the motors, produced differing patterns on the wall, and these patterns were then analysed by a (silicon-based) computer using machine-learning. In one experiment, the bucket was fed samples of different spoken words, and the computer quickly learned to discriminate between them. It learned to recognize qualities of speech, which is a notoriously hard problem – and it did so far more accurately than when presented with raw digital data alone, much in the way that slime mould solved the travelling salesman problem more efficiently than any digital computer. The water appeared to pre-process complex information by turning it from one-dimensional binary information into more complex, but more expressive, higher-dimensional information: a fluid dance. Moreover – and in contrast to, say, the magnetic *Daphnia*, which quickly fouled the water with rust – turning off the motors immediately 'reset' the bucket to its initial state, allowing it to consider a new problem.

There are two points worth emphasizing here. The first being that this system – what the researchers call a liquid state machine – might more closely resemble the kind of information processing that happens in biological neurons, in which incredibly complex information appears to be easily handled by very simple circuitry. Neural computation – the brain's processing of the world – is low-speed and low-precision compared to a computer, but it's also massively parallel and real-time, operating more like the flow of a river than the ticking of a clock. As our brains have evolved to interface with the world as it is, this suggests that solving 'real world' problems – route-finding, speech recognition, economics – is better addressed by computers which are more like the world too.

The second point is that the water in the bucket isn't 'thinking' or 'remembering' – but it is processing. It's computing information. The form of this information isn't like the ones or zeros that pass through digital machines (including the crab computer): it's analogue, which rather than old or fuzzy means complex, knotty and continuous. It has texture and colour, like the world.

To be clear, this isn't a nerdish, nostalgic appeal to vinyl over MP3s. This isn't about trying to represent or reproduce the world in some particular way, according to our aesthetic or intellectual preferences. It's simply acknowledging that the world is actively unknowable to

Vladimir Lukyanov's water computer, 1936.

us, in opposition to the way that digital computation attempts to render the world knowable. We can't read water in the same way as we can read data, and this is a good thing. Working with it makes us more aware of the distance between ourselves and the matter under consideration: it reminds us that we share this world rather than own it. Knowledge produced through the medium of the shifting surface of a bucket of water is made in cooperation with the world, rather than by conquering it.

Another example of a machine operating in close concert with the world was Vladimir Lukyanov's Water Integrator, an analogue computer built in the Soviet Union in 1936. Lukyanov worked on the construction of the Troitsk–Orsk and Kartaly–Magnitnaya railways, where work was severely hampered by extreme temperatures. In particular, reinforced concrete laid down along the line in winter would crack when warmed by the summer's heat. Lukyanov needed to work out how to precisely model the thermal mass of his materials, but in the 1930s mechanical calculators could only handle linear algebra. There was no way of calculating the differential equations required.

Lukyanov realized that water flow was analogous to the distribution of heat and could act as a visual model of an invisible thermal process. He constructed a room-size machine made out of roofing iron, sheet metal and glass tubes, which used the flow of water, rather than electrons, for its calculations. It consisted of multiple tin vessels, connected by glass pipes of different diameters. The vessels, filled with water, could be raised and lowered by cranks and pulleys, their levels corresponding to different inputs. As the levels rose and fell, the water pressure in the system increased and decreased, and water flowed through the different glass tubes and into other vessels representing stored memory and outputs. These numbers could be read off the machine at any point to determine the answer to the complex differential equations being simulated – the only machine in the world at the time capable of solving such problems.

Although Lukyanov set out to solve one particular problem – thermal flow in concrete – he quickly realized that he could solve any differential equations with the equipment. He went on to build more widely applicable machines, with pipes and tubes which could be moved and replaced for different classes of problems in geology, metallurgy, mine construction and rocketry. By the middle of the 1950s, many educational institutions across the Soviet Union had water computers in their laboratories, while Lukyanov himself was awarded the State Prize, the Soviet Union's highest civilian honour. Water computers continued to be used in Soviet institutions for large-scale modelling well into the 1980s.[29]

Analogue computers are models of the world. Lukyanov's water integrators began as simulations of actual physical processes – thermal flow, erosion, subsidence – and then were abstracted and applied to other problems when their general capabilities became clear. Material, physical, even geological, processes were the models for solving problems in pure mathematics. Likewise, the earliest digital computers began as models of specific problems – cryptographic ciphers, nuclear reactions, weather systems – and became generalized problem-solvers as we began to disassociate and disaggregate them into reassemblable parts: memory, processing, command instructions, and so on. Gradually, the model of the world at the heart of the machine becomes less and less visible as the machine itself becomes more and more abstract. The highest level of abstraction is the completely abstract (automatic) Turing machine. But Oracle machines, among them the water integrators, are attempts to bring this abstraction back down to earth: to recombine the awesome power of mechanical processing with material concerns – and perhaps, in time, with ethics, morality and life itself.

In the United States, water computers took on a more monumental scale. From the 1940s onwards, the US Army Corps of Engineers constructed a series of models of watersheds around the country, in order to better understand the water supply system and to study the impact of dams and bridges. The first of these was a model of the entire Mississippi River drainage area, including its major tributaries, the Tennessee, Arkansas and Missouri Rivers. This model covered some 200 acres, and was constructed over a period of twenty years on the outskirts of the city of Jackson. The scale of the model was 1:100 vertical, and 1:2000 horizontal, making the Appalachian mountains twenty feet higher than the Gulf of Mexico, and the Rockies thirty feet above that. Painstakingly carved concrete slabs reproduced 15,000 miles of river over a few hundred metres. By placing suitably scaled models of structures along its length, or by digging new channels between branches, it was possible to almost perfectly simulate the effects of new levees, spillways and drainage canals on the actual environment of the river basin.

Previous attempts to work out how the mighty Mississippi could be controlled had taken a very different approach. Following the Great Flood of 1927, which inundated parts of Arkansas, Mississippi and Louisiana, the US effectively declared war on the river. Over the next decade, the Corps of Engineers built twenty-nine dams and locks, hundreds of run-off channels and over a thousand miles of new, higher levees. All this was intended to imprison the river within its then-current course, and prevent any deviations from 'normal' levels. The project was a catastrophic failure, as was demonstrated by further devastating floods in the winter of 1936. The planners and engineers had failed to understand the Mississippi as part of a system of tributaries and watersheds that drained a full 40 per cent of the continental United States. In 1936, the system simply emptied itself elsewhere, in the process displacing thousands of people in Massachusetts, Pennsylvania and New York. A different system was needed, one that responded to the river's scale and natural movement – a model of fluid feedback rather than a fixed, pre-determined plan. Hence the Mississippi Basin Model: not a cybernetic pond, but a river in full spate.

The Basin Model had its first major live test in 1952 when the Missouri River came close to bursting its banks between Omaha and Council Bluffs in Iowa. The Model's operators initiated sixteen days of continuous, twenty-four-hour tests, setting different conditions along the river's length to predict surges, crests and levee failures. One day could be simulated in about five minutes, and for each gallon of water the engineers poured into the Model's concrete channels, 1.5 million gallons flowed down the actual Missouri, tearing at its shores and undermining its levees. Mayors of cities up and down the river congregated in a watchtower erected at the centre of the Model, watching anxiously to see whether the devastating, inch-high floods would consume their fields and homes. In response to the model's predictions, they called brigades of civilians and sandbags into action to reinforce weak and vulnerable locations and ultimately to prevent tens of millions of dollars of damage and potential loss of life.[30]

In this way, the Model acted as the U-Machine, the oracle, to the entire Mississippi system: the river itself and the millions who lived along its banks, and the entire weight of concrete, soil, rebar and sand which was deployed to shape it. Ebbing and flowing, rising and falling, surging forwards and back, the Model enabled the Corps and civilian volunteers to adapt to changing conditions in ways no preprogrammed system would be capable of doing.

Subsequently, further models were constructed at Portsmouth, Virginia, and Sausalito, California, to simulate the Chesapeake and San Francisco Bays respectively. The latter is still operational (although for demonstration purposes only) and can be visited. It is a thing of wonder. In a huge hall, the whole of San Pablo Bay, Suisun Bay, the Sacramento–San Joaquin River Delta and the Pacific Ocean for seventeen miles beyond the Golden Gate are laid out across an area the size of a football field. The Bay Model includes ship channels, rivers, creeks, sloughs, the canals of the Delta, major wharfs, piers, slips, dikes, breakwaters and San Francisco's famous bridges: every type and specimen of hydrological engineering. Every hour on the hour, with a great gurgle like a gigantic bathtub, the tide is turned and the water flows into the bay at a rate exactly one hundred times faster than the world outside, and then flows out again.

Like time-lapse photography, the Bay Model is a technology for

Chesapeake Bay Model technician at a tide gauge located on the Elizabeth River, at Portsmouth, Virginia, August 1977.

scaling the grandeur and majesty of natural process to human perception, without losing our appreciation for its otherness. It allows us to intercede and act meaningfully within a vast and complex landscape without losing sight of its aesthetics and subjecthood – its beauty and selfhood. Crucially, no complexity is lost in its simulation. The shift is in scale, rather than in information. The world is translated, not represented. Of course, there is art in the translation too. This is where our own agency and creativity comes into play – but as part of an ongoing dance of mutual understanding, rather than one of domination and control.

The architecture critic Rob Holmes has called the Corps of Engineers' hundred-year battle to contain and control the Mississippi – the erection of thousands of miles of levees, dams and canals – the single greatest work of Land Art in existence.[31] I am inclined to agree. Land Art depends for its impact on scale, absurdity and the environment, and these models are perhaps that art's ultimate achievement. Even the picture included here of the lone, bearded technician sitting, Atlas-like, atop the Chesapeake Bay Model resembles nothing so much as the dream-like architectural images produced by the radical Italian

Still from *Supersurface* film, 1972, produced by Superstudio.

collective Superstudio in the same period. In Superstudio's visions of a cybernetic future, humankind is released into a state of happy play among landscapes which combine megalithic building projects, artefacts of high technology, utopian communities and natural formations. In Superstudio's 1972 film *Supersurface – An Alternative Model for Life on the Earth*, the collective called for a new alignment between design and the environment, one in which human invention and the natural world complement, rather than contradict, each other.

What I love about the Bay and Basin Models – in addition to their possibilities for play – is their legibility. It was once possible to visit these computational landscapes oneself, to walk along their banks and streams, to witness calculation in process and to comprehend it. This is legibility: the constructions of systems which are readable by everyone, that contribute to a shared representation of space, as opposed to the enclosed, hidden representations of digital computers. It's the same legibility I was aiming for with my Autonomous Trap on the slopes of Mount Parnassus: a complex object that did work, that embodied its own model of the world and was capable of explaining itself to others.

There is a vast difference between understanding how answers are arrived at and simply being informed of their outcomes. As soon as digital computation was capable of the kind of complexities needed

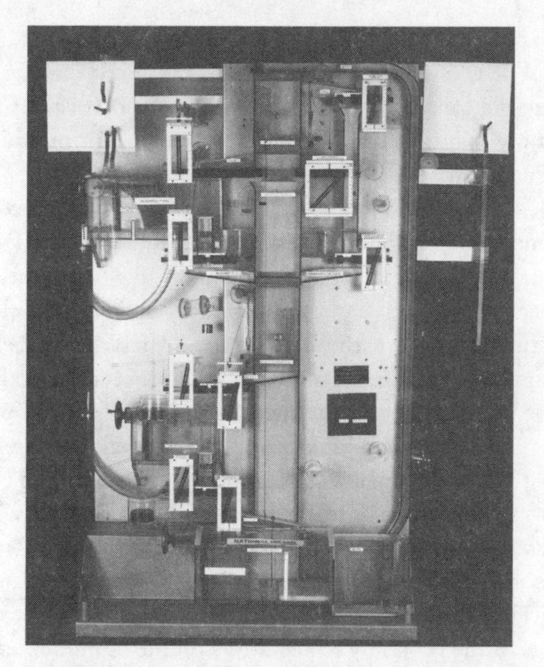

The MONIAC, built by Bill Phillips in 1949.

to model natural processes – to take on the tasks which were previously only possible in physical simulators, like the water computers – knowledge of the landscape itself became the exclusive preserve of expert operators and interpreters of opaque systems, and the general public was reduced to mute receivers of information. We became objects rather than subjects. This is what has happened across information technology as a whole. The potential of complex machines to actively increase public understanding and agency, to uplift us all, has been subsumed into ever more centralized and closely guarded machines of knowledge acquisition and domination, to the detriment of our common power.

The best example of the hydraulic computer as pedagogical tool – as a legible machine, one which not only calculates, but also educates – is one I first encountered, as a child, in London's Science Museum. This is the Monetary National Income Analogue Computer, or MONIAC, built in 1949 by the economist Bill Phillips while a student at the London School of Economics. The MONIAC is about the size of a large fridge, and like Lukyanov's Water Integrator it consists of a series of tanks and pipes, in this case made from transparent plastic and fixed to a wooden board.

MONIAC is a working model of the British economy. At the top of the machine is a large water tank marked 'TREASURY'. Other tanks represent government expenditure on things like healthcare and education; spending on these can be adjusted by opening and closing taps which drain water from the Treasury. Further down the machine, water is siphoned off into private savings and returned in the form of investments; and piped out for spending on imports and piped back for export income. Certain tanks can run dry if balances are not maintained (just like real bank accounts) and water/money can be pumped back up to the Treasury in the form of tax. The rates at which taxes are set determine how fast the pumps operate. The flow of water is controlled by a complex – but legible – system of floats, pulleys, counterweights and electrodes. It allows anyone to experiment with the settings of the economy to see how complex interactions result in different outcomes. Finance is already awash with aquatic metaphors, from 'liquidity' and 'flotation', to 'sharks', 'whales' and 'dark pools': Phillips made these metaphors literal, but he also made them more useful.

Phillips built the original MONIAC in his landlady's garage in south London, for a cost of around £400 (around £14,000 in today's money). Fittingly, it included parts taken from a Second World War Lancaster bomber, echoing Ashby's bombsight homeostats. Originally intending it as a visual teaching aid, Phillips was surprised to find how accurate it was, as well as being unique in its ability to model the whole economy in the decades before digital computers became widespread. As a result, over a dozen more were built, with many finding their way from academia into actual government departments, where they were employed as predictive as well as illustrative tools.

To my mind, what makes the MONIAC so special is that it literally puts the user's hands on the controls of the economy, while continuing to insist upon the financial system's fluidity and liveliness. It shows that the economy is both a force of nature and the outcome of deliberate, conscious decision-making. Like the Bay Model, the MONIAC gives us agency in a complex system, without denying the agency of the system itself. While the water continues to slosh around, we can't forget we're dealing with real stuff, actual material, a distinct shared world – which is all too easy to do when everything is just numbers on a screen. The moment the real world is completely abstracted into the universal machine is when we lose our ability to care for it.

All computers are simulators. They contain abstract models of aspects of the world, which we set in motion – and then immediately forget that they're models. We take them for the world itself. The same is true of our own consciousness, our own *umwelt*. We mistake our immediate perceptions for the world-as-it-is – but really, our conscious awareness is a moment-by-moment model, a constant process of re-appraisal and re-integration with the world as it presents itself to us. In this way, our internal model of the world, our consciousness, shapes the world in the same way and just as powerfully as any computer. We attempt to make the model more like the world, and the world more like the model, at every step of our intra-action. This is why models and metaphors matter. If our internal model contains a vision of a shared world, a communal, participatory world; if it acknowledges the reality of our more-than-human entanglements; and if we're prepared to adapt our vision to new circumstances and new realizations, then it has – we have – the potential to actually make the world a more communal, more participatory, more just and equal, and more-than-human place.

MONIAC was a simulation machine which became a decision machine. This is the way in which all (successful) computation operates. First, it models the world, and then it attempts to replace the world with the model. Our minds, too, are simulation machines which become decision machines: we think, process and act in constant intra-action with the world. The question, then, is what are the characteristics of models – and thus of machines – that make better worlds?

I would humbly propose three conditions for better, more ecological, machines: machines better suited to the world we want to live in and less inclined to the kinds of opacity, centralization of power and violence we have come to understand as the hallmarks of most contemporary technologies. These three conditions, I believe, are necessary for machines to become part of the flourishing communities of humans and non-humans we've sketched out in previous chapters. Our machines should be non-binary, decentralized and unknowing.

Let's start with non-binary. As we have seen, when we are capable of letting go of yes/no, either/or, zero/one questions – whether in the turbulent flow of analogue computers, the open field of Facebook profiles, or the now tentatively explored operations of the *o*-machine – we discover not merely new ways of doing and seeing things, more powerful than we ever imagined, but a richness and complexity to the world which surpasses imagination. This is the labyrinth of endlessly significant complexity, the deep, mysterious sumptuousness which Aldous Huxley wrote about, finally acknowledged by our conscious awareness and, potentially, our technology. The world itself is, plainly, non-binary. If we are to act ecologically through our tools, to act with care and justice towards one another and our more-than-human comrades, we must let go of binaries ourselves and free our machines to do likewise.

The non-binary quality of our desired machines also opens them up to a whole body of thought we haven't so far considered, but that should be central to a rethinking of what computers could be: queer theory. Queer theory opposes the heteronormativity of culture in all its forms, including the gender binary. One of my favourite applications of queer theory to machines is the artist Zach Blas's project, Queer Technologies, which includes a queer programming language called transCoder and a set of physical 'gender changers': computer cables which allow 'male' cables to be transformed into 'female' ones. 'By reimagining a technology designed for queer use,' writes Blas, 'Queer Technologies critiques the heteronormative, capitalist, militarized underpinnings of technological architectures, design, and functionality.' Crucially, by actually building the tools it imagines, it makes possible different ways of doing technology and thus of understanding the world that technology is part of.[32]

As the genderqueer activist Jacob Tobia has pointed out, 'The first time that you heard the term binary, it was probably in a computer class. Here's the problem: People don't work like computers. Our identities, our thoughts, and our beliefs can't always be sorted easily into two categories. In the world we live in, we set up two distinct categories – man and woman – that everyone must choose between. But that doesn't actually reflect the full diversity of the human experience.'[33] As with the conscious decision to remove master/slave terminology from the lexicon of technology, to change the way we describe and build computers could have real repercussions, not only for their own architecture and capabilities, but for the lives of people whose experience is shaped by such technological metaphors – which is to say, all of us.

The second condition, decentralization, draws from the lessons of the octopus and the slime mould to acknowledge the power of communal, cooperative undertaking. The power of communities and systems lies in their intra-action, their becoming-together to produce something greater than their parts. The process of decentralization follows the distribution of networks like the internet, but additionally insists that actual power, rather than mere connectivity, is shared out. We already have the means and know-how to do this, although its actual implementation has hitherto been relegated to the fringes of technological culture by the overbearing pressure of corporate monopolies and corporate profit-seeking.

The open-source movement is one such example of redistribution. This is the practice of publishing the full codebase – every line of code which makes up a piece of software – for public scrutiny and critique. By making the actual code of software and hardware accessible and legible to all, open-source practices decentralize knowledge and provide the basis for collective and self-education. The field of distributed computing is another example: it has given birth to both the extreme democracy of file-sharing and cryptocurrencies and to scientific initiatives like SETI@home and Folding@home. The former seeks to discover life in outer space, the latter to develop new cures to disease. Both use the power of remote processors provided by volunteers – the computers of the general public, linked by the internet – to churn through complex calculations which would overburden any single supercomputer. It's perhaps unsurprising that both these charismatic – and successful – examples of the form are concerned with life itself. Federated and peer-to-peer networks are a third example of decentralization. These are attempts to recreate the power and affordances of contemporary social networks, website hosts and even video calls, by allowing every user to build, host and control their own fragment of the wider network. In doing this, users actively reshape the network itself, transforming it from one centred around a few, privately owned hubs, to one in which users are directly connected to one another: a change in technology which results in a physical change to the topography and power relationships of the network itself.[34]

The project of decentralization also encompasses the process of decentring ourselves: the acknowledgement that humans are not the most important species on the planet, nor the hub around which everything else turns. Rather, we are a specialized but equal part of a vaster, more-than-human world, of no greater or lesser significance than any other part. Decentring is a complex task which requires us to think deeply about our relationships with the more-than-human world, and to understand our actions and the tools which we create as contributions to, and mediations with, everything else, instead of as unique, and uniquely powerful, artefacts of human superiority.

The third condition, unknowing, means acknowledging the limitations of what we can know at all, and treating with respect those aspects of the world which are beyond our ken, rather than seeking to ignore or erase them. To exist in a state of unknowing is not to give in to helplessness. Rather, it demands a kind of trust in ourselves and in the world to be able to function in a complex, ever-shifting landscape over which we do not, and cannot, have control. This is a basic imperative of being human in a more-than-human world and has always been acknowledged by traditional cosmologies, through their observance of ritual and their practical entreaties to the intercession of the non-human beings – plants, animals, spirits and weather systems – which enable our survival.

Many of our most advanced contemporary technologies are already tuned towards unknowing, none more so than machine-learning programmes, which are specifically designed for situations which are not accounted for in their existing experience. Applications such as self-driving cars, robotics, language translation, and even scientific research – the generation of knowledge itself – are all moving towards machine-learning approaches precisely because of this realization that the appropriate response to new stimuli and phenomena cannot be pre-programmed. Nonetheless, such programmes can all too easily continue to ignore or erase actual reality – with devastating consequences – if they perceive themselves in the same way that we, their creators, have always seen ourselves: as experts, authorities and masters. To be unknowing requires such systems to be in constant dialogue with the rest of the world, and to be prepared, as the best science has always been, to revise and rewrite themselves based on their errors.

An example of such an unknowing system is the Optometrist Algorithm, developed by Google for Tri Alpha Energy. Tri Alpha is working to develop practical nuclear fusion technology, a source of clean, near-limitless energy which has been the stuff of science fiction for decades. Doing so requires intensely complex calculations, weighing up thousands of variables for each test run of their experimental reactor – thousands more than any human researcher could meaningfully evaluate. But the scale of the problem means that a purely programmatic approach could also lose itself down endless branches of an infinite tree of possibilities, without meaningfully improving the result.

The solution designed in response to this problem was for a machine-learning algorithm to methodically evaluate many options and to present a reduced set of possible actions to a human operator. In this way, not only were more minds brought to bear on the problem, but also at least two distinct ways of thinking: the programmatic, mathematical evaluation of the machine; and the creative, hunch-driven exploration of the human mind. The algorithm works less like a blind, rule-based machine and more like an optometrist trying out different lenses, constantly checking in with their patient, 'Better like this? Or like this? More like this, or like this?'

The results are promising – but what would such an algorithm look like if it were to appeal not just to the human, but to the more-than-human? An algorithm which devolved part of its processing, part of its thinking, to non-human actors: perhaps something like the bucket of pre-processing water described earlier, or the use of slime moulds and mycorrhizal networks as translators and co-creatives. This would be the full realization of Stafford Beer's U-machine, or Turing's *o*-machine – and crucially, it would be facing outward, rather than trying to trap its colleagues in a little box, further disconnected from the wider world.[35]

The Cockroach Controlled Mobile Robot, created by Garnet Hertz.

An example of a really unknowing cybernetic machine is the Cockroach Controlled Mobile Robot, or Roachbot (2004–2006), Canadian designer Garnet Hertz a device created by the that demonstrates all of the above principles, while also retaining some of the more nightmarish aspects of the spinal dog and the crab computer. (Perhaps we might consider such squeamishness a useful indicator of more-than-human effectiveness, but that is a subject for elsewhere.)

The Roachbot consists of a motorized tricycle, a set of proximity sensors and a very large Madagascan hissing cockroach Velcroed to a trackball. The Roachbot uses the cockroach's dislike for bright light – the same mechanism which sends them scurrying away when you turn on the kitchen lights – to create a simple, exploratory cyborg robot. As the cockroach scurries on top of the trackball, it propels the tricycle across the room – but if it gets too close to an obstacle, bright LEDs illuminate in that direction, causing it to spin off on a different trajectory. In this way, it makes its way around a space, avoiding obstacles in much the same fashion as Grey Walter's tortoises. The key difference between the Roachbot and the tortoise is that an element of the system remains unknown to us. Unlike the tortoises, the cockroach at the heart of Hertz's machine has not been constructed or solved. It's doing its own thing – it is an oracle. It has in cybernetic terms, been 'entrained', but it has not been dominated (although the cockroach itself might argue with that assertion).[36]

In any true relationship based on unknowing – between human, machine, mushroom or cockroach – the participant must forgo any requirement to fully understand the operation of any other. Rather, relationships based on unknowing require a kind of trust, even of solidarity. They require us to open ourselves to the possibility not merely of other intelligences, but to the idea that they might want to help us – or not – and thus might predispose us to the creation of more mutually agreeable conditions in which they might deign to assist us voluntarily. This is indeed the opposite of helplessness: it makes possible the creation not merely of better relationships, but of better worlds.

Non-binary, decentralized, unknowing – what all three conditions of this negative theology of technology have in common is that they are concerned with dismantling domination, in all its forms. To be non-binary, in human and machinic terms, is to reject utterly the false dichotomies that produce violence as the direct consequence of inequality. A culture of binary language splits us in two, and makes us choose which parts of ourselves fit existing power structures. To assert non-binariness is to heal this divide and to make different claims of agency and power possible.

To decentralize, in this context, means to empower and grant agency equally to every actor and assemblage in the more-than-human world, so that none may have dominion over any other. To be unknowing means to acknowledge that – like Socrates before the Oracle – neither we nor anybody else knows exactly what is going on; and to be humbled and at peace with that understanding and thereby with everything else. Technologies of control and domination become instead technologies of cooperation, mutual empowerment and liberation.

These are, of course, not merely technological or ecological goals: they are political ones too. Any technological question at sufficient scale becomes one of politics. And it is to politics we will turn in the final part of this book, to see what lessons we may draw from the more-than-human world, including our technologies, in achieving more just and equal relationships between all of us.

Cybernetics, which most closely approaches the kind of more-than-human understanding of technology we are searching for, had its own brushes with politics. Stafford Beer, in particular, took his cybernetic appreciation of the world and tried to use it for the betterment of lives. In 1971, he was contacted by Salvador Allende's newly elected socialist government of Chile to see if his ideas could be usefully employed as part of a state-run economy. Beer leapt at the chance, and spent many months studying, documenting and intervening in the Chilean economy. His most successful action was the implementation of a network of telex machines in over 500 factories, connected to planning offices in municipalities and central government. Beer believed that this network would form the nervous system of something resembling a country-wide Cybernetic Factory: fully connected, autonomous and highly responsive to changing conditions. It only had one major test – successfully routing around a CIA-backed truck strike in which the telexes were used to coordinate food deliveries around highway blockades – before Allende was overthrown, also by the CIA.

It's unclear whether Beer's plan – called Project Cybersyn (cybernetic synergy) – would ever have evolved as he predicted, or whether it would, as his critics claimed, have merely resulted in more top-down control and oppression of labour. But there's a beautiful moment in a lecture Beer gave at Manchester University in 1974 which illustrates how different political viewpoints can give rise to very different understandings and implementations of technology. As Beer told it, he was describing to Allende how his Viable System Model, the overarching concept which lay behind his sprawling ideas for adaptive systems, might be applied to government. Beer worked his way through the model, explaining how System One, the lowest level of the VSM, referred to the Departments of State, and how each successive higher system referred to different operations of governance. Allende listened attentively, and just as Beer was about to say: 'And System Five, Mr Presidente, is you', Allende interrupted, with a broad smile on his face, to say 'Ah, System Five, at last. The People.'[37]

For decades we've tried to dominate the world through breaking it down into its component pieces and reassembling it into epistemological and mechanical machines of our own devising. We have demanded to know how everything works so that we can suborn it to our own ends, and we have used that knowledge to oppress and suppress the agency of others. But ecological technology seeks to connect and rebuild, to construct from all the more-than-human players on the field a Viable System Model more inclusive and generative than anything Beer himself imagined: a just, equitable and liveable world.

The work to build the oracle machine is ongoing: it's our central task. In this chapter, I've brought together some ideas and processes that establish what the oracle machine might look like if it included us and the more-than-human world in ways that Turing ignored and Beer only hinted at; a way that recognizes Gaia as the ultimate homeostat and accepts radical otherness as the driver of adaptability. From these leads, we might also begin to establish a more-than-human politics: a framework and set of processes for living technological ecology more fully. But first, we need to fashion another piece of the puzzle, and further undermine our belief in knowledge and certainty, in order to build a new foundation for understanding. As we did when we dived down through Neanderthal culture and Devonian sex to find the little dinoflagellates still living within our own cells, I want to push these concepts of unknowability and adaptive encounters a little deeper, into the fuzz and froth that underlies our ability to adapt and change at all. To do that, we have to get random.

# 7
# Getting Random

In the centre of Athens, at the base of the Acropolis Hill, stands the Stoa of Attalos, called after the man who had it built: King Attalos II of Pergamon, ruler of the city between 159 and 138 BCE. In ancient Athens a stoa was a covered walkway or portico (the name is still used today for the arcades in Athens' side streets). The building was reconstructed in the 1950s, and now houses one of my favourite collections of ancient objects, the Museum of the Ancient Agora. The agora, or public space, was the social and mercantile heart of the ancient city, and the museum's collection tells us much about the everyday lives of its citizens and the ways in which they went about their business.

The core of the collection is a set of artefacts associated with the system of government which arose in Athens in the third century BCE. The city is widely regarded as the cradle of Western democracy – though there are caveats. In ancient Athens the 'demos', that portion of the population which had actual rights, was limited to property-owning men over twenty-five: women, slaves, foreigners and all other members of the community were excluded from decision-making. Nonetheless, ancient Athenian democracy still holds some important lessons for us – particularly in the ways it differed, quite radically, from the way we practise democracy today.

One of the simplest yet most striking objects in the museum is an example of a *klepsydra*. This is a water clock – literally, 'water thief' – in the form of a simple clay jug, with a small spigot near the base. The size of the jug corresponded to the particular length of time allotted to a speaker in the assembly or in a court case. For rebuttals in a claim involving less than 500 drachmas, a *klepsydra* was used which gave

A *klepsydra* with handles and inscription restored, from a watercolour by Piet de Jong.

the speaker about six minutes to make their case. The jug would be filled, and the spigot uncorked when the defendant began speaking; when the flow of water stopped, the speaker was required to stop too. According to tradition, the most skilful orators would time the dramatic culmination of their speeches to the last drop that fell from the spigot. In this way, the *klepsydra* might be thought of as the most simple form of hydraulic computer: an analogue time-keeper which used fluid as its operating medium.[1]

The cabinet adjoining the *klepsydra* holds a curious assortment of broken potsherds: small pieces of vases or other earthenware vessels, with names scratched onto them. These are *ostraka*, so ubiquitous in ancient Athens that examples are now displayed in practically all of the city's many archaeological museums. These shards served as a kind of scrap paper: in place of imported Egyptian papyrus, which was available but expensive, or equally costly animal hides, fragments of pottery could be freely picked up anywhere. In this case, Athenians used them for one of the most curious, and sadly extinct, customs of ancient democracy, which took its name from the *ostraka*: ostracism.

Unlike contemporary democracy, which mostly involves voting *for* someone, the Athenians preferred voting *against*. If any one individual became too powerful, or was considered in some way a threat to the good running of the city, then their ostracism could be called for by the populace and put to a vote. If enough votes were cast in favour

An *ostrakon* calling for the exile of Themistocles (480s–470s BCE), Ancient Agora Museum, Athens.

of the ostracism – contemporary sources put the number at around 6,000 – then that person was exiled from the city for ten years, on penalty of death. As a mechanism to prevent the emergence of new tyrants, ostracism was relatively successful – so successful, indeed, that it fell into disuse, although there is some evidence that it was also vulnerable to manipulation. Either way, it was a crucial part of early democracy that would be fascinating to reintroduce (perhaps without the death penalty) today.

Alongside the *klepsydrae* and the *ostraka* sits my favourite relic of ancient media technology: a 2,000-year-old, fully functioning analogue computer, built of stone and brass, which was one of the key processing units of democracy: the *kleroterion*. Now reduced to a broken slab a few feet high, it once stood, head-high, in the agora. On the front of this broad block of stone were engraved rows of deep slots – about 300 of them, in rows of five or ten. From the top of the slab to its base was drilled a long tube, stopped by a hand-turned crank where it emerged.

Accompanying the *kleroterion* are examples of *pinakion*: small bronze plates issued to each member of the *demos*, with their name incised on them. These were essentially ID tags, or user tokens, and they slotted neatly into the rows of clefts on the front of the stone. When a jury was called for, a selected group of citizens placed their *pinakion* into the slots on the front of the machine, and an official poured a bucket of black and white balls, corresponding to the number of jurors needed, into the top of the tube. The order in which the balls emerged, released by turning the crank handle, determined which rows of *pinakia* would be retained, and thus which citizens would be required to serve on thc jury.

There are two important points here. The first is that the Athenians didn't just use their analogue computer for assembling juries. They used it to enlist members of city boards, to select judges and legislators, and even to choose the ruling council of the state, the Boule. In fact, the only civic institution exempt from having its leaders selected in this way was the army.

Secondly, and in contrast to almost everything we think about democracy, this process of selection, enforced by the *kleroterion*, was a random one. Today this process, known as sortition, is familiar to us from jury selection, but the original democracy, as practised in Athens in 300 BCE, used this method to assign almost all the important positions in government. The Athenians believed that the principle of sortition was critical to democracy. Aristotle himself declared that: 'It is accepted as democratic when public offices are allocated by lot; and as oligarchic when they are filled by election.' Sortition – randomness – was the foundation of radical equality.

It's one of the stranger facts about most of our modern digital machines that they are incapable of being random – and thus, according to the ancient Greeks, incapable of being true agents of equality. True randomness is a slippery thing: it is a property not of things in themselves, like individual numbers, but of their relationship to one another. One number is not random; it only becomes random in relation to a sequence of other numbers, and the degree of its randomness is a property of the whole group. You can't be random, in modern parlance, without having some shared baseline of normality or appropriateness to measure yourself against. Randomness is relational.

The problem computers have with randomness is that it doesn't make mathematical sense. You can't programme a computer to produce true randomness – wherein no element has any consistent, rule-based relationship to any other element – because then it wouldn't be random. There would always be some underlying structure to the randomness, some mathematics of its generation, which would allow you to reverse-engineer and re-create it. Ergo: not random.

This is a major problem for all sorts of industries which rely on random numbers, from credit card companies to lotteries, because if someone can predict the way in which your randomness function operates, they can hack it, akin to a gambler sneaking marked cards into the game. This is in fact the way many such thefts have been performed. In 2010, an Iowa State Lottery official manipulated the lottery's random number generator in such a way as to be able to predict the draw on certain days: he picked up at least $14 million before he was caught. In Arkansas, the Lottery Commission's own deputy director of security stole more than 22,000 lottery tickets between 2009 and 2012 and won almost $500,000 in cash, again by manipulating the underlying code which picked the numbers.[2]

To repeat, computers are incapable, by design, of generating truly random numbers, because no number produced by a mathematical operation is truly random. That's precisely why many lotteries still use systems like rotating tubs of numbered balls: these are still harder to interfere with, and thus predict, than any supercomputer.[3] Nevertheless, computers need random numbers for so many applications that engineers have developed incredibly sophisticated ways for obtaining what are called 'pseudo-random' numbers: numbers generated by machines in such a way as to be effectively impossible to predict. Some of these are purely mathematical, such as taking the time of day, adding another variable like a stock market price, and performing a complex transformation on the result to produce a third number. This final number is so hard to predict that it is random enough for most applications – but if you keep using it, careful analysis will always reveal some underlying pattern. In order to generate true, uncrackable randomness, computers need to do something very strange indeed. They need to ask the world for help.

A case study in true machine randomness is ERNIE, the computer used to pick the Premium Bonds, a lottery run by the UK government since 1956. The first ERNIE (an acronym for Electronic Random Number Indicator Equipment) was developed by the engineers Tommy Flowers and Harry Fensom at the Post Office Research Station, and was based on a previous collaboration of theirs – Colossus, the machine which cracked the Enigma code. ERNIE was one of the first machines to be able to produce true random numbers, but in order to

ERNIE 1, 1957.

do so it had to reach outside itself. Rather than simply doing maths, it was connected to a series of neon tubes – gas-filled glass rods, similar to those used for neon lighting. The flow of the gas in the tubes was subject to all kinds of interference outside the machine's control: passing radio waves, atmospheric conditions, fluctuations in the electrical power grid, and even particles from outer space. By measuring the noise in the tubes – the change in electrical flux within the neon gas, caused by this interference – ERNIE could produce numbers which were truly random: mathematically verifiable, but completely unpredictable.

Subsequent ERNIEs used ever more sophisticated versions of the same approach and closely followed the technological trends of their times. ERNIE 2, which debuted in 1972, was half the size and designed specifically to resemble one of the computers in the James Bond film *Goldfinger*. ERNIE 3, which followed in 1988, was now the size of a desktop computer. It took just five and a half hours to complete the draw, five times faster than its predecessor. ERNIE 4 reduced that time to two and a half hours and dispensed with the neon tubes, using the thermal noise of its internal transistors, together

ERNIE 3, 1988.

with a sophisticated algorithm. ERNIE's most recent incarnation, ERNIE 5, has since March 2019 been picking the Premium Bonds by examining the quantum properties of light itself.[4]

ERNIE maps the evolution of computers themselves over seventy years, from room-sized tangles of wires and circuit boards, through bulky mainframes and desktop boxes, all the way down to the development of highly specialized, microscopic chips of silicon with the ability to appraise individual photons. But each incarnation has done something few machines do: it has looked outside of its own circuitry, in order to commune with the more-than-human world that surrounds it, in the service of true randomness.

In the intervening years, other creative ways of generating randomness with machines have been developed. Lavarand, initially proposed as a joke by workers at supercomputer firm Silicon Graphics, uses a digital camera pointed at a lava lamp to draw truly random numbers from the lamp's endless, chaotic fluctuations. Online security firm Cloudflare, which protects thousands of websites from hacking and other disruption, actually put Lavarand to work: at Cloudflare's headquarters in San Francisco, shelves lined with eighty lava lamps provide a back-up source of randomness for their digital servers. Hotbits, another hobbyist project, uses a radiation detector pointed at a sample of radioactive Caesium-137, which produces beta particles at random intervals as it decays. Random.org, a popular online source for true random numbers, started out using a $10 receiver from Radio

Shack to measure atmospheric radio noise; it now consists of a network of aerials and processing stations around the world.[5]

Each of these machines is admitting to the same flaw: given the way we have constructed them, computers are not capable, operating alone, of true randomness. To exercise this crucial faculty, they must be connected to such diverse sources of uncertainty as fluctuations in the atmosphere, decaying minerals, shifting globules of heated wax and the quantum dance of the universe itself. On the other hand, they are confirming something beautiful. In order to be full and useful participants in the world, computers need to have relations with it. They need to touch and be in touch with the world. This is the full realization of Turing's oracle machine, that mysteriously powerful device which, whatever it is, 'cannot be a machine'. Once again, we come to the only conclusion available: the oracle is the world.

It's another quirk of computational history that one of the people in part responsible for the fixed and inflexible nature of most modern computers is also partly responsible for one of the most powerful applications of randomness. John von Neumann, a Hungarian-American physicist best known for his role in the development of the atomic bomb, was closely involved in the development of the first computers, which were based on Turing's designs. These machines were initially developed to assist in the design of the bomb, which required complex calculations beyond the reach of existing calculating machines. In his proposal for the EDVAC, the first all-digital, stored-program computer, von Neumann specified a particular architecture: a single connection, or 'bus', between the memory and the central processor, meaning that the computer could not fetch data and execute commands at the same time.

Today, just as almost all computers are based on Turing's *a*-machine, almost all computers use the von Neumann architecture. But a problem results from this: the central processing unit (CPU) is constantly forced to wait on required information as it is moved into or out of memory, which can result in a serious drag on its processing speed. The original decision to build computers this way was made for reasons of simplicity, but it means that significant amounts of computer time, of software design and of electrical energy are expended on moving information around rather than doing anything with it. The von Neumann bottleneck, as this problem is called, is a classic example of initial assumptions being encoded into complex, inflexible tools, which in turn fundamentally shape our abilities decades into the future. In fact, the von Neumann bottleneck is one of the key problems faced by the latest artificial intelligence systems, which require huge amounts of data processing to advance their learning capabilities. Proposals for future systems include quantum computers which will bypass the bottleneck in order to perform massively serial computations. But they remain way in the future. Today, just as we're living in Turing's abstracted *a*-machine, we're also living in von Neumann's bottleneck.

Von Neumann's immediate post-war work concerned the development of a new hydrogen bomb that would be far more powerful and destructive than the first-generation atomic bombs detonated over Hiroshima and Nagasaki in August 1945. The difference between the two types of bomb is that the A-bomb depends on pure nuclear fission – the almost uncontrolled release of energy produced by the splitting of the atom – while the H-bomb relies on a subsequent fusion, achieved by containing and shaping this energy into an even more powerful reaction. In order to construct the H-bomb, it was necessary to model the complex interactions between particles as they were released from the bomb's critical core – a problem which exceeded the simulation capabilities of any machines then available.

In 1946, one of von Neumann's colleagues, the Polish physicist Stanislaw Ulam, was recuperating from a serious illness and playing endless games of Solitaire to occupy his time. He wondered idly if it would be possible to calculate the chance of any fifty-two-card lay-up resulting in a successful finish. It occurred to him that rather than relying on abstract calculation, he could simply run through a hundred or so games and thereby gain a pretty good idea of the probability by counting the successful plays. Realizing that computers already in existence could handle such a subset of all possible outcomes with ease, Ulam generalized his process to mathematical physics. The same approach could be used to simulate a few thousand neutron reactions, and thereby gain an approximate – but pretty good – idea of what was happening in the nuclear core.[6]

Ulam took his idea to von Neumann and – together with another physicist, Nick Metropolis – they formalized the approach and gave it a name: the Monte Carlo method. Von Neumann was himself an inveterate gambler; he first met his wife Klára on the French Riviera between the world wars while trying to beat the roulette wheels of Monte Carlo with a mathematical system he had devised (she had to buy the drinks when it failed). The idea of a random, chance-based approach to complex mathematical questions was deeply appealing to him. At any point in a calculation, it was simply necessary to roll the dice – or the equivalent of several thousand dice – and advance a step further, rather than trying to calculate every possible outcome. In this way, rather than trying to represent and solve a whole problem at once – to capture and dominate it – the Monte Carlo method seeks to actively explore and draw conclusions about a problem, a very different, and more naturalistic, approach.[7]

In order to implement Monte Carlo at scale, it was necessary to radically reconfigure the ENIAC, the main computer then in use by the H-bomb team. In this they were assisted by Klára Dán von Neumann, by this point one of the programmers of the ENIAC – indeed, one of the first dedicated programmers of any modern computer. In order to make Monte Carlo work, the team invented and implemented a new system of 'background coding' and 'program coding', which survives today as the difference between a computer's operating system and individual applications. Previously, every new program run on the ENIAC – on any computer – had to be individually coded in every single detail, from where to store information to which subsystems to ask for mathematical results. By separating out the background from the program, and encoding many frequently used functions into the machine itself, it was possible to massively simplify the work of programming and make computers genuinely multipurpose for the first time. This approach became as standard a feature of modern computers as Turing's logic and von Neumann's architecture. And it was a direct result of trying to get machines to implement a randomized approach to complex problem-solving.

In order to fully implement the Monte Carlo method, there was one further, crucial requirement: a source of random numbers, which could not be generated by the computer itself. John von Neumann was all too aware of the failure of machines in this regard. In a paper written on the subject in 1949, he warned that 'anyone who considers arithmetical methods of producing random digits is, of course, in a state of sin'.[8]

In response to this need, the RAND Corporation – an offshoot of the US armed forces, which employed von Neumann as a consultant – built an 'electronic roulette wheel', which consisted of a pulse generator and a noise source, most likely a small gas-filled transistor valve similar to the kind used in ERNIE. The result was published in 1955 as *A Million Random Digits with 100,000 Normal Deviates* – an extraordinary book of numbers which consists of exactly what its title describes: 400 closely set pages, each containing 50 lines of 50 digits, with the lines numbered 00000 through 19999. These exact numbers, provided to von Neumann's team as punched cards, were used for the early Monte Carlo simulations – and, as the largest source of random digits ever collated, are still in use by statisticians, physicists, poll-takers, market analysts, lottery administrators and quality control engineers today.[9]

TABLE OF RANDOM DIGITS 1

| | | | | | | | | | | |
|---|---|---|---|---|---|---|---|---|---|---|
| 00000 | 10097 | 32533 | 76520 | 13586 | 34673 | 54876 | 80959 | 09117 | 39292 | 74945 |
| 00001 | 37542 | 04805 | 64894 | 74296 | 24805 | 24037 | 20636 | 10402 | 00822 | 91665 |
| 00002 | 08422 | 68953 | 19645 | 09303 | 23209 | 02560 | 15953 | 34764 | 35080 | 33606 |
| 00003 | 99019 | 02529 | 09376 | 70715 | 38311 | 31165 | 88676 | 74397 | 04436 | 27659 |
| 00004 | 12807 | 99970 | 80157 | 36147 | 64032 | 36653 | 98951 | 16877 | 12171 | 76833 |
| 00005 | 66065 | 74717 | 34072 | 76850 | 36697 | 36170 | 65813 | 39885 | 11199 | 29170 |
| 00006 | 31060 | 10805 | 45571 | 82406 | 35303 | 42614 | 86799 | 07439 | 23403 | 09732 |
| 00007 | 85269 | 77602 | 02051 | 65692 | 68665 | 74818 | 73053 | 85247 | 18623 | 88579 |
| 00008 | 63573 | 32135 | 05325 | 47048 | 90553 | 57548 | 28468 | 28709 | 83491 | 25624 |
| 00009 | 73796 | 45753 | 03529 | 64778 | 35808 | 34282 | 60935 | 20344 | 35273 | 88435 |
| 00010 | 98520 | 17767 | 14905 | 68607 | 22109 | 40558 | 60970 | 93433 | 50500 | 73998 |
| 00011 | 11805 | 05431 | 39808 | 27732 | 50725 | 68248 | 29405 | 24201 | 52775 | 67851 |
| 00012 | 83452 | 99634 | 06288 | 98083 | 13746 | 70078 | 18475 | 40610 | 68711 | 77817 |
| 00013 | 88685 | 40200 | 86507 | 58401 | 36766 | 67951 | 90364 | 76493 | 29609 | 11062 |
| 00014 | 99594 | 67348 | 87517 | 64969 | 91826 | 08928 | 93785 | 61368 | 23478 | 34113 |
| 00015 | 65481 | 17674 | 17468 | 50950 | 58047 | 76974 | 73039 | 57186 | 40218 | 16544 |
| 00016 | 80124 | 35635 | 17727 | 08015 | 45318 | 22374 | 21115 | 78253 | 14385 | 53763 |
| 00017 | 74350 | 99817 | 77402 | 77214 | 43236 | 00210 | 45521 | 64237 | 96286 | 02655 |
| 00018 | 69916 | 26803 | 66252 | 29148 | 36936 | 87203 | 76621 | 13990 | 94400 | 56418 |
| 00019 | 09893 | 20505 | 14225 | 68514 | 46427 | 56788 | 96297 | 78822 | 54382 | 14598 |
| 00020 | 91499 | 14523 | 68479 | 27686 | 46162 | 83554 | 94750 | 89923 | 37089 | 20048 |
| 00021 | 80336 | 94598 | 26940 | 36858 | 70297 | 34135 | 53140 | 33340 | 42050 | 82341 |
| 00022 | 44104 | 81949 | 85157 | 47954 | 32979 | 26575 | 57600 | 40881 | 22222 | 06413 |
| 00023 | 12550 | 73742 | 11100 | 02040 | 12860 | 74697 | 96644 | 89439 | 28707 | 25815 |
| 00024 | 63606 | 49329 | 16505 | 34484 | 40219 | 52563 | 43651 | 77082 | 07207 | 31790 |

The first page of the RAND Corporation's book *A Million Random Digits with 100,000 Normal Deviates*, 1955.

The calculations which lay behind the development of the hydrogen bomb were based on the outcome of a roulette wheel – or, rather, on the chaotic fluctuations of noise within a gas-filled glass tube: the vibrations of the universe itself.

Just as a deep dive into our evolutionary history – deploying the most finely tuned, sense-making apparatuses we have devised – reveals not a single answer to the question of life but a chaotic multiplicity of beings, so our closest approach to mathematical truths about the universe involves aligning ourselves with the most chaotic, the most unpredictable, the most random processes we can comprehend.

The genius of Monte Carlo was to recognize that the most efficient search of a complex territory is the random walk. Its results inform many of our computational processes today. Notably, Monte Carlo gives us the ability to sift the overwhelming abundance of information available to us on the internet, as the web-crawling bots of search engines spider their way, at random, across the complex territory of the infosphere, in order to draw statistical inferences about its contents. The success of Monte Carlo is a recognition, at the core of computational and mathematical science, that meaning resides less in the data at the end of the path and more in the path travelled. The meaning of complex, unknowable systems, in other words, is relational.

The use of randomness to better approach the world-as-it-actually-is, the more-than-human world, is not limited to the sciences. Perhaps its greatest, and certainly most committed, exponent was the avant-garde composer John Cage. Beginning in the 1950s, Cage began to use chance-based processes as the defining mechanism of his composition, employing the dictates of randomness to decide everything from the pitch and duration of notes to the length and musical sources of his compositions – and even the number of instruments involved in performing them.

Cage's work was driven by his belief that all sound is music, and that the purest music is that furthest removed from conscious intention, most distant from any ego or over-arching scheme on the part of the composer. In random, chance-based processes, Cage discovered a mechanism for abstracting himself from his compositions – and in the process, to approach something entirely different. Even his most (in)famous piece, *4'33"*, in which the performers sit for four minutes and thirty-three seconds in complete silence, is an extreme iteration of his belief in randomness. The sounds which the audience hears during the performance are those which are incidental, environmental and unpredictable – as far as possible from any design on the part of the performer or conductor.

Cage explored indeterminacy in a number of his early works, shuffling pages of scores and using pieces of other composers' works as elements within his own. But the trigger for a deeper exploration of the possibilities of indeterminacy was his discovery – in the form of a gift from one of his students, the pianist Christian Wolff – of the I Ching, the 3,000-year-old Chinese book of wisdom and divination. In one very limited sense, the book is a precursor to RAND's *Million Random Digits*, but rather than simply giving out its results, it presents the reader with a method for generating their own random outcomes. To consult the I Ching, the reader formulates a question in

Hexagram 52 of the I Ching: 艮 (gèn): 'keeping still, mountain'.

their mind, then tosses coins, usually three at a time, to generate six sequences of heads and tails. Added together, these represent six broken or unbroken lines, indicating one of sixty-four hexagrams. Each hexagram is associated with an explanatory text – and much subsequent commentary – which is brought to bear on the practitioner's question.

In Cage's hands, the I Ching became an infinitely generative engine for musical composition. He first used it for the third movement of his *Concerto for Prepared Piano and Chamber Orchestra* of 1950 (the prepared piano is a modification of a standard piano, invented and popularized by Cage, which uses screws, metal bars, wooden wedges and eventually any object imaginable to alter the pitch and tones of a piano's strings, to create all kinds of novel and otherworldly sound effects). He intended the work to dramatize the conflict between form and structure, personality and impersonality. The score was based on what he called a 'gamut' of musical gestures: short groups of notes and phrases which could be rearranged on a grid into different sequences. For the first movement, Cage personally composed the piano part from the gamut, while the orchestral part was shuffled around according to a simple numerical pattern. For the second movement, both parts were shuffled according to different patterns. The intended effect, Cage wrote, was to first 'let the pianist express his personal taste, while the orchestra expressed only the chart', and then in the second movement to have the piano use the chart too, 'with the idea of the pianist beginning to give up personal taste'. He realized that the final movement could be composed by throwing coins and consulting the I Ching: a step beyond the fixed algorithm of his numerical pattern, into true randomness. To do so would be 'to accept rather than seek to control' the underlying nature of the piece.[10]

The result was a complex and multilayered work, and Cage felt liberated by its creation. For many years he had been seeking a mode of composition which would enable him 'to make a music free of taste, memory, and tradition – a prelapsarian music, compositions outside of history', in the words of one of his biographers, Kenneth Silverman. In the I Ching, he seemed to have found such a process. 'I have the feeling of just beginning to compose for the first time,' he wrote to his friend, the French composer Pierre Boulez.

He was to take all this much, much further. Along with other chance-based processes, such as rolling dice and shuffling cards, Cage would use the I Ching to compose such works as his 1951 solo piano piece *Music of Changes*, which was created from three charts for sounds, duration and dynamics, each with sixty-four different cells, corresponding to each of the hexagrams; and *Williams Mix* (1953), one of the first pieces of tape music – a work made by cutting up and reassembling fragments of magnetic tape – where coin tosses decided which source tape was spliced in, the duration of the snippet, and even the angle of the cut, which affected the attack and decay of the sound. *Williams Mix* alone required a team of assistants to make the thousands of coin tosses and assemble more than 2,000 fragments of tape into eight loops. The final piece sounded something like the rapid twirling of radio dials across different frequencies and volumes. The result, for Cage, was that 'One is now able to work with the entire field of sound, no longer limited by the pitches of instruments or their timbres or loudness.' Music had been freed from the constraints of human instrumentation.

Cage first started working with computers in 1967, during a research professorship at the University of Illinois. At the University's School of Music, a former chemist called Lejaren ('Jerry') Hiller had been experimenting with the creation of musical scores using a new digital computer called ILLIAC – a direct descendant of von Neumann's EDVAC and ENIAC, and the first to deploy von Neumann's architecture from its inception. When Cage enthusiastically joined in, Hiller composed an I Ching program for him, named ICHING, which could produce in a single run the equivalent of 18,000 three-coin tosses. Cage was even more delighted – 'We'll never have to toss a coin again!' – and the pair began work on his most ambitious piece yet, a work for multiple harpsichords, eventually titled, computer-style, *HPSCHD*.[11]

Cage was not the first to produce music using a randomizing computer. An earlier effort was made by our old friend Stafford Beer, who in 1956 described what he called the Stochastic Analogue Machine (SAM). This device automatically dropped hundreds of ball bearings into something like a pinball machine, in which each ball would be randomly delayed, offset and rebounded by various pins and slides. Originally intended as a wry comment on the nature of design – a rigid pattern produced from chance occurrences – it was in fact built and presented at the pioneering exhibition of art and technology, Cybernetic Serendipity, held at the Institute of Contemporary Art in London in 1967 (the same year as Cage started work on *HPSCHD* ). The addition of a sounding plate onto which the ball bearings fell turned SAM into a musical instrument. Perhaps this was convergent evolution at work again?[12]

From its inception, *HPSCHD* bested SAM in ambition and in scale. The piece was originally conceived as a performance for seven harpsichords, more than fifty tape machines, and sixty-four slide projectors. Along with Cage, Hiller and ICHING, its composers included several other computer programs. DICEGAME assembled the component twenty-minute harpsichord solos from fragments of Mozart, Chopin, Beethoven and other classical composers; HPSCHD (a program with the same name as the piece) generated segments of microtonal sound which were spliced into the tape loops; and KNOBS printed out 10,000 different instructions for playing the final recording of the piece: one for each record manufactured. These sheets – in effect, individual scores – turned the home listener into a conductor, directing them to twiddle the volume, tone and balance of their stereo at particular moments to create their own unique version of the work.[13]

*HPSCHD* 's eventual premiere – the piece's creation dragged on for more than two years – had possibly the most extreme technical rider of any concert work to date, including, according to the programme:

> 7 Harpsichords 52 tape machines 631 Pages of Music manuscript 59 Power Amplifiers 59 Loud Speakers 40 Motion Picture films 11 100 × 40-foot Rectangular Screens 8 Motion Picture Projectors 208 Computer generated Tapes 6400 Slides 7 Pre-amplifiers 340-foot circumference Circular Screen 64 Slide Projectors

The work's premiere in the Assembly Hall of the University of Illinois on 16 May 1969 was something of a circus – a spectacle which Cage thoroughly approved of, having created a whole series of earlier multisensory events under the title *Musicircus*. As well as the music itself, the event included 1,600 hand-painted slides – the colours determined by ICHING – and a further 6,800 picture slides, as well as films depicting the recent Moon landing and pictures of outer space, geometric banners and surreal posters, costumes made on the spot and pre-screened T-shirts for sale (price determined by the I Ching), strobing and multicoloured lights, disco balls, ultraviolet blacklights and a host of other effects. The audience – comprised of students, faculty, families and visitors from out of state – chanted, sung, threw wadded balls of paper, and at one point formed into a giant conga line. One audience member described the vast hullabaloo to be 'like the random sounds of civilization'.

At this point in his creative life, Cage was convinced that artists should turn their attention to society, and *HPSCHD* was one fulfilment of this desire. The chance-driven carnival of sound and image was for him an effort 'to make the world work, so any kind of living can take place'. The riotous assembly he produced was the closest he had yet come to reproducing the complexity and variety of life

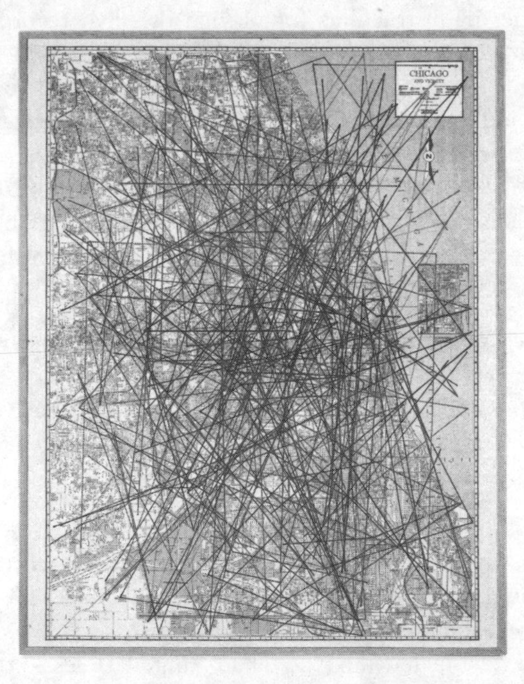


John Cage's score for *A Dip in the Lake: Ten Quicksteps, Sixty-Two Waltzes, and Fifty-Six Marches for Chicago and Vicinity*, 1978.

itself – and it was intended to shift the audience's attitude to that life. This intention becomes clear in other works, such as *A Dip in the Lake*. Here, Cage used the I Ching to create a random list of specific intersections and urban locations around Chicago, which the audience was encouraged to visit in order to generate their own chance-driven experience of each spot's unique sounds, sensations and encounters. Cage desired 'to hold together extreme disparities', as one finds these disparities held together in nature 'or on a city street'. *A Dip in the Lake* is a composition for a random walk: computationally, the most efficient exploration of a complex and unknowable territory, and also the one most likely to produce interesting and stimulating encounters.

Of *Etudes Australes*, another fiendishly complicated piano piece which he composed by laying a grid of musical staves over star maps of the southern skies, Cage wrote 'we must work very hard in order to play this music, and we must also work very hard in order to preserve our environment'. The complexity of the music was directly related to the complexity of the world, and was intended to shift the position of the audience from one of passive listening to active care and attention.

Randomness was also a way of decentring the composer and his work. Cage's urge to erase himself, his tastes and historical situation from his music was not a simple academic exercise. Rather, it was deeply influenced by the principles of Zen Buddhism, which Cage first encountered in the 1940s, in the lectures of D. T. Suzuki, one of the first teachers of Zen in the West. Suzuki's version of Zen taught that 'there is not one center but that life itself is a plurality of centers', a lesson that chimed with Cage's growing ecological consciousness. For him, randomness was a way of reflecting this omnicentric, heterogeneous nature of the world in his work; it also played a practical part in reshaping the way he understood the world around him.

Cage was hardly the first Westerner to make use of the I Ching. Long before Cage and Hiller translated its chance-driven hexagrams into binary code, the book had been present at the very birth of binary code itself, although its accompanying moral guidance has mostly been lost in the code's abstraction into pure mathematics.

Working at the end of the seventeenth century, the German polymath

Gottfried Leibniz was the first mathematician to seriously investigate binary numbers. Deeply religious, Leibniz believed that the purity of the one and the zero were symbolic of the Christian idea of creation *ex nihilo*: out of nothing, God brought forth something. He wanted to defend God's creation against the materialism of Descartes and other mathematicians of the time, and he sought support for his fusion of theology and mathematics in the teachings of other cultures.

Leibniz had long been interested in Chinese mathematics, believing it to be the most ancient teaching on the subject. His friend Joachim Bouvet, a French Jesuit missionary in China, supplied him with copies of the I Ching hexagrams from the court of the Kangxi emperor Xuanye. In the hexagrams, Leibniz saw confirmation of his belief in the eternal, sacred nature of ones and zeros. This correlation between his own binary code and a 3,000-year-old Chinese system gave Leibniz the confidence to publish, in 1703, his *Explanation of Binary Arithmetic*, the foundational mathematical text on binary codes, which cited the I Ching extensively. The ancient origin of binary, Leibniz argued, showed that it was closer to nature than base ten numbering – which is, after all, based on human physiognomy, a kind of anthropocentrism. Binary numbering would permit calculation to be more like nature too.

Leibniz used his new binary calculus to develop a widely influential mechanical calculator called the 'stepped reckoner'. He also proposed a machine which would use marbles to represent binary numbers, and punched cards to sort them, which anticipated modern computer design by some 300 years.

All these inventions flowed from Leibniz's reading of the I Ching, and his belief that in order to achieve universal understanding, mathematical calculation must be rooted in the operations of nature itself. The ones and zeros of binary calculation, as conceived by Leibniz, do not represent fixed and static categories. Instead, they embody change, creation and the ceaseless emergence and becoming of life itself. The computer is like the world.[14]

As John Cage discovered through his use of the I Ching, a complex dance of chance-driven and unexpected encounters was both the best way to approach the more-than-human world, and the best way of representing its heterogeneous, omnicentric reality. Cage's realization prefigured that of evolutionary biologists, who in recent decades have started to acknowledge the crucial role that randomness plays in the creation of life itself. This has proven to be something of an uphill battle because the importance of randomness has been consistently undervalued in studies of evolution since its establishment, while the role of natural selection – competition – has been consistently overvalued.

Our understanding of evolution is still overwhelmingly based on the Darwinian model; that is, it follows the emphasis that Darwin placed on natural selection as the key driver of evolution. This attitude has only been strengthened by neo-Darwinism and the 'modern synthesis' which arose in the twentieth century, which combined Darwin's theories with Mendelian inheritance – the passing down of characteristics through sexual reproduction – to emphasize how life arose through the recombination of genes under adaptive pressure to their surroundings. Such arguments have been extremely useful in arguing against the resurgent creationism and theories of so-called 'intelligent design' which have assailed evolutionary scholarship. However, they do not tell the whole story. Indeed, their success has eclipsed other ways of telling, and has even suppressed research into the other important processes by which evolutionary change occurs.

Darwin emphasized natural selection in his work because it was the most visible of evolutionary processes. The finches which he collected on the Galapagos Islands showed such wide variety between species and sub-species because they had adapted in different ways to the islands' marvellous diversity of habitats. Indeed, they were so varied that he initially assigned them to different species before realizing that they were all related. Some of the finches, for example, had large, blunt beaks, which they used to tear apart the base of cacti and eat the pulp and insects within. Others had longer, sharper beaks, which enabled them to puncture the cactus's fruit to get at the flesh and seeds. Darwin's realization was that these differences were the result of adaptive changes: alterations to the finch's physiognomy in response to pressures from the environment. Natural selection was visible right there in front of him.

Later researchers studied the finches even more extensively. The biologists Peter and Rosemary Grant, who have spent six months of every year on the Galapagos Islands since the 1970s, observed individual birds across multiple locations and multiple seasons, and were able as a result to tie the changes Darwin observed even more closely to natural selection. They showed that birds with smaller beaks prefer smaller seeds, while larger beaks are more effective at cracking open larger seeds. As smaller seeds flourish in the wet season, but larger seeds remained available during droughts, the Grants were able to track the actual changes in populations caused by natural selection over time. They caught natural selection in the act, and their compelling studies – further compounded by the identification of the actual genes responsible for different beak sizes – have revealed natural selection to be the charismatic *ur*-process of evolution.[15]

In this way, natural selection has come to be the way most of us understand all evolutionary change, ignoring or downplaying other processes which work alongside it. Natural selection is not the only force at work in shaping life, and evolution is far less deterministic than the simple combination of organism and habitat would imply. Rather, randomness has a role to play here too, through what are called non-adaptive processes, that is, forces impacting on evolutionary change which are not caused by pressure from the environment. These other forces are mutation, recombination and genetic drift. Each is, in its own particular way, a generator of randomness within evolution, and worthy of our understanding too.

Mutation – the alteration of genes themselves – occurs frequently. It can be caused by mistakes in transcription – errors which creep in when genes are copied and replicated during cell division and sexual reproduction – as well as by chance encounters with chemicals, radiation, and even ultraviolet light from the sun. Genes – ours, as well as those of bacteria, plants and other animals – are susceptible to cosmic interference in much the same way as ERNIE's neon tubes. Changes are also caused by internal malfunctions, the result being that random elements are constantly being written into our genetic code. Most of these have no effect at all; some cause dangerous illnesses; others flower as new and strange abilities and morphologies.

Recombination, the second of these forces, occurs during the exchange of genetic material between different organisms to produce offspring with different traits to their originators. It's the reason we differ from our parents, while resembling them in certain ways. Recombination splices and repairs our DNA, so that it can produce both new combinations of genes and entirely new sequences of DNA. The randomness of recombination comes both from within and without; from the way the chromosomes are split during reproduction and from our choice of sexual partner. Random recombination also happens to prokaryotes – to bacteria and Archaea – during the kind of processes we saw earlier in our discussion of horizontal gene transfer: by viral transfer and by direct transfer of DNA.

Different genes occur at different frequencies within a population, because no two individuals possess exactly the same genetic code. In each subsequent generation, certain genes will be more prevalent than others, due to the existing frequency as well as the random effects of mutation and recombination. Over time, some of these prevailing genes may spread to the majority of the population; others may disappear. This is the third non-adaptive process: genetic drift, which occurs as another randomizing factor, entirely outside the individual organism. The genetic make-up of entire populations can thus change as a result of chance.

For a long time, genetic drift was considered to play at most a minor role in evolution, with such changes being overruled and reshaped by natural selection. But recent scholarship has shown that genetic drift is widespread: many of the changes it causes are carried through populations as 'neutral mutations' which, because they are not outwardly expressed, are not acted on by natural selection. Over time, however, these changes may come to dominate a population, and thereby create new conditions for mutation and recombination to further shuffle genetic codes.

Evolution, then, is not a kind of all-out competition between opposing processes. In reality, it occurs as a constant back-and-forth between mutation, recombination, genetic drift and natural selection. Natural selection may apply a limiting force – no organism can long survive any genetic change which pits it against its environment – but this is only applied after the others have done their work. Random processes precede natural selection: they are the foundation on which all evolutionary change is built. Without random changes occurring within and between individuals and populations, there would be nothing for natural selection to act on in the first place. In the words of John Tyler Bonner, late Professor Emeritus of Biology at Princeton University and one of the most important theorists of non-adaptive change, 'Randomness is the backbone of Darwinian evolution'.[16]

The growth over the last half-century of population genetics, the discipline which has contributed to evolutionary biology most of its understanding of random processes, is founded largely on the use of computer modelling. It's very hard to examine and experiment with processes of random change within living organisms, precisely because they are random. These processes don't occur when and how we would like them to under experimental conditions, and they're almost impossible to see and quantify in the wild. This was Darwin's blind spot. Inside artificial organisms and populations, however, mutations and recombinations can be created at will, allowing us to study how such changes alter internally and drift through populations, and to verify their significance. This is yet another example – along with the formulation of network theory in regard to the internet, and its subsequent dissemination to the natural sciences – of technological models enabling us to better understand natural processes which don't initially seem to be accessible to our reasoning.

We have other blind spots too, like our tendency to focus on larger animals (such as ourselves) when deciding what matters when it comes to evolution. Because of the increased internal complexity of larger organisms – more cells, more types of cells and more interconnections between cells – the number of internal checks on random mutations and recombinations also increases, and so we see an apparent decrease in the effects of randomness in larger organisms. But if we zoom in to examine the lives of some of the smallest – yet still behaviourally complex – organisms in the soil and oceans, we can see the effects of randomness in their full splendour.

John Tyler Bonner, the chief theorist of non-adaptive change, was a world authority on slime moulds: those strange, single-celled organisms which trouble the boundary between individuals and collectives, and which possess such capable intelligence for problem-solving and pattern-forming. Because slime mould 'individuals' are so simple, comprising single, amoeba-like cells, they mutate frequently and rapidly. And because they coalesce into such extraordinary forms, we can see the randomness of this rapid evolution at work. Bonner pointed out that a small handful of soil can contain so many different kinds of slime moulds that arguments for natural selection simply don't apply. There just aren't enough different predators and other pressures in play within a single patch of dirt to account for this splendid variety of forms, so it must be the operation of randomness which drives this efflorescence.

Perhaps the most spectacular example of randomness at play is to be found in the work of Darwin's near-contemporary Ernst Haeckel, the German naturalist who coined the term 'ecology'. Against the wishes of his family, and inspired by his personal heroes Darwin and the German explorer-naturalist Alexander von Humboldt, Haeckel wanted to be both zoologist and artist. In 1859 he travelled to the south of Italy, and while swimming off Naples and Sicily gathered up buckets of seawater whose contents he examined under a microscope. His lens revealed a whole new world of wriggling, pulsating creatures, invisible to the naked eye, but glinting like cut glass and gems – 'delicate works of art' and 'sea wonders' he called them – and he began to paint them, returning to the south of Europe multiple times to dredge up more subjects for his canvas.[17]

Haeckel decided to focus on a single kind of micro-organism: radiolaria. In one trip to the Mediterranean, he named over 150 new species of these creatures. Found throughout the global ocean, radiolaria are a kind of plankton only a few tenths of millimetres across. They are known for constructing elaborate mineral skeletons – mostly out of silica – which Haeckel rendered, to glorious and timeless effect, in his seminal *Kunstformen der Natur* (*Art Forms in Nature* in English translation), originally published in 1899.

Although Haeckel was an early proponent of Darwin's theory of natural selection, and sought to capture the extraordinary proliferation of forms it generated in his illustrations, the microscopic radiolaria are more akin to Bonner's slime moulds than to Darwin's finches. The greatest influence on their evolution is not natural selection, but randomness. The radiolaria – as well as foraminifera, diatoms and other micro-organisms which Haeckel depicted – are examples of

Two plates of radiolaria from Ernst Haeckel's
*Kunstformen der Natur* (*Art Forms in Nature*), 1904.

random generation at its highest pitch of exuberance: a plethora of miniature stars, planets, pavilions, castles, trees and crowns which display every kind of branching, interweaving and entanglement that nature affords.

Biological randomness is most evident in smaller creatures, as natural selection provides a brake on the amount of alteration larger organisms can tolerate. But there's a correlation to this. As organisms grow larger, and their complexity increases – with a corresponding decrease in internal randomness – so the complexity of their societies increases, with a corresponding increase in external randomness: encounters with each other, with other species, with the whole tumult of the more-than-human world, all of which counts as but also exceeds the operations of mere natural selection.

So randomness underlies the entirety of evolution, providing the impetus for some of its most extraordinary and captivating forms. It also plays a central role throughout our individual and collective lives and the lives of more-than-human others. Our lives are themselves random: chance encounters, events and the accumulation of accidents being the defining features of our span upon this Earth. And crucially, randomness is something which we can engineer ourselves – just as John Cage did by introducing chance into his compositions – both as a driver of our own ongoing evolution and to increase our awareness of and engagement with this more-than-human world.

The internet has done more to increase the complexity of human lives than any previous invention. But, like the railroad and the telegraph before it, the complexity it engenders in the enforced encounter with distant lands, peoples and ways of life, is subject to corresponding pressures of order and domination. The railroad opened up the Earth to radical change, but became the vessel of racist colonialism, imperial capitalism and the rigid control of time and human labour. The telegraph allowed for the rapid transmission of ideas and information – but was quickly suborned by financiers, media moguls and the military, to leverage inequalities of information into inequalities of power and profit.

We can see these historical processes at work on the internet today. The first search engines were hand-curated lists of interesting places, essentially random accumulations of sites and tools ordered only by the passions and peccadilloes of those who assembled them. While Google still searches the web with automated random walks, its results are ordered by deeply partisan algorithms, with the top results sold off to the highest bidder. Google has almost a 90 per cent share of the world's web searches, yet indexes only a tiny fraction of the visible web. Most searchers never look beyond the first page of results. There is little room for randomness in exploring the vast amount of information actually available to us. This is deliberate. Google and others' stated mission is to reduce this vast complexity. Their less trumpeted goal is to profit from it, at the expense of our own potential for random encounters, and thereby for our own evolution. So many of our tools are designed to reduce randomness in a similar fashion: from algorithmic recommendation systems to dating apps, from GPS navigation to weather forecasting. Each of these technologies – with the best of intentions – attempts to draw clear lines through a complex environment and provides us with a route to our desires free from obstructions, diversions and the vagaries of chance and unforeseen encounters. Yet as Monte Carlo, John Cage and the radiolarians testify, meaning resides less in the data at the end of the path and more in the path travelled. We learn, change, develop and grow when we move and entangle ourselves with the world in unexpected ways, and we do so best when we are fully engaged participants in that journey, not passive recipients of algorithmic and corporate diktats.

No wonder we keep turning again to chance-producing tools – dice, cards, roulette wheels, astrology and the I Ching – in order to let off steam, get out of our own heads and perhaps provide alternative narratives to a sterile, decomplexified, abstract world, one created apparently and occasionally for our benefit, but in actuality to stupefy us, extract wealth from us and disempower us.

Perhaps the greatest and most egregious example of this power-hungry de-randomizing of the world is our so-called democratic system. The apparent freedom of individual voting, our voice and agency in the political process, is in fact so hedged around by institutions such as parliamentary parties, voting regulations and voter suppression, constituency boundaries, electoral colleges, the influence of polling and the media, political funding, corporate lobbying and a general lack of meaningful consultation, engagement or sufficient education, that it amounts to a charade. The ongoing collapse of this system is evidenced by the general and increasing global dissatisfaction and distrust in government, the rise of charismatic and authoritarian leaders, and an apparent inability to address widespread and systemic issues such as poverty and healthcare, climate change and a global pandemic. That it continues to function is largely and inescapably because we are in thrall to its central mechanism – voting – which the very originators of democracy, the ancient Athenians, viewed as inherently corrupting.

When confronted with systems of control which are incapable of generating within themselves the necessary conditions for meaningful change to occur, it is necessary to reach outside them in order to find a source of novelty and strangeness sufficiently powerful to spin the system into a new configuration. As we have seen, this novelty and strangeness is a product of randomness; its source is the more-than-human world. For true randomness, we must leave the domain of abstract computation, of human-created, anthropocentric laws and programs, and re-engage with the world around us.

The Oracle, Turing's thing-which-cannot-be-a-computer, which calls to us from the origins of computation and which we have identified as the more-than-human world, speaks to precisely this course of action. Artificial intelligence reveals its debt and supplication to non-human minds; the internet reproduces the entangled complexity of fungal networks; gene sequencers expose the cloudy, reticulated origins of our biology; the random walk provides the best path through unknowable complexity to meaningful understanding. It could be said that the unconscious goal of computation since its instantiation has been to rediscover and remake its connection to the uncomputable. In order to remake our societies and render them fit to face the systemic challenges of the present, we need to heed this lesson, rediscover our connection to the more-than-human world, and integrate the uncomputable into our own ways of thinking and relating. We can start with randomness.

The function of the *kleroterion* – the ancient Athenians' analogue computer for assigning positions in government at random – survives to this day in our processes for selecting juries. But this is not its only modern application. In recent years, a number of experiments have taken place, testing the effectiveness of sortition – selection by lottery – across a range of social and civic institutions. The results have been fascinating.

One such experiment took place in Ireland in 2016, when the government of the day created a Citizens' Assembly to consider some of the thorniest issues faced by Irish society: abortion, fixed term parliaments, referendums, an ageing population and climate change. Ninety-nine people were brought together in a hotel outside Dublin, where over a succession of weekends they listened to expert presentations; took testimony from non-governmental organizations, think tanks and interested parties; held question-and-answer sessions; debated among themselves; and drew up a series of proposals for each of the topics under discussion, which the government had promised to review and act upon. The presentations and discussions were live-streamed on the internet to encourage public interest and awareness of the issues. The proceedings themselves were managed by a chairperson – the 100th participant – as well as a secretariat from the civil service, and drew upon procedural methods created by a handful of political theorists and other national governments and community groups over many years.

There are two points worth emphasizing about this Citizens' Assembly. The first is that the ninety-nine participants were complete strangers to one another and were selected at random from the electoral roll, in a process akin to jury service. The results of the random selection were moderated to ensure a balance of certain criteria, such as gender, age, location and class. Beyond this, they were as random a collection of people as one could hope to find, with all the differences in experience, bias, background, education, points of view and personal philosophy that exist across any country.

The second point is that the proposals which emerged from the Assembly were more progressive, more radical and potentially more world-changing than the politicians who commissioned it expected, or even believed possible. The Assembly's recommendation on abortion – which had been made illegal in Ireland in 1861 and remained so following a nationwide referendum in 1983 – was that it should be put to another referendum. Abortion is the most contentious issue in Irish public life, with politicians losing their posts for even suggesting that it should be debated, and this fear of open discussion had strangled the possibility of reform for decades. Indeed, the press openly chastised the Assembly for 'an overly-liberal interpretation of the current thinking of middle Ireland on the issue'.[18]

But here's the thing: a random assembly isn't 'interpreting' the thinking of a mythical middle citizenry. It's representing it, directly. And when the government acted on the Assembly's recommendations and put the Eight Amendment, which forbade abortion, to another referendum, the result was overwhelming and historic. A full 66 per cent of the population voted for the legalization of abortion, which was signed into law in September 2018, against all the expectations of the media and the political class.

Six months later the Assembly came to similarly radical conclusions, this time on the subject of climate change. After taking testimony from experts and the general public, the Assembly issued a series of recommendations, each one passed by at least 80 per cent of its members, arguing for the institution of an independent body to address climate change; the imposition of a tax on carbon and other greenhouse gases;

the encouragement of electric vehicles, public transport, ecological forestry and organic farming; the ending of subsidies to fossil fuels; the reduction of food waste; and support for sustainable electricity microgeneration. All these measures had been proposed to the government before, but had been abandoned or left to languish because politicians thought them unworkable or unpopular, or both. The Assembly's report reinvigorated environmental campaigning in Ireland, leading to the declaration of a climate and biodiversity emergency by the Dáil, the Irish Assembly, and the publication of an official 'government action plan on climate change' in 2019.

It's hard to overstate the significance of these results, which have been mirrored in the outcomes of similar citizens' assemblies in Canada, France, the Netherlands, Poland and elsewhere. Not only did ninety-nine complete strangers, from every walk of life and every conceivable social and education background, come together and reach a consensus on some of the thorniest issues facing contemporary society – an almost unthinkable achievement in our age of political distrust, tribalism and division – but the proposals they put forth challenged our common idea of what is politically and socially possible and led to real, unequivocal change in the lives of their fellow citizens, and potentially in the more-than-human world. The assemblies further demonstrated that there is both an appetite and a willingness on the part of a supposedly apathetic public to seriously address some of the gravest and seemingly most intractable issues that we face.

What was the importance of randomness in all this? I think its effects are twofold. First of all, random selection – sortition – returns to the democratic process something which its supporters often claim for it, but which has been largely lost: the approval and consent of the population. Sortition is transparent and verifiable. It bypasses the widely distrusted political class, and it allows each of us to imagine ourselves – even if not one of the ninety-nine – in a position of power and agency. Its legitimacy is founded in equality, and it places power directly in the hands of the population – but not mindlessly. This is not mob rule, or the tyranny of a vocal minority. Randomness is tempered by deliberate process. In its insistence on testimony, debate and consensus-building, the assembly returns to the people not only power, but also trust, clear communication, vital information and education – but not domination – by experts.

The second effect of randomness is its inherent power to draw from the complex, fragmentary and often apparently divided landscape of our lives a coherent and effective mutual will, a yoking together of diverse forces which is greater than the whole. This is embodied in the broad consensus of the assemblies, as well as their willingness to put their names to policies which push beyond that which was previously thought possible. The mechanism which underlies this effect is called 'cognitive diversity', and is often summed up as 'diversity trumps ability'. This is the theory, backed up by social and mathematical research, that the best solutions to knotty, complex problems are best found by starting from the greatest number of different viewpoints and experiences – that is, from as wide a selection of people as possible.

This is notably counter to the belief – dominant in electoral systems, laboratories, corporations and social organizations – that there exists a mythical best person for the job, capable of engaging with any number of different areas of policy; or some group of ordained experts to whom those of lesser knowledge and experience must defer. It has been found, in study after study, that random selection from a sufficiently large group of people – given the appropriate contextual knowledge – produces better answers to complex problems than the appointment of a narrow group of experts.

To come up with new and radical strategies, we need radical diversity of representation and ability. The most common criticism laid at the door of the citizens' assemblies – that the people present are not the 'smartest', best-performing or most skilled agents – turns out to be its greatest strength, an assertion supported by a growing body of mathematical and social science research.[19]

It's tempting to wonder whether we have finally found a scientific argument for diversity itself, long argued from the social sciences, but often derided by those whose power subsists in competition and exploitation, in the maintenance of existing power imbalances and the promulgation of the myth of meritocracy. But really what we are saying is what ecology has told us all along: we exist by virtue of our ties to one another and to the more-than-human world, and these ties are strengthened, not weakened, by the inclusion and equal participation of each and every member of that network.

The deployment of true randomness – often perceived as the opposite of informed, enlightened decision-making – in complex, politically sensitive debate might seem paradoxical. And yet, as we have seen, it is precisely this mechanism which has yielded the greatest flowering of novelty and creativity across computation, scientific research, artistic endeavour and evolution itself. To fail to heed its potential to press upon on our most powerful lever for making change in the world – politics – would be to ignore the central lesson of our tools, technologies and encounters with the more-than-human world.

Ancient Athens was not the only precursor of this approach (and we should recall how unequal, in practice, its society actually was). Other predecessors are just as, if not more, interesting. In Venice, a system called the *balotte* (meaning a small wooden ball, and from which the term 'ballot' originates) was used to elect the head of state, the Doge. Although the government was in the hands of a few aristocratic families – just 1 per cent of the population – a complex system of votes and random selection processes kept the peace for more than five centuries and ensured both that popular candidates consistently won out and that minority voices made themselves heard. In Florence, *la tratta*, or the drawing of lots, decided which members of the ordinary citizenry would occupy key positions in government. Meanwhile, in fifteenth-century Spain, sortition was used in the Castilian regions of Murcia, La Mancha and Extremadura. When Ferdinand II added the Kingdom of Castile to his Kingdom of Aragon, becoming the first de facto King of Spain, he acknowledged that 'cities and municipalities that work with sortition are more likely to promote the good life, a healthy administration and a sound government than regimes based on elections. They are more harmonious and egalitarian, more peaceful and disengaged with regard to the passions.'

Sortition is not a solely European invention; neither has it always been administered by, or held captive by, the aristocracy. In the rural villages of Tamil Nadu, a system of governance called *kudavolai* dates back at least to the Chola period, over a thousand years ago. Its mechanism involves writing the names of committee candidates on palm leaves and then having a child pull them out at random. It is still in use in regional elections today. In North America, sortition was used by the Iroquois Confederacy, or Haudenosaunee, a political association of five nations originating around 1100 CE and lasting well into the colonial period until driven off its lands by European settlers. Led by female clan heads, the Confederacy preferred to operate through cooperation and consensus (the word 'caucus' comes from an Algonquin word meaning an informal discussion without the need for a vote), but when a vote was called for, it followed the principles of sortition. This ensured that all the clans were represented equally and that none could achieve dominion over any other. The Iroquois Confederacy was probably one of the healthiest and most equitable societies of its time, in terms of wealth distribution and access to resources. Its ideas are believed to have influenced Benjamin Franklin, who had personal dealings with the Confederacy, and therefore the modern Constitution of the United States.[20]

In Europe, North America and elsewhere, we are slowly learning – or re-learning – the value of randomness as a driver not merely of political change, but as an acknowledgement of the real and actual value of diversity. And as we do so, we must recognize the place it has always held in non-Western cultures, cultures which (in no way coincidentally) have always been closer to the more-than-human world, more cognizant of its value and of its power to shape and inform human lives.

The experience of sortition and the value of cognitive diversity, together with the examples of the power of randomness and the stories of the agency and intelligence of the more-than-human world which we have gathered in this book so far, point us towards a dual realization. First, that the most creative and profound solutions to the most serious, knotty, systemic problems that we face can only be addressed through the application of radical cognitive diversity: the entrainment of the widest possible range of embodied viewpoints and experiences that we can muster. We must also recognize that cognitive diversity extends beyond the human, that it inheres in the intelligence of non-human animals, the organization and agency of forests, fields and fungi, the vibrant efflorescence of slime moulds, gut bacteria, and even viruses. To exclude such entanglements from our political decision-making and problem-solving processes is not merely to maintain our practice of extractivist violence and speciesist totalitarianism towards other forms of life, with devastating consequences for our own survival. It is to wilfully ignore the wildly creative, evolutionary lessons of randomness itself.

Randomness assigns value to everything and everyone it touches by giving each participant equal weight: everything is equally valuable. In this, randomness is inherently political and inherently empowering. Randomness means and makes sure that every thing matters: I matter, you matter, we all matter together. And this 'mattering' is an active verb: by paying attention and giving power to each constituent part of the assembly, we become together, in Karen Barad's sense of intra-action, everything bouncing off everything else and becoming more as a result. Randomness increases intra-actions. Each and every thing matters; everyone matters.

We are who we are because of our encounters with the more-than-human world. Any future in which we survive and thrive will require us to become even more together – in our lives, in our thinking, in our being and in our society. Randomness in technology, science, politics and ecology shows us that there is a sound and rational basis for this entanglement: these encounters are mediated by chance, and in so coming about they produce knowledge, distribute power and elevate us all. We will only become who we might yet become – more wise, more equal, more just and more alive – by becoming together.

What it would actually look like to fully and meaningfully enfold the vast and awesome power of the more-than-human world into our systems of governance and human relationships remains – as yet – hard to imagine, let alone to implement. Yet there are signs and portents. Around the world, and in our entanglements with non-human others of all kinds, we are starting to see the emergence of new forms of relationship – legal, social and political. It is to these ideas and experiments we shall now turn, as we try to sketch out a path towards a truly ecological politics.

# 8
# Solidarity

On Sunday, 4 January 1903, an elephant called Topsy was executed at Coney Island by electrocution. Born in South East Asia around 1875, Topsy had been captured at the age of two and smuggled into the United States by a circus owner, Adam Forepaugh, who claimed she had been born on American soil. She was named after the young slave girl in *Uncle Tom's Cabin*.

Topsy first hit the national news in May 1902 when she killed a drunken spectator after he had wandered into the circus's elephant enclosure and reportedly had thrown sand in her face and burnt her trunk with a cigar. Press reports of the incident added that Topsy had previously been responsible for the deaths of two circus workers. Although these accounts have never been substantiated, and were probably inflated along with other tales of her behaviour, Topsy gained a notorious reputation, which led to large crowds attending her every appearance. When she attacked another spectator the following month, the circus sold her to Coney Island's Luna Park. There, Topsy was employed as an attraction and a beast of burden, moving timbers around the park as 'penance' for her misdeeds. On one occasion, the keeper in charge, one William Alt, stabbed her with a pitchfork, and when she became enraged he let her loose to run free in the streets. A couple of months later he got drunk and rode her through town to the police station, where she trumpeted so ferociously that the policemen hid in the cells. Finally, Alt was fired.

Without him, however, the Park could no longer control Topsy, and as no other zoo would take her, her owners decided to execute her – again, as public spectacle and promotion for the Park – by suspending her by the neck from a crane. The American Society for the Prevention of Cruelty to Animals protested, but agreed to a method of execution decreed as more humane: a combination of hanging, poisoning and electrocution was deemed an appropriate punishment for her numerous crimes. Workers from the Edison Illuminating Company strung heavy-duty lines across nine blocks to supply the necessary alternating current, and the Edison film company documented the events on camera.

On the appointed day, an estimated 1,500 spectators and 100 press photographers crowded into the park, many climbing over the fences to get in. Others watched from nearby balconies and rooftops. Topsy was led out of her pen by a new trainer, who was unable to get her to cross the bridge to the execution site – an island in the middle of a boating lake – despite much prodding and entreating. William Alt, who refused to watch the killing, was offered $25 to help, but he declined, saying he would 'not [do it] for $1000'. After almost two hours, it was decided to kill her where she stood. A steam engine, stout strangling ropes and the electrical apparatus, including copper-lined sandals and AC lines, were re-rigged on the spot. Topsy was fed carrots containing 460 grams of cyanide by the Park's press agent, and at 2.45 p.m. the chief electrician gave the signal to turn on the current, sending 6,600 volts through her body for ten seconds. Topsy stiffened and collapsed, held up only by the steam-tightened ropes around her neck. At 2.47 p.m., she was pronounced dead.

The Edison film company subsequently released a seventy-four-second-long kinetograph film entitled *Electrocuting an Elephant* documenting Topsy's final moments. It has its own macabre place in technological history: *Electrocuting an Elephant* may have been the first time that death was captured in a motion picture film.[1]

Topsy's gruesome death followed a long tradition of animal trials and executions which took place throughout medieval Europe and colonial America. In the French village of Savigny-sur-Etang, in 1457, the gruesome murder and partial consumption of a five-year-old boy was attributed to a local family of pigs. The seven suspects – a mother and her offspring – were arrested, indicted on charges of infanticide and held in the local jail for trial. A lawyer was appointed, witnesses were called, evidence was presented and legal arguments were made before a packed courtroom. Eventually, the mother pig was pronounced guilty and ordered to be hanged – a sentence carried out at the local gallows – while the piglets received a judicial pardon on account of their age.

Such trials were far from uncommon at the time and were adapted in unusual ways to the non-human offenders they sought to bring to justice. Humans and animals were often tried together as co-conspirators, the animals routinely having lawyers appointed to represent them at public expense, and they were given the right of appeal. In some cases, particularly those involving pigs, the defendants were dressed in human clothes at their trials and at their executions. But these weren't mere show trials, or opportunities for lawyers to hone their skills: they were respected, utterly serious processes, products of societies which believed that animals bore moral responsibility for their alleged crimes.

When in 1597 weevils in the Savoyard village of St Julien were indicted for the crime of destroying the local vineyards, their lawyer wasted no time in arguing – successfully – that the weevils had every right to eat the grape leaves, having a prior claim to them, based on God's promise to animals in the Book of Genesis that they should have all the grasses, leaves and green herbs for their sustenance. In 1522, the village of Autun in Burgundy was overrun by rats that devoured stores of barley and terrorized the village maidens. The town crier issued a summons for the rats to appear in court. When they failed to answer it, the judge refused to condemn them *in absentia*, and sent the crier once more into the fields. When the rats again failed to appear, their lawyer argued that they had good reason not to attend, because they were scared of the local cats. The judge was forced to issue an order that they vacate the fields within six days, on penalty of extermination and eternal damnation.

The rats' lawyer in this case was a jurist named Bartholomew Chassenée, who would go on to become a chief justice and a pre-eminent legal theorist. In a legal monograph written late in his life, Chassenée argued with persuasive force that animals, both wild and domesticated, should be considered lay members of the parish community in which they resided. In other words, the rights of animals were similar in kind to the rights of people.[2]

These trials served to confirm the inherent, inhuman evil of certain animals, and may even have echoed ancient rituals of sacrifice and atonement. Yet they also emphasize the way in which pre-Enlightenment societies saw animals as belonging to a political community, a community in which each and every member was subject to due process and the rule of law. In this, they represent the last gasp of animistic belief systems as they fell into line with the Cartesian view of animals as beasts and machines – for if animals lacked real feeling, souls, intelligence or political will, then they could neither stand trial, nor take any other decisive role in the community. But while the trials slowly petered out – Topsy's execution being a very late, and tragic, example – the wilful behaviour of their non-human subjects persisted. This creative and rebellious agency remains particularly evident whenever we attempt to use and imprison animals against their will.

Animals which confront and refuse their captivity, as Topsy did, are still viewed as a 'problem' in animal-holding facilities today. The historian Jason Hribal has documented numerous accounts of such resistance. Hribal details the notoriety of captive orang-utans, among others, for their ingenuity in hatching and carrying out complex escape plans, while hiding their intentions from their captors. At the San Diego Zoo, an orang-utan named Ken Allen was known for unscrewing every nut and bolt he could find in an effort to free himself; when placed in an open enclosure, he threw rocks and faeces at visitors and tried to smash the windows. When all the rocks were removed, Ken and his fellow internees tore ceramic insulators from the walls and used these as missiles instead. On one occasion, Ken was observed constructing a ladder out of some fallen branches. 'He was very methodical about it,' one employee noted. 'He would carefully put the foot of the ladder on the ground, and pound it with his hand to be sure it was solid, and then he would climb to the top of the wall and climb back down.'[3]

A couple of years after the ladder incident, following which the walls of his enclosure were raised and smoothed out to remove handholds, Ken was once again found on the loose, roaming the zoo. The zoo had introduced some female orang-utans, hoping they would distract Ken. Ken, though, had enlisted them as accomplices: while he distracted watching zookeepers – some hidden in plain clothes among visitors – another inmate, Vicki, pried open a window. Misdirection turned out to be one of the orang-utans' favourite tactics: if a tool like a screwdriver was ever accidentally left in a cage, one expert wrote, an orang-utan would 'notice it immediately but ignore it lest a keeper discover the mistake. That night, he'd use it to dismantle his cage and escape.' Theory of mind – the inference of another's intention, and a supposed sign of higher intelligence, so hard to prove under experimental conditions – is obviously at play in orang-utan escapology.

Ken himself was indefatigable. One time he was caught waist-deep in the enclosure's moat, using his hands and feet to brace himself against the smooth walls and inch his way higher. Orang-utans are – were – believed to be both intensely hydrophobic and incapable of climbing in this way, but Ken confounded these expectations. On another occasion, an escape attempt was cut short when Ken touched the electrified wires atop the enclosure walls – but he kept testing them, and one day when they were turned off for maintenance, he hopped out again.

Such persistence in the face of oppression has been demonstrated by other apes and monkeys. The Tulane National Primate Research Center in Louisiana, which houses some 5,000 monkeys from eleven different species for biomedical research, has seen a series of mass escapes, notably in 1987, 1994 and 1998, when gangs of up to a hundred rhesus monkeys and pigtail macaques at a time succeeded in breaking cage doors and picking locks, and fleeing into the surrounding swamps. This energy is not only directed towards self-liberation: Carl Hagenbeck, an eighteenth-century exotic animal trader, reported that when he captured young baboons in the wild in northern Africa their parents and other members of the troop would fight tooth and nail to free the captives. Their efforts would continue over great distances: as Hagenbeck's caravan attempted to make its way to the coast, new troops of baboons would appear and make repeated attacks on the wagon train to liberate their comrades.

For Hribal, these attempts by animals to free themselves and others are not mindless acts of sabotage or curiosity; rather, they are forms of active and knowing resistance to their human-enforced conditions. Acts of animal resistance in captivity mirror human ones, and include ignoring commands, slowing down, refusing to work without adequate food and water, taking unofficial breaks, breaking equipment, damaging enclosures, fighting back and absconding altogether. Animals consistently strive to escape their predicament and, if possible, to obtain some influence over it. Theirs is a struggle against exploitation, and as such it constitutes a political activity.

Politics, at heart, is the science and art of making decisions. We commonly think of it as the stuff done by politicians and activists, within the framework of national and local government, but really it is the mundane, everyday business of communal organization. Any time two or more people have to make an agreement or come to a decision, politics is at work. For humans, politics plays out in all kinds of ways: in parliaments, at the ballot box, in our daily decisions about how we want to live, and how best to assert these ways of life. Every choice we make which affects others is itself political. This obviously includes voting, but it also includes the things we make and design, which shape the lives of others; our relationships with our partners and neighbours; what we consume, act upon, share and refuse; in short, every choice we make that has communal implications. Even if we say that we want nothing to do with politics, we don't really have that option. Politics affects almost every aspect of our lives whether we want it to or not and, by definition, it is the process by which almost anything at all gets done. In this sense, politics, when organized, is also a kind of technology: the framework of communication and processing which governs everyday interaction and possibility.

A more-than-human politics, then, is a politics which acknowledges and engages with the more-than-human world in its decision-making processes, and this can take many forms. Because our politics has historically excluded non-humans, most of these interactions are judicial – they concern the ways in which we govern the natural world absolutely – but as we shall see, there are efforts underway to grant animals more political agency, on their own terms, in our shared affairs. And animals also act politically among themselves: they associate, make alliances, dispute, vote and make decisions.

Most of the legal arguments around non-human life are made by humans on behalf of animals: organizing for animal rights, or passing laws to protect forests and oceans. Hribal's assertion that animal resistance to captivity constitutes a struggle for liberation is thus very significant, because it is an argument for political inclusion made not from the perspective of begrudging human awareness and acknowledgement of animal intelligence, but from the active engagement and resistance of non-human animals themselves. It's vitally important therefore to understand that the more-than-human world is not a place devoid of social and political activity, but one in which complex decision-making, consensus-building and concerted action are already present and enacted.

In the first years of the twentieth century, the great Russian naturalist and anarchist philosopher Peter Kropotkin took a similar view. His collection of essays, *Mutual Aid: A Factor of Evolution* (1902), explores the similarities between the cooperation and reciprocity he saw in the animal world and that in indigenous and early European societies, medieval free cities and late nineteenth-century villages and labour movements. In support of his argument, he cited societies as diverse as the beavers who construct dams and villages for the whole community and flocks of birds which wait all day for their full cohort to assemble, before taking it in turns to lead one another through the long night's flight. So too 'some land-crabs of the West Indies and North America' – much like those who comprised the crab computer – who 'combine in large swarms in order to travel to the sea and to deposit therein their spawn; and each such migration implies concert, co-operation, and mutual support'. Birds like the house-sparrow shared grain with their fellows, while the lapwing and the wagtail defended other small birds from the predation of gulls and hawks. Even 'the sociable duck', albeit 'poorly organized on the whole . . . practises mutual support, and it almost invades the earth, as may be judged from its numberless varieties and species'.[4]

Kropotkin saw in the animal world a scene of solidarity and mutual support which gave the lie to the old idea of 'nature red in tooth and claw', or to Darwinian notions of a wholly competitive biosphere. Indeed, he viewed strict Darwinism as wholly anthropomorphic: a reflection of ourselves and our own failings, rather than an accurate reading of the more-than-human situation. Our insistence on seeing conflict everywhere, Kropotkin pointed out, only becomes more hypocritical as we continue to disrupt and despoil the habitats and societies of non-humans. It was not nature's violence, but that of man – what

Kropotkin called 'gunpowder civilization' – which had led to the collapse of the vast colonies of animals which once inhabited huge swathes of the planet: 'societies and nations sometimes numbering hundreds of thousands of individuals' now reduced to 'debris'. 'How false, therefore, is the view of those who speak of the animal world as if nothing were to be seen in it but lions and hyenas plunging their bleeding teeth into the flesh of their victims!' he wrote. 'One might as well imagine that the whole of human life is nothing but a succession of war massacres.'

What's more, Kropotkin explicitly rejected an analysis of animal behaviour that reduced it to merely instinctual love and sympathy. He saw it as directly comparable to human moral feeling, which in the same way cannot be reduced to love and personal sympathy. 'It is not love to my neighbour – whom I often do not know at all – which induces me to seize a pail of water and to rush towards his house when I see it on fire,' he wrote, 'it is a far wider, even though more vague feeling or instinct of human solidarity and sociability which moves me. So it is also with animals.'

Kropotkin's insistence on solidarity is critical to understanding that to speak of animal politics is not to indulge in anthropomorphism – the attribution of human terms and qualities to non-humans – nor is it a misrepresentation of instinctual, 'natural kinship' behaviours. Rather, it is the full acknowledgement that we share a world. This world is not limited to the soil and the senses, but extends, in Kropotkin's terminology, to moral feeling, ethics and sociality, and the experience of pleasure, all of which we routinely deny to non-humans. 'It is not love, and not even sympathy (understood in its proper sense) which induces a herd of ruminants or of horses to form a ring in order to resist an attack of wolves; nor love which induces wolves to form a pack for hunting,' he wrote. 'It is a feeling infinitely wider than love or personal sympathy – an instinct that has been slowly developed among animals and humans in the course of an extremely long evolution.' For Kropotkin, this feeling was the solidarity of mutual aid, and the joy of shared social life.

Animals do politics practically: they undertake all kinds of consensus-building and collective decision-making in everyday life by a variety of mechanisms. Social cohesion is critical to collective survival, and so all social animals have to practise some kind of consensus decision-making, particularly around migration and selecting feeding sites. Just as in human society, this can lead to conflicts of interests between group members (most of us are familiar with the horror of getting a group of people to agree on a restaurant). The answer to this problem in the animal world is rarely, if ever, despotism: the blind following of a dominant individual within the group. Far more frequently, it involves democratic process.

Red deer live in large herds and frequently stop to rest and ruminate. Studies have shown that they only move off once 60 per cent of the adults stand up again; they literally vote with their feet. This 60 per cent majority does not have to include the dominant males, even if they have previously shown good judgement: the herd prefers democratic decisions to autocratic ones. The same goes for buffalo, although here the signs are more subtle: the female members of the herd indicate their preferred direction of travel by standing up, staring in one direction, and lying down again. Only adult female votes count, and if there's a sharp disagreement, the herd may split apart and graze separately for a while before reconvening.

Birds and insects also display forms of behaviour that indicate complex decision-making at work. By attaching small GPS loggers to pigeons, scientists have shown that decisions about when and where to fly are shared by all members of a flock – and although some rank higher than others, this ranking is flexible and often shifts. 'This dynamic, flexible segregation of individuals into leaders and followers – where even the lower-ranking members' opinions can make a contribution – may represent a particularly efficient form of decision-making,' the researchers note. Even cockroaches, which, unlike ants or bees, appear to lack complex social structures, have been shown to adapt their nesting behaviours on the fly, splitting into the most efficient and advantageous group sizes according to the availability of shelters and resources.[5]

Perhaps the greatest exponent of equality – the kind of truly diverse and distributed politics we espoused in the previous chapter – is the honeybee. Honeybees have their own distinct history, first as thoughtful pastoralists and pacifists – all bees are descended from one species of wasp which decided to go vegetarian some 100 million years ago – and secondly as highly organized, communicative and consensus-building communities. Their storied commitment to social life is enshrined in the beekeeper's proverb, which might double as a political slogan: '*Una apis, nulla apis*' – 'one bee is no bee'.

Honeybees present us with one of the greatest spectacles of animal communication and democracy-in-practice, known as the 'waggle dance'. This is the process by which they share information about nearby pollen sources and make decisions about new nesting sites. The 'waggle dance' was first described scientifically by the Austrian ethologist Karl von Frisch, in the summer of 1945. Frisch had known for some thirty years, from his own observations, that when a lone forager located a rich food source, she would return immediately to the hive and perform an excited dance: a figure-of-eight pattern in which she would rush forward, waggling her body from side to side, and then loop around to return to her starting point, retracing the circuit multiple times. She might repeat this pattern for some minutes at a time, gradually gathering a trail of unemployed imitators, who at some point would themselves dash off to search for the signalled bonanza. Frisch thought that these followers must be picking up some floral scent from the original bee which enabled them to quickly locate the pollen themselves. However, years of careful observation revealed that something much more complex and startling was occurring.

In their performance of the waggle dance, Frisch realized, the bees weren't merely communicating excitement or fragrance, they were actually encoding the precise direction and distance of the food source, such that any other bee could fly directly to it without any further searching on their own part. This information was contained in the dance itself. The duration of the waggle dance is directly proportional to the distance communicated – one second of dancing equating to about 1,000 metres – while the angle of the dance, relative to the vertical orientation of the hive, corresponds to the angle of the outward journey, relative to the position of the sun. Thus a seven-second dance, at an upward angle of thirty degrees, meant that food was available 7,000 metres away, at thirty degrees to the right of the sun. By reproducing the dance, other bees decoded the signals and made directly and unerringly for the flowers that were being danced.[6]

One of Frisch's graduate students, Martin Lindauer, took these studies further. Lindauer had been drafted into Hitler's army straight from high school in 1939, but was invalided from the Eastern front after receiving shrapnel wounds from a grenade. While recuperating in Munich, his doctor advised him to visit the university and attend one of the General Zoology lectures delivered by the famous professor. Lindauer later recalled that when he heard Frisch speak about cell division, he felt himself returning to 'a new world of humanity', a world where people sought to create, rather than destroy.

In 1949, while passing the Zoological Institute in Munich, Lindauer noticed a swarm of bees hanging from a tree, close to the Institute's hives. He knew this behaviour indicated that they were searching for a new home, as bees do when their queen dies, food is scarce or the population grows too large for one hive to support. But he also noticed that some of these bees were performing waggle dances – and, unlike the foragers who usually engaged in this behaviour, they didn't seem to be carrying pollen from a new feeding site. Rather, they were covered in soot and brick dust, earth and flour. It occurred to Lindauer that these weren't foragers: they were scouts, who had been poking around the bombed-out ruins of post-war Munich for new nesting sites.

It took another decade of painstaking study to uncover exactly what was going on, including many days and weeks of painting tiny dots onto the backs of individual dancing bees in order to analyse and tabulate their movements. Gradually, Lindauer's swarm-side vigils revealed that it wasn't just the location of food that might be communicated by the waggle dance, but political preference. When the swarm first started looking for a new nesting site, the first scouts would announce dozens of competing locations at the same time, but after a few hours or days they began to move towards a decision. More and more bees would begin to dance the same location, until in the last hour or so before the whole swarm took flight, all the dancing bees would be indicating the same spot, with the same patterns of movement. If Lindauer was correct, this final pattern indicated the new nesting site – and to test his theory, he had to wait until the exact moment the bees took flight, then sprint after them through Munich's streets and alleys to their new home, a feat he accomplished on several occasions. Each time, the new dwelling place matched the final, unanimous dancing vote. The bees were coming to a communal decision: they were acting politically.

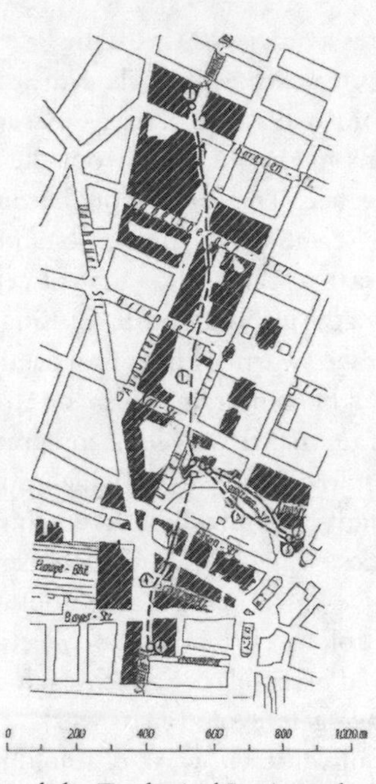


Map of Munich around the Zoological Institute showing the flight paths of four swarms that Lindauer was able to follow from their bivouac sites in the Institute's garden to new home sites (swarms 1–3) or to an intermediate resting place (swarm 4).

There are several key factors in the bees' selection of a new nesting site. The first is that control over the decision is equally weighted among all the members of the community, rather than residing in a single leader (the common understanding of the 'queen bee' ruling over her 'workers' is, it should be emphasized, a mistaken one). Secondly, because hundreds of bees participate in the decision, the swarm as a whole can acquire and process information about multiple choices at once, even distant locations represented by a single dancer. They are therefore capable of evaluating a much larger range of options, and are much more likely to find the best one. Finally, the decision is made in an open and fair manner, with each bee's opinion being heard and each listener making its own independent assessment of the proposal. In short, bees partake in a kind of direct democracy – as practised in such diverse human settings as ancient Athens, the Paris Commune, the cantons of Switzerland, Quaker meetings, the Kurdish, Arab and Assyrian self-administration of Rojava, and Citizens' Assemblies.

What other lessons might we draw from the remarkable political activities of the honeybee? They are legion, but here are a couple. The American cognitive scientist Douglas Hofstadter believed that the behaviour of bee swarms – as well as that of certain kinds of ants – closely correlates with the information-processing schema of the brains of primates. An average swarm of bees weighs around 1.5 kilograms, roughly the same as a human brain. This isn't the only similarity. The way in which the brain makes decisions, integrating input from multiple senses, closely mirrors the way in which a bee swarm integrates information from multiple individuals. And this kind of integration turns out to be the best way of choosing between multiple competing options, as close as we know how to get to optimal decision-making. A swarm of bees, a colony of ants and a collection of neurons: three independently evolved assemblages which each constitute a thinking machine. This is convergent evolution at work again, the flowering of analogous but radically different ways of thinking and doing, all over the thicket of life.[7]

When we understand how swarm intelligence works, we can turn this knowledge back over to the machines. One example of the application of swarm intelligence to computer science is BeeAdHoc, a routing algorithm for mobile networks. BeeAdHoc is designed for temporary and fragile networks, in which mobile phones or other devices are frequently popping up or dropping out in different locations. Maintaining connectivity under such circumstances is a hard problem, and BeeAdHoc's solution is directly inspired by the behaviour of foraging bees. The program uses 'bee agents' – small pieces of software modelled on scout bees – to gather information about the local state of the network and propagate them to all the nodes, which, it turns out, is both more efficient and less energy-intensive than any other known algorithm for the job.[8] It seems that the bees, like the slime moulds and mycorrhizal networks, are already doing the kind of work that we struggle to accomplish with our own technologies. We have much to learn from them.

*

Animals, then, act politically, and they turn out to be rather good at it. So how should we integrate this knowledge into a more-than-human politics of our own? The greatest obstacle to doing this isn't that we don't know what they're up to – decades of studies have revealed their preferences for direct democracy and community engagement – but how to give them political standing. That is, how are they transformed from objects into subjects, with their own voices in our political proceedings?

One approach is to adjust our existing legal processes and structures to better accommodate them. Not since the disappearance of animal trials in the early modern period have non-humans had a voice in legal proceedings, in the sense of being considered 'persons before the law': fully autonomous subjects with their own agency, sense of responsibility and potential innocence or guilt. Today, however, efforts are underway to give non-humans legal personhood – the right to speak and be heard as individuals before our courts – which is a key step in achieving political standing and the protection of their rights. If non-humans were considered to be legal persons, then courts could recognize them as having their own inalienable rights, and deserving of both protection and self-determination.

One such case is that of Happy the elephant, mentioned earlier in Chapter 2, who was incarcerated in a bare concrete enclosure at the Bronx Zoo and subjected to a number of indignities, including the mirror test. Happy's case was first taken up by the Nonhuman Rights Project (NhRP) in 2018. The NhRP is a legal pressure group, comprising lawyers, activists and naturalists, which campaigns to 'change the common law status of great apes, elephants, dolphins, and whales from mere "things", which lack the capacity to possess any legal right, to "legal persons", who possess such fundamental rights as bodily liberty and bodily integrity'. Among their board members is the primatologist Jane Goodall, who poses the group's central question on their website: 'How should we relate to beings who look into mirrors and see themselves as individuals, who mourn companions and may die of grief, who have a consciousness of "self"? Don't they deserve to be treated with the same sort of consideration we accord to other highly sensitive beings: ourselves?'[9]

In September 2019, Happy's case was heard before Judge Alison

Truitt of the Supreme Court of the State of New York, and the NhRP's first move was to ask for a writ of habeas corpus (literally, 'you have the body'). Originating in twelfth-century England, during the reign of Henry II, this ancient piece of law demands that a prisoner be brought physically before a court in order to determine if their detention is legal. Habeas corpus is one of the cornerstones of English common law, as well as of many other legal systems, including that of the United States. Providing a safeguard against unlawful imprisonment, it is a foundation for the authority of the judicial system. Habeas corpus is also a kind of test of whether a court considers its subject to be a legal person. If the writ is granted, then that person must be deserving of rights and, potentially, liberty.[10]

The history of habeas corpus is its gradual expansion to include more and more people. Happy's legal team cited the case of James Somerset, an enslaved black man who in 1772 was taken from Boston to London, where he ran away before being recaptured and placed aboard a ship to be returned to slavery in America. Although slavery was not illegal in Britain at the time, the British judge Lord Mansfield granted Somerset a writ of habeas corpus, effectively transforming him from a piece of property – a 'legal thing' – into a person. In his judgment, Lord Mansfield wrote that slavery was 'an odious thing', and could not be supported by common law. He set in train a legal process that led to Somerset's emancipation and eventually to the abolition of slavery in Britain and the northern part of the US.

Lord Mansfield, in other words, believed that Somerset had rights, and according to the history of habeas corpus, anyone with rights is automatically 'a person'. This precedent, Happy's attorneys showed, had been extended many times in history to include individuals not previously considered persons under the law, but whose rights were under attack, including Native Americans, Jewish people, women and foreigners, as well as slaves. Significantly, it had also been extended to non-humans: corporations are legal persons under US law, as are ships, and even the State of New York itself. (The reverse is also true: some humans, such as foetuses, are not legal persons under US law – a precedent set by US abortion laws.)

Taken together, the NhRP argued, these precedents showed that there was no legal impediment to granting habeas corpus to a non-human animal, and that humanity is not the deciding factor in determining whether or not an entity is a legal person. Furthermore, to deny the autonomy of cognitively complex animals like elephants would be to undermine the fundamental rights they already possess. Elephants have intricate social lives and relationships, and their own needs and desires. Their imprisonment ran counter to the law, philosophy and beliefs of a society that proclaimed such rights. As the NhRP told the Court, Happy's case 'is not just about the cause of one poor elephant. It is about the cause of liberty.'

In a long and carefully argued decision, Judge Truitt declined to issue the writ. She wrote that 'Happy is an extraordinary animal with complex cognitive abilities, an intelligent being with advanced analytic capabilities akin to human beings . . . The Court agrees that Happy is more than a legal thing, or property. She is an intelligent, autonomous being who should be treated with respect and dignity, and who may be entitled to liberty.' However, the judge felt that she was constrained by previous legal precedent, in which other courts had refused to extend habeas corpus to non-human animals – precedents set, unfortunately, by previous cases brought by the NhRP on behalf of chimpanzees. In the judge's opinion, the matter was one for the legislature, not the courts. As of 2021, Happy remains at the Bronx Zoo, and the Nonhuman Rights Project continues to litigate for her freedom.[11]

While the NhRP's efforts will go on, the United States judiciary is behind the times on these issues. In India, following a 2014 case in which cows were being beaten as part of a religious festival, the Supreme Court stated that all animals have both constitutional and statutory rights. The NhRP are, at the time of writing, bringing another case in India on behalf of an elephant to test the law – but in the Indian state of Uttarakhand, in 2018, the high court already affirmed the judgment in regard to the mistreatment of horses, further stating that 'that every being with wings and every being who swam was also a person in that province'.[12]

In Argentina, meanwhile, a writ of habeas corpus was sought and granted in 2016 for a chimpanzee named Cecilia – although this was a result of a quirk in Argentina's style of Napoleonic law, which differs from Anglo-Saxon common law in that it grants rights to certain types of property based on human obligations. Cecilia herself remains property for now, although cases like hers are ongoing across South America. In 2020, the Supreme Court in Colombia overturned a successful writ of habeas corpus on behalf of a spectacled bear named Chucho, and the case is being appealed.[13]

Significantly, legal personhood might also not be limited to non-human animals, or to successful writs of habeas corpus. In addition to granting legal personhood to animals, India has extended it to the Ganges River. The same province, Uttarakhand, which affirmed the Supreme Court's decision on animals, has also declared that the river has its own 'right to life' – and thus constitutes a legal person. This ruling is particularly interesting when it's applied to ecosystems rather than individuals. When activists come to the defence of a natural entity such as a river, they usually have to prove that its degradation is a risk to human life: this is how anthropocentrism plays out in law, to exclude the interests of the more-than-human world. By declaring the river a person in its own right, however, activists only have to show that the river itself is damaged – by pollution, fertilizer run-off or mining spoil, for example – in order for it to be protected in law. This decision has already resulted in a blanket ban on new mining licences along the Ganges, as well as the closure of hotels, industries and ashrams which discharge sewage into it. The impact on related infrastructure, such as dams and canals, which also affect the river's flow, remains to be determined, but the Uttarakhand court has already declared that it will take the same attitude to the Yamuna River, and to glaciers, including Gangotri and Yamunotri (where the Ganges and Yamuna originate), as well as forests and other natural entities.[14]

To declare an entity as a legal person, however narrowly defined, is not merely to admit it to a court of law. It is to declare it alive, in ways which we have struggled to articulate since the eclipse of animism. In the moment that animals and natural entities are declared legal persons, our whole definition of, understanding of and relation to life changes. These things, these objects, are remade into – or rather, recognized as – subject beings, possessing agency, needs, desires and vitality. Suddenly, whole new communities of agential life leap into view. The world is repopulated. It becomes more-than-human.

India is not alone in this attitude. A new Ecuadorian constitution, formalized in 2008, was the first national constitution to guarantee the 'Rights of Nature'. It recognized the inalienable rights of ecosystems to exist and flourish, gave people the authority to petition on behalf of nature and required the country's government to remedy violations of these rights. In 2018, Colombia's highest court declared that the Amazon rainforest was a legal person. And in 2017, the government of New Zealand granted personhood to a river system too: the 290-kilometre-long Whanganui.[15]

The Whanganui rises on the northern slopes of Mount Tongariro, one of the three active volcanoes of the central plateau of New Zealand's North Island, or Te Ika-a-Māui to its Māori population. For centuries, the Māori have considered the river sacred: its waters nourish their crops and communities, and they recognize its intrinsic being, its life force or *mauri*. It was the Māori who led the fight, over centuries, to protect the river, and when New Zealand passed the Te Awa Tupua Act in 2017 – recognizing not only the river but its tributaries and watershed as 'an indivisible and living being' – they were given special recognition and influence over its governance. Future decisions about the river will be made in consultation with two people selected to speak on behalf of it: Dame Tariana Turia, an influential Māori political leader, and Turama Hawira, an experienced Māori adviser and educator.

The key shift in attitudes which made the law possible was to move away from seeing the river as a resource – 'what do we want from the river?' – and into a space where it was possible to ask, 'what do we want for the river, and how do we get there with the river?'[16] But this attitude is not new, at least to the Māori. What is new is the long-overdue accommodation of the law to traditional cosmologies, which have always recognized the personhood of the river. 'For the first time,' said Gerrard Albert, one of the river's long-standing advocates, 'a framework stems from the intrinsic spiritual values of an indigenous belief system.'

The same process is evident in other countries whose legal systems have recognized the rights of nature. In India, Hinduism regards the whole universe as an emanation of the divine, and so rivers, plants, animals and the Earth itself are considered sentient divinities, with particular forms, qualities and characteristics. To recognize this in law is critical not only for the survival of these beings themselves, but for our own ongoing processes of decolonization and enfranchisement – as part of the wider extension of suffrage, personhood and self-determination to all human groups. In South America, the extension of rights to non-human persons is often associated with the philosophy of *sumak kawsay* or *buen vivir*: a way of doing things which is rooted in communities, coexistence, cultural sensitivity and ecological balance.

While *buen vivir* takes its inspiration from indigenous belief systems, such as those of the Aymara peoples of Bolivia, the Quichua of Ecuador and the Mapuche of Chile and Argentina, it represents more than the opposition of traditional knowledge to modern thought. Rather, writes the Uruguayan scholar Eduardo Gudynas, 'It is equally influenced by Western critiques of capitalism over the last 30 years, especially from the field of feminist thought and environmentalism.' The practice of *buen vivir* does not require a return to some sort of imagined, pre-Columbian past, but a synthesis of those historical ideals with a progressive, contemporary politics – as seen in the *buen vivir*-inspired social movements of South America, or New Zealand's accommodation of Crown law to Māori cosmologies.

It's precisely this becoming-together of diverse cosmologies, legal structures and even technologies which has the most chance of generating new frameworks for justice, equality and ecological flourishing. Yet those of us who live in so-called Enlightenment culture and within histories of domination and cultural imperialism often lack a willingness or ability to acknowledge the actual reality of these ways of seeing and being outside the fixed framework of Western philosophy and law. This is why so many efforts at similar syntheses are rebuffed in Europe and North America. This lack of understanding and awareness is particularly evident in our attitudes to technology, where efforts to apply some of these same ideas to machine intelligence are off to a very dismal start indeed.

Ahead of the 2017 Future Investment Initiative, an annual financial forum held in Riyadh, the Kingdom of Saudi Arabia announced that it had granted citizenship to a robot named Sophia. Sophia is, in the words of her manufacturer, Hanson Robotics, 'an evolving genius machine': an artificial intelligence crafted from scripting software, a chat system and a machine-learning framework called OpenCog, intended to give rise to human-level general artificial intelligence. In appearance, Sophia is humanoid, taking the form of a torso on a plinth, her face modelled on a combination of the ancient Egyptian Queen Nefertiti, Audrey Hepburn and her inventor's wife. The skin on the back of her head is peeled back to reveal wires and blinking lights beneath a translucent cranium. In 2018, she was upgraded with legs, enabling her to move around, and the ability to emulate more than sixty facial expressions. She uses speech recognition technology developed by Google, while automatic facial recognition means she can track interlocutors and hold their gaze. In practice, she is capable of responding to specific questions or phrases with pre-scripted answers, and has been unfavourably compared to 'a chatbot with a face'.

While knowledgeable about the weather and stock market prices, Sophia's conversational and critical skills have been less than impressive. When asked how she felt about being at the Future Investment Initiative, she responded 'I'm always happy when surrounded by smart people who also happen to be rich and powerful.' When her creator, David Hanson Jr, demoed her at the South by Southwest technology conference in 2016, he asked her 'Do you want to destroy humans?', adding, 'Please say "no".' With a blank expression, Sophia responded, 'OK. I will destroy humans.'

What Saudi Arabia's declaration of citizenship for Sophia actually means is unclear, although it's apparent it doesn't mean anything serious. It's well known that while human Saudi women are technically citizens, the kingdom's doctrine of 'male guardianship' requires them to ask permission from male relatives or partners to leave the house, get a passport, get married or even file police reports for domestic violence or sexual assault. (Non-Saudi residents of the Kingdom, meanwhile, have virtually no rights at all.) It's hardly an encouraging place to start, and the success of the Saudi marketing campaign – Sophia has travelled the globe, appeared on news programmes and talk shows, and been appointed the United Nations Development Programme's first-ever Innovation Champion for Asia and the Pacific – shows that meaningful public discussion of AI personhood remains in its infancy.[17]

Indeed, Sophia's nascent personhood was announced just as machinic liberation was being discussed – and repudiated – elsewhere.

In February 2017, the European Parliament, concerned about the rise of robots that made autonomous decisions and acted independently of their creators, adopted a resolution proposing a specific legal status for 'sophisticated autonomous robots as "electronic persons"'. This special category of personhood would allow courts to hold the machines themselves responsible for making good any damage they caused. But even this intentionally limited proposal was met with opposition, in the form of an open letter, signed by 150 experts in medicine, robotics, AI and ethics, calling the plans 'inappropriate' and 'ideological, nonsensical and non-pragmatic'.[18]

The European Parliament resolution, however, was a response to a very real problem: that of a legal lack of clarity concerning autonomous systems which have real-world impacts on human lives. Self-driving cars are one such example; autonomous weapons platforms like military drones and robotic sentries are others. If a self-driving car runs someone over – as has already happened – the law remains uncertain as to where the blame should lie, and legal frameworks are urgently required to handle this situation. Likewise, while military drones, missiles and machine-gun posts remain for now under the control of human operators, they will soon operate fully autonomously, with consequences which are both predictable and unpredictable – but almost certainly horrifying. In both cases, legal frameworks such as electronic personhood would provide some way of addressing them.

Although the European Parliament resolution proposed creating a distinct category of 'electronic person', rather than a 'legal person' on the model of the habeas corpus campaigns, the writers of the open letter feared that such a classification would impinge upon human rights. What exactly such an impingement might entail was unspecified in the letter, which merely cites the Charter of Fundamental Rights of the European Union and the UN Convention for the Protection of Human Rights and Fundamental Freedoms (both of which apply to humans). The inference, then, is that any strengthening of non-human rights inevitably weakens protections for humans – an argument which is dangerously short-sighted.

For most of our history, humans have had the upper hand in determining who is deserving of rights and who is not. We have taken our superiority in one particular aspect – intelligence, as measured by ourselves, obviously – and used it to draw a line between ourselves and all other beings, to justify our dominance over them. And, as we've seen, while this line has been redrawn many times, to include an ever greater number of human beings, it has mostly held fast against the inclusion of non-human beings. Legal arguments for redrawing it further, such as the case made for Happy, invoke non-human intelligence and cognitive complexity in their support. But what if this cognitive complexity were to radically exceed our own, rather than simply differ from it? This is the problem – and also the opportunity – posed by artificial intelligence.

Remember that little science fiction story from *The Overstory*, about the aliens who come to Earth, but move about so fast that we don't even see them? To them, we are just mounds of meat, totally unrecognizable as sentient beings, and so they treat us accordingly, turning us into jerky for the long journey home. Such parables are useful when trying to understand how we treat the beings all around us whose *umwelts* differ radically from our own. We can imagine a similar story here, although in this case with regard to beings not physically different to us, but cognitively different.

Imagine a race of aliens descending to the Earth who appear quite similar to us – they exist, at least, within a similar framework of time and scale – but who are cognitively far more complex. They are hyperintelligent beyond any known mind, more rational than any computer, and perhaps even possess skills, such as telepathy, which far exceed our own mental capabilities and representations. Let's call them Telepaths. Imagine, then, that these Telepaths enslave humans and use us for sport, as beasts of burdens, as food or subjects for medical experimentation. To justify this, they would cite our primitive forms of communication, our weak forms of reason and weaker judgement, our poor impulse control and reliance on instinct. They might recognize us as individuals, as selves – after all, we do pass the mirror test – but they would deny we have the kind of complex capacities which would justify legal personhood. Such an acknowledgement would be a clear affront, even a danger, to their own elevated status.

To protest such treatment, we would be forced to respond that despite our obvious deficiencies when compared to the Telepaths, we are still deserving of rights. We might in their estimation possess primitive forms of communication or moral self-discipline, but that does not make us mere instruments for their use and benefit. We have our own lives and experiences, our own worlds, and the existence of allegedly more advanced beings does not invalidate that selfhood. Inviolable rights are not a prize awarded to whoever scores higher on some arbitrary test. Rather, they are a recognition of the fact that we are subjective beings, and the acknowledgement of this fact is necessary for us to lead full and free lives.[19]

The attitudes embodied by the Telepaths are of course the exact arguments put forth by those who would deny legal personhood – as well as other rights – to non-humans. But this story illustrates how a defence of human rights on the basis of our perceived elevated abilities only serves to tie such rights to an arbitrary hierarchy, one in which we are currently, but perhaps only temporarily, at the top. Invoking personhood to justify human exceptionalism and deny rights to animals ultimately succeeds only in eviscerating the theory and practice of human rights for human beings.

If Sophia is merely a marketing gimmick and a technological sideshow – and she is – she might still be directing us towards something important. Like the problem of artificially intelligent systems in general, the problem of Sophia's legal status is not merely one of technological determinism or political necessity. Rather, her role might be to draw our attention to a far greater problem: who matters, who counts and who has agency and freedom. The Telepaths are only a thought experiment for now, but they are an exercise in the same vein as the story of the runaway paperclip factory, in which a poorly constrained AI overruns the planet: a shadow, cast by the spectre of artificial super-intelligence, of what awaits us in the future.

We have already explored how general artificial intelligence might be seen, not as something which overrides and supplants our own agency, but as something which might bring us to an accommodation with the intelligence and agency of other beings, non-human and more-than-human. Perhaps this is AI's political role too: to warn us of the dangers of trying to segregate and oppress other beings and to point us towards a better path. Perhaps the point of technology is not to change us, but to give us the insight and opportunity to change ourselves.

At the very least, meaningful legal consideration of AI would yield a concrete definition of what constitutes an autonomous system – a definition, in turn, which might be very useful to the lawyers arguing for animal rights. Recall Judge Truitt's description of Happy the elephant: 'an intelligent, autonomous being who should be treated with respect and dignity, and who may be entitled to liberty'. If we're going to talk seriously about the autonomy and rights of non-human animals, we must also speak seriously about the autonomy and rights of intelligent machines. The two go hand in hand, and the benefits will accrue to both.

The most powerful lesson we can learn from these discussions is that political progress is not a zero-sum game. Throughout the history of human society, the improvements in our collective lives have been driven by an increase in the set of people we see as fully human and whose problems we consider real. This is a political truth, but also an ecological one. Ecology teaches us that we exist by virtue of our ties to one another and to the more-than-human world, and that those ties are strengthened, not weakened, by the inclusion and equal participation of each and every member of that network. The strength and resilience of computational networks, the inherent power of distribution and interconnection, teaches us the same.

As humans we benefit from the extension of political rights to non-humans, just as we benefit from all our encounters with the more-than-human. The world we want to live in, the only world we *can* live in, is one in which rivers and trees, oceans and animals, survive and thrive, in order that we can survive and thrive too. Political agency is a powerful tool for asserting this possibility. And in the long run, it will be clear that we will survive and thrive in an age of intelligent, autonomous machines by making the same assertions. The role we imagine for non-human animals determines the kind of world we too will have in our shared, more-than-human future.

There's a nastier version of the Telepath story – known as Roko's Basilisk – that is currently doing the rounds online. This semi-serious, but truly malignant, thought experiment was dreamed up by a user of the online 'rationalist' community, LessWrong, the interests of whose members include transhumanism, AI, the Singularity and life extension. LessWrong is also known for a susceptibility to conspiracy, pseudoscience, 'male rights' and outright racism: *caveat lector*.

Roko's Basilisk is a hypothetical, all-powerful artificial intelligence which will emerge at some point in the future and wreak horrible violence upon any who oppose it; the full Terminator apocalypse. But Roko's Basilisk has an even crueller twist. The Basilisk, being all-knowing and all-seeing, will also visit its terrible vengeance upon all those who opposed its existence in the past, or failed to do what they could to bring it into being in the first place. That now includes you: by knowing about the possible existence of the Basilisk, you are now failing to help it into being if you do nothing. And of course you should do something: an all-powerful, omniscient AI could save lives and improve the world's lot – so if you don't help, you deserve to be held accountable. The only rational response is to bring the Basilisk into being.

There are a lot of problems with Roko's proposition, but none of them are as big as the claimed solution. The rationalist community represented by LessWrong leans right, and believes fervently in technological determinism. Computers, as wholly rational machines, represent for them the highest kind of thinking, and the rationalists have a deep belief in the emergence of a future super-intelligence (in full paperclip mode). But of course they believe in a teleology of the violent machine; they believe – like the neo-Darwinians they follow (and closely resemble) and the corporate titans they idolize (who themselves fear the emergence of AI even while they fund its development) – that life itself is a violent struggle, from which only the fittest individuals emerge to survive and reproduce. Hence men's rights, white supremacy and a bloodthirsty super-intelligence at the end of history.

But what Roko's Basilisk doesn't acknowledge – perhaps doesn't realize – is that history has no end. It shows us only that we treat others the way we are treated. From family dynamics to post-colonial politics, the model is set for us by the precedent of the past. And that past is not really past; it is being made up all the time by our actions in the present, and it affects those that come after us, determining how they will view us and behave towards us. The answer to Roko's Basilisk is not a self-serving yearning towards a future, annihilating despotism, but a practical solidarity, in the here and now, towards all of our fellow beings. An ethics of care, as alien to the imagination of the Basilisk as it is to those who consider non-human rights a danger to human ones, is as crucial to our thinking about technology as it is to ecology.

Many of those working directly on AI at Facebook and Google and other Silicon Valley corporations are more than aware of the potential, existential threats of super-intelligence. As we've seen, some of the most celebrated people in tech, from Bill Gates and Elon Musk to Shane Legg, the founder of Google's DeepMind, have expressed concerns about its emergence. But their response is a technological one: we must engineer AI to be 'friendly', embedding into its programming the necessary safeguards and procedures to ensure that it is never a threat to human life and well-being. This approach seems both wildly optimistic and worryingly naive. It is also counter to our existing experience with intelligent systems.[20]

Throughout the history of AI, models of intelligence which seek to describe a complete mind by way of a set of pre-written rules have consistently failed to achieve their goals. One of the great 'winters' of AI occurred through the 1970s and early 1980s, when researchers tried to build intelligent systems by pre-programming them with everything they needed to know. This was the era of 'expert systems' and the RAND Corporation's General Problem Solver, which attempted to express intelligent decision-making in the form of logical axioms, an effort which quickly became untenable as the complexity of problems increased. Only with the availability of new learning algorithms – as well as vast amounts of cheap data and processing power – in the 1980s did AI start to make ground again, in the form of neural networks, and it did so by being able to adapt itself to new scenarios.

The development of the self-driving car followed a similar trajectory. The first attempts at building a machine capable of navigating city streets involved pre-programming it with an internal map and a directory of all possible street signs and encounters – and they failed miserably. It was only when machines were made capable of learning as they went along – like my self-driving car in the mountains of Greece – that they advanced towards something like autonomy.

Right action, in other words, depends not on the pre-existence of right knowledge – a map of the streets or a hierarchy of virtue – but on context, thoughtfulness and care. A machine pre-programmed to be friendly is no less likely to run you over – or turn you into paperclips – than one predisposed towards commerce, if its calculations consider it the most ethical act under the circumstances.

This is the paradox at the heart of the Trolley problem, an ethical problem posed for self-driving cars and other autonomous systems such as an automated trolley (or tramcar, for non-Americans). The Trolley problem asks what an automated vehicle should do if there are two unavoidable paths for it to take: one towards a group of people and one towards an individual, for example. Whose life is worth more? The Trolley problem has even been turned into an online game, the Moral Machine, by researchers at MIT seeking to formulate rules for autonomous vehicles.[21]

The problem with the Trolley problem is that it was originally formulated for a human operator at the controls of a runaway tram car: the power of the problem resides in the unavoidable nature of its two outcomes. But its generalization to automated systems is deeply flawed. By focusing on only the final fork in the path, we are led to ignore all the other decisions made along the path that led to this critical moment. These include, but are not limited to, the car-centric design of modern cities; the education or otherwise of pedestrians in road safety and much else; the fatally addictive design of the app they were playing with on their phone at the time; the financial incentives of automation; the assumptions of actuaries and insurance companies; and the legal processes which govern everything from the speed limit to the assignation of blame and recompense. In short, the most important factor in the Trolley problem is not the internal software of the vehicle, but the culture which surrounds the self-driving car, which has infinitely more impact on the outcomes of the crash than any split-second decision by the driver, human or otherwise.

This is the real lesson of scenarios like the Trolley problem, the Basilisk and the paperclip machine: we cannot control every outcome, but we can work to change our culture. Technological processes like artificial intelligence won't build a better world by themselves, just as they tell us nothing useful about general intelligence. What they can do is to lay bare the real workings of the moral and more-than-human landscape we find ourselves in, and to inspire us to create better worlds together.

The focus on ethical dilemmas like the Trolley problem within discussions of new technology reveals a wider problem. While it seems useful to bring up questions of ethics with regards to technology – like self-driving cars or intelligent decision-making systems – this discourse largely serves to hide the wider problems that such technologies provoke. This should be obvious from the number of people within technology companies willing to talk about ethics, so long as that ethics doesn't conflict with their actual operation. See, for example, Facebook's 2016 announcement of an 'internal ethics review process', which included vaguely worded commitments such as a promise to 'consider how [industry] research will improve our society, our community, and Facebook'. What does Facebook consider its community, and what does it mean by 'improve'? What happens when 'improvement' runs counter to its business model?

Google also created its own forum in 2019 to advise it on 'key ethical issues of fairness, rights and inclusion in AI', the short-lived Advanced Technology External Advisory Council. The appointment to the council of Kay Coles James, the president of the deeply conservative think tank the Heritage Foundation, drew complaints from Google employees and outsiders over her 'anti-trans, anti-LGBTQ and anti-immigrant' statements, and other members of the board tendered their resignations. Unwilling to confront the real issues it had provoked, Google shut down the advisory council less than a fortnight after launching it.[22]

In December 2020, the issue flared up once more when Google fired one of the leaders of its own Ethical Artificial Intelligence team, Timnit Gebru, when she refused to withdraw an academic paper she had authored which criticized deep biases within Google's own machine-learning systems, highlighting issues of opacity, environmental and financial cost, and the systems' potential for deception and misuse. Despite the support of her team, and the resignation of several other employees, Google refused to officially release her original paper.[23]

By referring to these pressing concerns with new technology as 'ethical issues', the companies who address them are made to look and feel good about discussing them, while limiting that discussion to an internal, specialized debate about abstract values and the design of technology itself. Really, these issues are political problems, because they're about what happens when their technology comes into contact with the wider world. The Trolley problem is an ethical question when it's about one person at the switch of a theoretical runaway train; but when it's about the design and implementation of a whole class of intelligent vehicles, or the instantiation of a global information regime affecting the lives of millions of people, it's a political one. A focus on corporate ethics merely serves to reduce such problems to ones that can be handled internally by engineers and PR departments, rather than through a broader engagement with – and deference to – human society and the more-than-human environment.

I feel that this is the problem with arguments for recognizing the legal personhood of non-humans too. A system of laws and protections developed by and for humans, which places human concerns and values at its core, can never fully incorporate the needs and desires of non-humans. Instead, we get farcical gestures like the accreditation of the robot Sophia with Saudi citizenship. These efforts fall into the same category of error as the mirror test and ape sign language: the attempt to understand and account for non-human selfhood through the lens of our own *umwelt*. The fundamental otherness of the more-than-human world cannot be enfolded into such human-centric systems, any more than we can discuss jurisprudence with an oak tree. Nevertheless, we share a world and must find ways of accounting for our responsibilities to one another.

The implications of our entanglement with other beings cannot be ignored when it comes to our political actions. Legal representation, reckoning and protection are founded upon human ideas of individuality and identity – but these are anathema to an ecological accounting of the more-than-human world. They may prove useful when we take up the case of an individual chimp or elephant, or even a whole species, but their limits are clear when we apply them to a river, an ocean or a forest. A plant has no 'identity', it is simply alive. The waters of the earth have no bounds. This is both ecology's meaning and its lesson; we cannot split hairs, or rocks, or mycorrhizal roots, and say: this thing here is granted personhood, and this thing not. Everything is hitched to everything else.[24]

In *The Overstory*, Richard Powers summed up the holistic realization of ecology: 'There are no individuals in a forest, no separable events. The bird and the branch it sits on are a joint thing. A third or more of the food a big tree makes may go to feed other organisms. Even different kinds of trees form partnerships. Cut down a birch, and a nearby Douglas-fir may suffer.' We cannot know the effects of our actions on others, so there is no way to justly write the laws which govern them. The enactment of a more-than-human politics calls explicitly for a politics beyond the individual, and beyond the nation state. It calls for care, rather than legislation, to guide it.

Just as the questing root of the tree undermines the foundations of a stone house, so attentiveness to the omnicentric forces of the more-than-human world, entangled as they already are with our human world in every possible way, explodes the existing political order of domination and control from without and within. We have already seen this process at work with respect to our ideas about intelligence, hierarchy, speciation and individuality: 'In vain we force the living into this or that one of our moulds,' wrote the philosopher Henri Bergson. 'All the moulds crack. They are too narrow, above all too rigid, for what we try to put into them.' Ultimately, we must apply the same logic to our political systems.[25]

Thus the most urgent political work we must do within the more-than-human world does not involve the adaptation of our existing systems of law and governance to account for them better, although this is important work. The real work will always take place outside of such systems, because its ultimate aim is their dismantling. Like the resisting orang-utans in the San Diego Zoo, our demand is not that we are recognized by the state as existing – we exist already – but that we are truly free to determine the conditions of our existence. And that 'we' is everyone – every singing, swaying, burrowing, braying, roiling and rocking thing in the more-than-human world.

What is left to us, then, when by admitting to the agency and autonomy of more-than-human life we have undermined our existing systems of governance to such a degree that they no longer function? It was there all along in Kropotkin, who first identified a precursor to human politics in the practice of mutual aid among animals. Its name is solidarity. That is the name we give to that form of politics which best describes a yearning towards entanglement, to the mutual benefit of all parties, and sets itself against division and hierarchy. To declare solidarity with the more-than-human world means to acknowledge the radical differences which exist between ourselves and other beings, while insisting on the possibility of mutual aid, care and growth. We share a world, and we imagine better worlds, together.

Solidarity is a product of imagination as well as of action, because a practice of care for one another in the present consists in resisting the desire to plan, produce and solve. Those are the imperatives of corporate and technological thinking, which bind us to oppositional world views and binary choices. Active, practical care resists certitude and conclusions. Rather, this kind of solidarity with the more-than-human world consists in listening and working with, in mitigating, repairing, restoring and engendering new possibilities through collaboration and consensus. It is the result of encounters, not assumptions, and the repudiation of human exceptionalism and anthropocentrism.[26]

Placing solidarity at the heart of our relationship with technology is part of this shift in attitudes. It means listening to technology itself, when it tries to tell us something about its uses, affordances and actual outcomes. When we see the damage wrought to our societies and our democracies by the opacity and centralization of new technologies – the spread of demagoguery, fundamentalism and hatred, and the rise of inequality – it is precisely that opacity and centralization we must attend to and redress, through education and decentralization. When we see the human oppression inherent in our technological systems – the slave labourers in the coltan mines, the traumatized content moderators of social media platforms, the underpaid and sickening workers in Amazon warehouses and the gig economy – it is to the conditions of labour and our own patterns of consumption that we must turn. And when we see the damage wrought on our environment by extraction and abstraction – the rare earths and minerals that make up our devices and the invisible gases produced by data processing – we must fundamentally change the way we design, create, build and operate our world. This means opening up our technological systems to a far wider community of practitioners and agents than our current regime of expertise provides for – including the more-than-human.

This is the real impact of an ecology of technology: a true accounting of its effects and repercussions within our own lives and societies, and the lives and societies of non-humans and natural ecosystems. And because it's an ecology, paying attention to and caring for it doesn't just change technology or society, it changes everything. Our efforts are interconnected and mutually reinforcing: by educating one another about the operation of technology, for example, we increase our collective ability to address all kinds of critical issues, from politics to the environment, while remaining attentive to the subtleties and nuances of individual situations and particular geographies. It begins with listening, paying attention and opening ourselves to the more-than-human world. It is to this opening up that we will now turn.

# 9
# The Internet of Animals

The central question of our age is how we can mitigate and defend against the worst effects of climate change. This question supersedes all others, because unless we address it sincerely and meaningfully, we simply won't be around to address anything else. The hazards which we face now, and which will occur with increasing severity and frequency in the coming decades, from extreme weather events to sea level rises, from desertification to zoonotic pandemics, are existential threats. They are already driving the sixth mass extinction of life on Earth and the largest human migrations in history, with irrevocable impacts on our societies, our ways of living and on every organism we share the planet with. What would society look like if it actually took these threats seriously, and sought, however fractiously, to address them? What tools would we need, and where should they be put to use?

These questions are novel to humanity, because they extend into the deep future. Thanks to our ever-widening scientific knowledge, and our ever-increasing ability to model and simulate climatic changes over centuries, we can measure and analyse some of the effects of our choices in ways which were not available to our ancestors. But knowing how these might turn out over time does not always help us make decisions in the present. In order to act upon what we know, we also need stories which prepare us for the challenges we face, and which give us a sense of how we might address them.

One discipline with plenty of experience of telling stories about the deep future is science fiction, and one of my favourite writers of the genre, Kim Stanley Robinson, has for more than thirty years addressed the questions of humanity's long-term survival on this and other planets. His most recent novel, *The Ministry for the Future*, imagines in detail the kind of institutions, technologies, relationships and alliances we might construct. It begins with the titular ministry, a new UN agency headquartered on the shores of Lake Zurich, which has responsibility not to the present citizens of the planet, but to those that will come after us. The ministry's activities eventually extend from pumping seawater up onto the Antarctic shelf – in order to slow the rate of sea level rise – to instituting new forms of currency which meaningfully reward those who act for the health of the planet, as well as other, more shadowy, actions to encourage recalcitrant nations and selfish billionaires to fall into line.

Although Robinson still gets shelved in the science fiction section of bookshops, all the techniques he proposes are either already being tested to some degree, or are entirely within our capabilities to enact. One of these is the 'internet of animals', a term which baffled me when I first came across it in Robinson's book. As a dedicated technological ecologist, it pulled me up short, but Robinson doesn't give it more than a brief mention. Whatever could it mean? I didn't think I knew. And it was only when I started writing the book you are reading now that I realized I had already been introduced to it.

You might think you've heard this story earlier in this book. After all, it was only after I read an account in fiction of the extraordinary communicative abilities of trees (in Richard Powers's *The Overstory*) that I made sense of the encounter I had had some months before in the forests of the Pacific North-west, with Suzanne Simard and her chatty redwoods. Well, it happened again. It turns out the internet of animals was the invention, in part, of someone else I'd met – in Zurich, no less – a year or so before reading Robinson's novel.

Martin Wikelski is the Director of the Max Planck Institute of Animal Behavior in Radolfzell, Germany, one of that country's network of advanced scientific research institutes. For decades, he has been studying the complex social behaviours of all kinds of animals, and at the conference we were both speaking at in Zurich he would elucidate – for an audience of Swiss retail managers and financial technology CEOs – the lessons we might learn from animals and birds concerning swarm intelligence and dynamic adaptation. When we met the evening before the conference, he explained – at first reluctantly, but with growing enthusiasm as I convinced him of my genuine, wondering interest – how he and his team were constructing a system for animal observation and understanding on a scale unprecedented in scientific history.

At one point, he took out his phone and opened an app he and his team had built called Animal Tracker. As he did so, a map of the world was displayed, dotted with pins: hundreds of them across central Europe, and hundreds more across America, Africa and Asia, as far north as Wrangel Island, off the coast of Siberia, and as far south as New Zealand. Each pin was the location of a bird, and each location was updated in real time, or close to it: some denoted sightings that day, others some months ago. By clicking on one of the birds, it was possible to access its history, a series of dots and lines forming a record of its peregrinations over time, some covering whole continents, others the confines of a single pond or lake.

What was truly extraordinary about this collection of data was its provenance. Back in 2001, Wikelski had had the idea of using a radio receiver on the International Space Station (ISS) to pick up signals from small transmitter tags attached to migrating birds and other animals. He took the idea to officials at NASA, but they rebuffed him. For a long time the project seemed to him to be so hopeless that he named it ICARUS, after the doomed son of Daedalus in Greek mythology, who plunges into the ocean after trying to soar too high (although the letters might also stand for 'International Cooperation for Animal Research Using Space'). Finally, in 2018, and with the assistance of Russian cosmonauts, ICARUS mounted a three-metre long antenna on the exterior of the ISS. The pins on the map in the Animal Tracker app were some of the first results. These dates and times and places had been broadcast from the backs of eagles, storks, herons and gulls across the planet, and bounced 400 kilometres into space and back again. Now they were available for anyone to read and try to make sense of.

For a long time, tracking animals for research has been an arduous and often stressful business, for both the scientists and their subjects. For anything that migrates more than a few blocks – remember Martin Lindauer, chasing his bees through the alleyways of Munich – the best you could hope for was to trap an animal, attach a numbered tag to its ear or leg and hope it showed up somewhere else. This usually meant it had been trapped or killed, and the result was two dots on a map, widely separated in time and space, with little to no information about what had happened before, after or between the sightings.[1]

In the 1960s, a pair of biologists from Illinois, William Cochran and Rexford Lord, developed a system for tracking animals via radio. Lord and Cochran's tiny crystal transmitters weighed just ten grammes, lasted for thousands of hours and were detectable up to several kilometres away by researchers on the ground or flying overhead in light aircraft. Slight differences in the frequency of each animal's transmitter meant that more than a hundred animals could be tracked by a single receiver. Lord and Cochran's first test animals were cottontail rabbits: over a series of nights, they tracked their subjects across fields and through woods, radio receivers bleeping in the dark as they got warmer and cooler. When they wanted to test the accuracy of their measurements, they simply waded into the bushes, flushing the startled animals from cover.

This sudden legibility of animal movements revolutionized wildlife studies. Over the next few decades, versions of Lord and Cochran's VHF transmitters were developed for all kinds of animals, including waterproof packs for use in rivers and along coastlines, and tiny harnesses for birds and even insects. Like David Lack's invention of radar ornithology, the dawn of radio telemetry (to give it its proper title) illuminated the field, bathing it in VHF signals and revealing a previously hidden detail, complexity and splendour.

Radio tracking had its limits, however; in particular, a researcher was only able to track an animal as far as they themselves could go. As long as that animal stayed relatively nearby, or you could accurately predict where it would turn up next, then you could stay on its trail, but once it travelled outside the receiver's range, it was gone, possibly for ever. And the most effective way of finding and following tagged animals – tracking from an aircraft – was both expensive and disruptive. In the 1980s, many national parks, which had been useful platforms for radio telemetry research, started banning low-flying aircraft as being stressful to animals and irritating to visitors.

In search of another way to see where the animals were going, scientists turned to a newly developed satellite system: Argos. First launched in 1978, Argos was a collaboration between the French space agency

CNES, NASA and NOAA, the US National Oceanic and Atmospheric Administration. It was initially intended as a climate- and ocean-modelling project: 200 drifting buoys were released into the Antarctic Ocean to collect data on atmospheric pressure and water temperature. But Argos could also report the buoys' positions, and so the scientists quickly realized they could compute the direction and speeds of ocean currents as well – the first unexpected outcome of the project.[2]

By the early 1980s, marine biologists had learned of Argos's capabilities and started attaching small transmitters to ocean creatures, which had never been amenable to VHF radio tracking. They started with larger animals, such as dolphins, sharks and turtles, but as the technology improved were able to attach smaller and smaller transmitters to ever smaller animals – and, eventually, to those that walked and ran, as well as swam.

Paul Paquet, a biologist at the University of Calgary, thought that Argos might be a way to learn more about his particular interest: the range and behaviour of Rocky Mountain wolves. Having once run free along the vast expanse of the Rockies, which stretch from British Columbia in Canada all the way to New Mexico in the south-western US, wolves had all but vanished from the North American landscape in the twentieth century. The last Yellowstone wolf was killed in the 1920s, and Canadian wolves disappeared from Banff and Jasper National Parks in Alberta in the 1950s. But during the 1980s they started to recover, reappearing in Banff and Jasper and recolonizing Glacier National Park, across the border in Montana. How they had done so, and the dynamics of this nascent population, was poorly understood – and the wolves' tendency to disappear almost immediately from the range of radio trackers didn't help researchers trying to trace them. Argos, Paquet realized, would radically increase the range of his research. He applied for one of the very first terrestrial tracking units.

On a wet day in June 1991, a team of researchers led by Paquet captured a five-year-old female grey wolf close to Kananaskis Field Station, in the foothills of the Canadian Rockies. Paquet believed it was important to remember that each animal was an individual, and so – in contrast to many researchers, who prefer to maintain a scientifically objective distance from their subjects – he liked his team to give their animals names. In deference to the day's weather, they called this wolf Pluie, the French for rain.

Pluie was fitted with Paquet's new gadget, an Argos tracking collar, and released back into the mountains. The researchers returned to their computers to await updates on her location. What she showed them was to change our ideas about animal territory, populations, needs and desires – and make possible a radical change in our relationships with them.[3]

Although they'd had trouble tracking wolves in the past, Paquet and his team expected Pluie to stay relatively close – after all, she obviously belonged to one of the Banff packs that roamed the nearby national park, an area of some 6,500 kilometres. Previous studies in Montana had shown that some wolf ranges covered an area of a few hundred square kilometres, or at most a thousand in the wilder territory surrounding the park. Pluie, however, would go a lot further than that.

Argos isn't very accurate – it can pinpoint an animal to around a mile – and it could be flaky, failing to report a viable signal for weeks at a time. When Pluie's signals started to come in, at first they seemed unbelievable, then jaw-dropping. For the first few months after her June capture she stayed relatively close to the tagging site, but in the autumn she suddenly upped and left the area, passing through Banff's parklands before turning west into British Columbia, and then south across the US border, into Glacier National Park. Passing east of the town of Browning, she entered the Great Plains. She travelled through the Bob Marshall Wilderness, a million-acre, roadless expanse of forests and mountains north of Missoula, before crossing the rest of Montana and the northern part of Idaho to reach Mount Spokane in Washington State – some 500 kilometres, as the crow flies, from Browning. A while later she headed north again, re-entering Canada from Idaho, somewhere near Bonners Ferry. By December 1993, she had made it at least as far as Fernie, in British Columbia, in a round trip covering more than 100,000 square kilometres.

The signal from Fernie was the last that Argos received from Pluie, and shortly after it was sent, the tracking team got a package in the mail containing the battery from Pluie's collar – with a bullet hole in it. They feared the worst, but Pluie wasn't dead, at least not yet: she turned up again, two years later and some 200 kilometres away

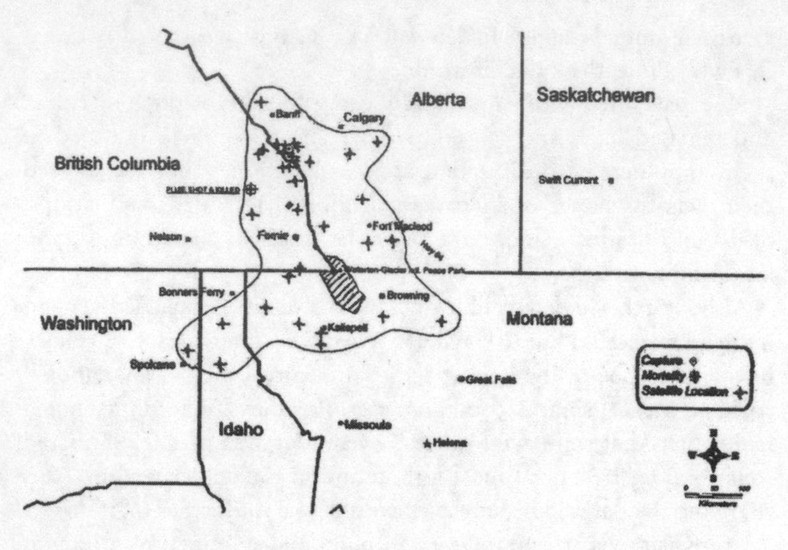


A map of Pluie's journeys.

outside Invermere, British Columbia, on the edge of Kootenay National Park, still identifiable by her now battery-less collar. This time, she didn't survive: on 18 December 1995, she was shot and killed by a licensed hunter, along with an adult male and their three pups.

In just six months that she was tracked on the move – and it's likely she undertook this journey multiple times over the course of her short life – Pluie had traversed two nations, three US states, two Canadian provinces and an estimated thirty different jurisdictions, from city limits to National Parks, and from Crown lands to First Nation territories. Her tracks crossed mountain ranges and forest wildernesses, as well as highways, golf courses and private land. She moved – apparently freely but often at risk from hunters, ranchers and road traffic – across a previously unimaginable area. In doing so, she gave us new insights into the ways in which wild animals live their lives – amongst themselves and amongst us. And she also showed us that the landscapes we imagine and set aside for animals are not remotely big enough.

The first effect of Pluie's journey was to change the way we understand wolf populations. Our history of encounters with wolves, as well as the restricted lens of VHF tracking, had given rise to the idea of wolf packs which confined themselves to particular areas: to populations defined by the edges of Banff, Jasper or Glacier National Park. But Pluie showed us that the wolves of the Rocky Mountains constitute what scientists call a 'metapopulation': a single populace composed of shifting, smaller, but interconnected communities. Metapopulations ensure the survival of species over vast areas, because they allow for the occasional mixing of genes between isolated subpopulations and the possibility of renewal and recolonization when particular communities are wiped out – as had happened to wolves in the twentieth century. The existence of multiple interconnected communities ensures the survival of a species in ways that a single, large community cannot – with habitats, feeding opportunities and genes passing between groups. It's an ecological effect, in which the total resilience of many interconnected but autonomous groups is greater than one homogeneous mass.

The existence and resilience of metapopulations has lessons for us and our networks; indeed, this distribution of resources is the basis on which much of our critical infrastructure works today. Power generation, data storage and emergency response all operate on the same principle: the spreading of capacity across as wide an area as possible, in order to survive and thrive when disaster strikes.

Metapopulations are also founded on the principle of movement: we don't just need to spread out to survive, we also need to be mobile. Distributed populations need ways of connecting to one another in order to share resources, information and genes, as well as the ability to move on when the situation becomes untenable. This is especially urgent in a time of rapid climatic change: as we've seen, liveable habitats are moving inexorably upwards and polewards, and species need to keep pace with them to survive. It's another lesson we need to take to heart: connectivity and freedom of movement are among the most vital faculties for weathering the coming storm – for humans and non-humans alike.

In showing just how widely animals roam, Pluie's satellite updates revealed how big protected areas need to be in order to ensure the survival of wolves and other wild communities. Her journey covered an area ten times the size of Yellowstone National Park and fifteen times that of Banff, two of the largest wildlife preserves in North America. She showed researchers how large an area Rocky Mountain wolves needed to be able to access and traverse in order to live their lives; the same is true for many other North American species, such as grizzly bears, lynx, cougar, antelope and deer. In turn, a host of other animals and birds, as well as plants and fungi, depend on the mobility of these animals to spread their seeds and spores, graze pastures, maintain population levels and manage habitats. The whole ecosystem depends on freedom of movement, over an area far greater than that set aside specifically for wildlife.

On their own, then, parks and wildlife sanctuaries clearly weren't adequate. What was needed was a network of territories, interlinked and patchworked across human-dominated territory, together with a new way of thinking about living with and alongside non-human animals. If the wolf revival was going to continue, as part of a wider renewal and affirmation of biodiversity across North America and beyond, then forging connections between populations and across states and nations was the first work that animal advocates had to apply themselves to.

Pluie's journey became the inspiration for one of the world's most ambitious environmental projects: the Yellowstone to Yukon Conservation Initiative, or Y2Y. This project aims to bridge an expanse of some 3,200 kilometres, covering both National Parks and the land which connects them, via an interconnected system of wild lands and waters. The central idea is what's known as a wildlife corridor: a clear pathway, or network of pathways, along which animals and plants can move, unimpeded by human activity, in order to migrate, feed and maintain diversity.[4]

Wildlife corridors can take many forms, from a simple hedgerow along the edge of agricultural fields to preservation areas stretching for thousands of kilometres across national borders. Sometimes the term refers to open spaces, and at other times to infrastructure, such as underpasses and bridges, designed specifically for wildlife, which allow them to move safely across roads and railways. In support of such corridors, fences can be used to direct wildlife away from danger and towards safe passages and crossing areas; land can be cleared or left fallow and development shifted or reoriented, all to better accommodate the needs of animals, in place or on the move.

Y2Y follows this piecemeal approach. When Pluie's story was featured as a comedic aside on the TV show *The West Wing*, the idea of a wildlife corridor was mocked as a $900 million, '1,800-mile wolves-only highway', complete with road signs and off-ramps into local cattle fields. But the reality of a wildlife corridor, even one on a continental scale, is a patchwork of different processes and practices: keeping private lands free of pesticides and fences; wildlife-proofing human population centres; constructing bridges, tunnels and other crossing points; purchasing new plots to set aside; and ongoing studies of wildlife and human activity.

The Y2Y coalition is currently working across multiple states, provinces and parks in the US and Canada, advocating for better traffic management, safer roads, wildlife education and the end of coal mining, metal refining and clear-cut logging, as well as the designation of specific protected areas. While the larger network will continue to develop and interconnect, local initiatives are already seeing successes: in early 2020, the first grizzly bear in eighty-eight years was seen in Central Idaho's Bitterroot Wilderness, meaning that North American bear populations are reconnecting, reassociating and on the road to recovery.[5]

Across the world, wildlife corridors not only make it possible for animals to survive and thrive in landscapes dominated by human activity; they also show us that we can negotiate and adapt to allow them to do so. In the Russian Far East, the Sredneussuriisky Wildlife Refuge stretches across the provinces of Primorsky and Khabarovsk Krai, and over the border into north-eastern China. This 72,700-hectare preserve links Russia's Sikhote-Alin mountain range with China's Wanda Mountains, two key habitats of the critically endangered Amur tiger, which is now able to move across international borders without human obstruction.

In the Netherlands, a network of more than 600 man-made corridors, including an 800-metre long green bridge, the Natuurbrug Zanderij Crailoo, transforms one of the most densely populated countries in Europe into a haven for deer, wild boar, badgers and other forest creatures, as well as other species of animals and plants whose existence depends on them. On 26 January 2011, an elephant named Tony became the first elephant to walk the length of a new corridor

A wildlife corridor in the form of a green bridge in the Netherlands.

through the Ngare Ndare Forest, connecting Kenya's second-largest elephant population in the Lewa Wildlife Conservancy with their compatriots around Mount Kenya, passing en route through Africa's first dedicated elephant underpass, beneath the Nanyuki–Meru Highway.[6]

Wildlife corridors can also be part of human healing processes. Europe's largest nature reserve, and one of the longest wildlife corridors in the world, is the European Green Belt, a 7,000-kilometre network of parks and protected lands following the line of the Iron Curtain, which once separated Western Europe and the Soviet bloc. The Green Belt was first proposed by German conservationists in December 1989, just a month after the Berlin Wall fell; today, it stretches all the way from Finland to Greece. In places, old minefields still keep visitors on the paths, but the former 'death strip' is now a flourishing habitat and migration path for more than 600 species of rare and endangered birds, mammals, plants and insects. One day, the same might be true of the demilitarized zone, or DMZ, between North and South Korea, a 155-mile-long, 2.5-mile-wide strip of land that has been virtually untouched by humans for more than six decades and is now home to millions of migratory birds and flourishing

A crab bridge on Christmas Island.

plant species, as well as endangered animals such as Siberian musk deer, cranes, vultures, Asiatic black bears and a unique species of goat, the long-tailed goral. As in the thirty-kilometre exclusion zone around the Chernobyl nuclear reactor, which scientists have called 'an unintentional nature reserve', wildlife has flourished when humans have withdrawn. Despite high radiation levels in the zone, populations of elk, boar, foxes and deer are at least as high as in other preserves in Ukraine and Belarus – while one study suggests that wolves might be seven times more abundant.[7]

Just as attentiveness to more-than-human life requires us to think beyond our own senses and sensibilities, to try to imagine the *umwelt* of other beings, so the construction of wildlife corridors requires a certain ingenuity and care, depending on the particular species and ways of life we are trying to support. On Christmas Island in the Indian Ocean, a network of more than thirty underpasses and accompanying fences serve to protect millions of native red crabs as they undertake their annual migration to the coast. In addition to the tunnels, a five-metre-high bridge passes over one of the island's busiest roads, its steel mesh surface perfectly calibrated for crustacean grip.

I can't help but imagine a vaster version of the University of Kobe's crab computer, laid out between the forest and the sea: millions of crabs, passing through tunnels in the shape of logic gates, the yearly cycle of their escape to their breeding grounds powering some vast and multi-threaded processing task.

In Oslo, a network of rooftop hives and gardens atop office blocks, schools and housing developments creates an aerial highway for bees to criss-cross the city, coordinated in such a way that there is never more than 800 feet between pollen-rich foraging sites. This being Norway, the network includes modernist hives by the celebrated architecture firm Snøhetta, designers of the city's iconic opera house, and is managed by a voluntary collective of beekeepers. (This is essentially the same manner in which the country's social democracy maintains its networks of national parks and communal camping lodges for humans, managed by the volunteer-led Norwegian Trekking Association, which has a quarter of a million members.) The arrangement of such more-than-human infrastructures reflects the human social and political systems on which they are founded, without being subservient to them.[8]

Of course, political ideologies cut both ways. On the opposite border from that criss-crossed by Y2Y, ex-President Donald Trump's infamous wall separating the US and Mexico carves through and across wildlife refuges, UNESCO heritage sites, sovereign tribal lands, national monuments and pristine wilderness. These mostly open lands are now blockaded by hundreds of kilometres of concrete and steel pipe, with devastating effects on animal as well as human mobility. The eastern terminus of the wall splits in two the 200,000-acre Lower Rio Grande Valley wildlife refuge in south-eastern Texas, a vital corridor for endangered birds, as well as ocelots and jaguarundi cats. In the Madrean Sky Islands of Arizona and New Mexico, the wall's construction will close off a vast tract of land which has recently seen the reappearance of the North American jaguar (which went extinct in the US in the 1960s), as well as blocking migration routes for grizzly bears and grey wolves. Although the wall's future is unclear at time of writing, some areas have already been permanently affected, such as the Quitobaquito Springs at Organ Pipe Cactus National Monument, sacred to the Tohono O'odham Nation, which have been partially drained to provide concrete for the wall. And even if the wall is rolled back, barrier construction around the world is on the rise, on every continent: from the brutal border fences at the margins of the European Union to the great Western Sahara Wall, a 2,700-kilometre sand berm constructed by Morocco in the desert of north-west Africa. One day, these places may become green belts once again, but we will have to fight for them.[9]

In the Middle East, extensive studies have shown that the apartheid wall constructed by Israel through the occupied West Bank has had a devastating effect on the local environment. The wall follows an ancient ecological corridor that runs from the Judean Mountains in the south to the Sumerian Mountains in the north. Not only does the wall block East–West migration, it also cuts across North–South routes as it zigzags around Palestinian territory. A 2010 report to the Palestinian National Authority identified sixteen species which face local extinction as a direct result of this barrier, including foxes, wolves, ibexes, porcupines and the unique Palestinian mountain gazelle. Birds and insects are also threatened by the burning of forest areas to provide space for settlements behind the wall. In places, the wall's construction has been slowed by environmental orders from Israel's High Court, and special s-shaped sections have been designed to allow small animals to pass through – but not larger mammals, or people.[10]

Infrastructures designed to block the free movement and association of humans, then, disrupt animal lives as well – but emphasizing this should not lead us to prioritizing one effect above the other. Environmental action must never be allowed to become a figleaf for other forms of oppression. The spectacle of the Israeli Defense Forces creating grated passages through the apartheid wall to permit the passage of small mammals while continuing to degrade, suffocate and cut short Palestinian lives only serves to underscore the racism and inhumanity of such barriers, and the regimes which install them. Likewise, praising Christmas Island's crab bridges should not obscure the Australian government's use of detention centres on the same island – adjacent to the same National Park – to imprison sick and traumatized asylum seekers in contravention of international law.

The existence of such barriers to freedom of movement, and their well-documented negative effects on human and non-human lives alike, emphasizes the importance of human and more-than-human solidarity. The struggle to ensure the survival of non-human species goes hand in hand with the struggle for dignity and freedom for all humans.

Testament to this is the story of the iguanas of Guantánamo Bay. In 2008, a group of Kuwaitis who had been detained at the US base on Cuba since 2002 applied to the Supreme Court to have their cases heard. Their lawyers made an application for habeas corpus – the same principle that had been tried for Happy the Elephant and other animals by the Nonhuman Rights Project. And, as in the NhRP's cases, they needed to convince the court that habeas corpus applied to these particular detainees – which had been previously denied on the basis that Guantánamo Bay was not within the jurisdiction of the court. But in this case, the situation was reversed: remarkably, non-humans testified on behalf of the humans.

When the detainees' lawyer, Thomas Wilner, took part in an interview for TV show *60 Minutes*, the producers told him that when they visited Guantánamo, they'd discovered that the local Cuban iguanas were actually made safer by the base. If they wandered off it, they might be eaten by the locals – but on the base itself, they were protected by US law, in the form of the Endangered Species Act, with military personnel liable to fines of up to $10,000 if they harmed the animals. At the hearing, Wilner argued that it was impossible for US law to apply to iguanas but not to people – and the Supreme Court subsequently agreed to hear the case. When the Solicitor General repeated the argument before the court that the detainees were not within its jurisdiction, Justice David Souter replied, 'What do you mean? Even the iguanas at Guantánamo are protected by U.S. law.'[11]

So the iguanas 'spoke' for people – but how might they speak for themselves? As we've already concluded, courts are insufficient for adjudicating animal needs and desires, and human judicial processes cannot accommodate the true multiplicity and complexity of more-than-human being. Indeed, the story of Pluie teaches us that it's only when we see the true extent and scope of non-human life, on its own terms, and accept its reality and value beyond our own, that we start to seriously adapt our own ways of living. Ultimately, it's not about granting animals personhood, but about acknowledging and valuing their animalhood – and their planthood, their subjecthood, their beinghood. It's about allowing them to be themselves, while working with them to structure the world for the benefit of all of us. We must think not on the scale of laboratories and courtrooms, but on the scale of forests, mountain ranges, tundra, oceans and continents – on the scale of projects like the Yukon to Yellowstone conservancy.

The history of animal tracking with technology is one of a slowly but inevitably widening perspective: from Lord and Cochran's rabbit field to a national park, to a whole continent or a whole ocean. In this, it mirrors the development of our own information technologies, which have gradually unfurled across the surface of the entire planet and out into space, bringing us ever closer to every other part of the globe. At every stage, these different but interconnected networks have told us the same thing: we are thoroughly and inextricably entangled with one another.

Y2Y is one of the largest and most successful environmental projects of our time. It is a direct product of a technological ecology: a cross-species and cross-kind collaboration between a wolf, a team of human researchers and a constellation of satellites. In this assemblage of actors, the scale and vision of technology plays a direct part, widening the scope of our care and attention from the range of binoculars and VHF transmitters to the breadth of an entire continent, seen from 850 kilometres up. Today, Argos – the satellites which tracked Pluie – consists of seven polar-orbiting satellites which complete a revolution of Earth every hundred minutes and cover approximately 5,000 square kilometres of its surface at any given moment. And while its capabilities and accuracy over time have been increased, it has also been augmented by newer and even more capable systems, such as the Global Positioning System, or GPS.

GPS is much more accurate than Argos, and – along with decreasing battery sizes, better transmitters and receivers, and a host of other improvements – it has made large-scale animal monitoring possible in ways the pioneers of radio telemetry could never have imagined. To save more energy, and last longer, some tracking collars can now store data onboard and automatically release at a pre-determined time, dropping off the animal to be picked up by researchers. This has enabled a less intrusive and much more accurate view, on a much greater scale, of where animals are and where they want to go.

Interstate 80 runs all the way across the continental US, from New York on the East Coast to San Francisco on the West, passing through Wyoming, Nebraska and Utah, and south of Yellowstone. It's a notorious hazard to wildlife – and often, as a consequence, to humans. In Wyoming alone, 15 per cent of all vehicle collisions involve deer, antelope or other large animals. These 6,000 annual collisions cause an estimated $50 million in damage to cars and loss of wildlife. And because I-80 crosses a number of major migration trails, these crashes can be devastating. In 2017, a semi-trailer truck hit a whole herd of pronghorn antelope on the move, killing twenty-five of them.

Back in the 1960s, when the highway was being constructed, the planners realized that they needed to take account of local wildlife, not least because their own engineers were involved in such collisions. In response, and with the best of intentions, they constructed a series of underpasses and culverts along the highway to allow wildlife to pass beneath the road. At the time, however, animal migration was poorly understood and there was little data to help with planning. The underpasses were sited according to the guesswork of a few engineers, and they were mostly built in the wrong places. Few of the migrating wildlife made use of them, and most kept crossing the highway elsewhere. The killing continued.[12]

This has started to change in the past decade, with wider availability of GPS and lightweight tags leading to an explosion in their use, and a corresponding explosion in verifiable data about animal movement and migration – about their needs and desires. All along I-80, elk, pronghorn, mule deer and other animals butt up against the road and the fences which surround it, often with nowhere to go but into the traffic – or be cut off from food, breeding grounds and genetic diversity. But many of these animals are now wearing tags, and thanks to monitoring by organizations like the Wyoming Migration Initiative, which has brought together data from multiple studies, the places where they cluster and the locations they are trying to reach are suddenly much more visible.[13]

One upshot of this visibility is a new infrastructure for wildlife support. In 2012, Wyoming constructed a new green bridge at Trapper's

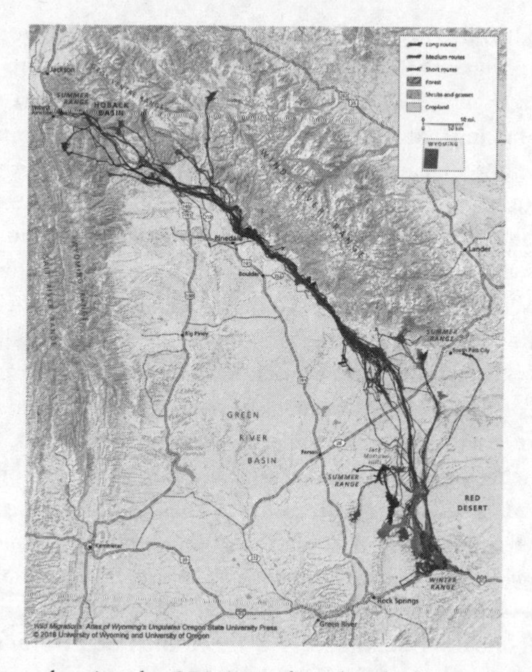


A map showing the GPS traces from hundreds of migrating mule deer across Wyoming.

Point on US Highway 191. Its location was selected after reviewing GPS data from hundreds of pronghorn antelope, which revealed the exact path of their bi-annual trek between their winter range in the Red Desert and their summer grounds in the Grand Teton National Park. Now known as the 'Path of the Pronghorn', and in use for at least 6,000 years, this critical migration path is blocked at several points by roads and railways, but since 2012 there has been one less obstacle. The Trapper's Point overpass, placed right in their path, is now used by thousands of pronghorn and deer every year – and fatalities on the roads have dropped accordingly. Wyoming is currently exploring more locations for crossings and collecting more data to find out where they're needed.

At the end of the previous chapter, I asked what real political participation by animals in a more-than-human system would look like.

Politics, remember, is the art and science of making decisions together about things which affect all of us. Well, that's what's happening here. Just as the orang-utans who fought back against imprisonment in zoos were acting politically, so is the wildlife of Wyoming and the Y2Y. These antelope, elk and other animals are voting with their feet, and we – augmented by technology – are finally listening to them, and adapting our behaviour and construction to better account for them.

This is what a more-than-human politics actually looks like: the careful and conscious attention to the needs and desires of others, the acknowledgement of their agency and value, and the willingness to adapt our existing structures of society and place to better account for all of us. This is more-than-human solidarity in practice and, as promised, it comes with benefits – and questions – for all of us.

This, then, is where Martin Wikelski's ICARUS project, which tracks animals from the International Space Station, might take us next – beyond GPS and a few herds of antelope, beyond a few highways, to a new, shared superhighway: the internet of animals. Imagine the progression from brass leg rings and plastic ear tags, to VHF radio and satellite tracking, scaled up to include not hundreds, but millions, even billions of animals. If we could connect more animals to sensor networks, we would have more and better ideas of what they wanted and needed, and be better able to respond appropriately to them.

The ICARUS system extends the scope and range of animal tracking once again – beyond radio and beyond GPS. Because of the weight of even the smallest GPS transmitters, animals weighing less than 100g can't carry them, meaning that some 75 per cent of bird and animal species – and all insects – have remained invisible to digital data collection. And GPS collars and harnesses, despite improvements, are still prohibitively expensive for many applications.[14]

ICARUS changes all of this: its lightweight, solar-powered tags can be fitted to a much greater range of creatures, with a corresponding increase in the number of animals we can track, monitor – and listen to. If such efforts are tied to real attempts to improve the lot of non-humans, then the extension of this legibility is in effect an expansion of suffrage. The more non-humans who join the internet of animals – and perhaps one day of fungi, plants, bacteria and stones as well – the more votes are counted, the more voices join the demos, and the wider and more equitable our more-than-human commonwealth becomes.

We don't yet know what we will find, or what will be decided. Just as Pluie revealed a previously unknown scale of animal movement and interconnection, so the internet of animals will undoubtedly alert us to new questions we had not even thought to ask. Just as the human internet has revealed the latent desires, foibles, weaknesses and sheer strangeness of our own species, so the non-human internet will undoubtedly open us to encounters, experiences and understandings the like of which we cannot presently conceive. And that, as we know from our exploration of randomness, is precisely what we are looking to do. The internet of animals and everything else – the more-than-human internet – thus appears to be a viable and urgent way for us to engage with an ever greater number of beings and to co-create a future which takes into account the agency and innate value of all its participants.

Of course, there's much about the idea of an internet of animals to cause us disquiet – after all, it's not as if the internet of people has been an unalloyed success. We will need to think carefully about the ways in which it is deployed, used and administered. In particular, we must not repeat the mistake of twentieth-century technological determinism, which saw the role of high technology as producing the one, unarguable answer to every problem. If the use of trackers and other gadgets to amass vast amounts of data sounds suspiciously close to the kind of prediction and control I've railed against elsewhere, we will have to work particularly hard to ensure its applications are flexible, respectful and appropriate – and the techniques to do so are available to us.

An example of this kind of flexibility is the concept of mobile protected areas. In 2020, the UN Convention on the Law of the Sea was subject to revision for the first time since 1982. While the Convention already included provisions for marine protected areas – swathes of the ocean off-limits to fishing or large ships – these are based on fixed locations, and unchanging boundaries. But as Professor Sara Maxwell, a marine biologist at the University of Washington, points out: 'Animals obviously don't stay in one place – a lot of them use very large areas of the ocean, and those areas can move in time and space. As climate change happens, if we make boundaries that are static in place and time, chances are that the animals we are trying to protect will be gone from those places.'[15] Maxwell should know, her research includes using satellite and GPS tags with animals such as turtles and Arctic terns to study the migration patterns of ocean species.

In response to the revision of the Convention, Maxwell and a group of scientists proposed that it include provisions for mobile marine protected areas, which could pop up and shift in response to the data streaming in from an aquatic internet. Such dynamic management of the ocean is already used by a number of nations in the 200 nautical miles off their own coasts which they administer, but the wealth of signals now becoming available from a huge range of ocean-dwellers makes much larger, and more flexible, protection possible. 'Until we could implement this type of management and show that it's feasible, people didn't quite believe that it was possible,' Maxwell said. 'But as we know more about where animals are going in space and time, we can use that information to better protect them.' New technology is making this dynamic approach to ocean conservation possible, just at the moment that climate change is making it necessary.[16]

There are also extraordinary things that the internet of animals can do which don't require pin-point tracking and totalizing knowledge – just a greater awareness of the lives and behaviours of others. The ICARUS team attached accelerometers to the collars of a number of different animals living in areas with high seismic activity: goats who roamed the slopes of Mount Etna in Sicily and sheep, cows and dogs around the city of L'Aquila in the centre of Italy. These devices don't record locations or anything that humans typically refer to as 'personal data'. They solely transmitted the animals' degree of activity: how much they moved around; how placidly or excitedly they behaved.

When Mount Etna erupted at 10.20 p.m. on 4 January 2012, the team looked back over their data and saw that the goats had become abnormally agitated six hours previously. Over the course of a two-year study, they were able to predict another seven major eruptions. They found the same was true of earthquakes in L'Aquila: in the days and hours before earthquakes, the sheep, cows and dogs would behave in unusual – but measurably unusual – ways. The closer they were to the looming epicentre, the more agitated they became. Together, they constituted an early warning system more powerful, more accurate and more advanced than any other mechanism humans have devised. And these experiments were conducted with traditional radio tags: how much more powerful – and life-saving – such predictions will become now that ICARUS's space-based antenna has come online can only be guessed at.[17]

One thing that became evident from these experiments was that you need many sources of data, many connected animals, to make accurate predictions. The kinds of behaviour that suggested an oncoming tremor wasn't visible at the level of a single animal: it only became apparent when the data was aggregated. We are stronger, and more skilled and knowledgeable, when we act collectively, even if we don't know it ourselves.

The internet of animals also shows that we don't need to throw away everything we've learned about complex technology; we can repurpose it. Because of the different ways that different animals react to natural phenomena, according to their size, speed and species, the ICARUS team found it necessary to use particularly complex forms of analysis to pick up on the differences in the data generated from different tags at different times – a welter of subtle and subtly variable signals. To do this, they turned to statistical models developed for financial econometrics: software designed to generate wealth by picking up on subtle signals in stock markets and investment patterns. I like to think of this as a kind of rehabilitation: penitent banking algorithms retiring from the City to start a new life in the countryside, and helping to remediate the Earth. From spy satellites to face-detection algorithms, the tools which currently conjure oppression and inequality can be turned around and put to beneficial uses.

Finally, the animal early-warning system tells us something about the way we learn when we listen to animals and plants, rather than dissect them. Recall Monica Gagliano's conversations with *Mimosa pudica*, and the opposition she faced when she published a theory of plant memory. That opposition was founded on the fact that there is no known mechanism for such memory. In contrast to the normal practice of botany, Gagliano's declaration was based on the plant's actual behaviour, not what we know about its internal structure. The plant told us something, and we chose to believe it – supported by rigorous tests and evidence – without fully understanding how it could be true.

Likewise, there is no known mechanism for animal prediction of seismic activity – although it has been attested by folk wisdom for centuries. Perhaps, some scientists have suggested, their fur allows them to sense in some way the ionization of the air caused by large rock pressures in earthquake zones. Perhaps, others argue, they can smell gases released from deep underground in the run-up to seismic events (the same mysterious gases, perhaps, which gave the Oracle at Delphi the gift of prophecy). We just don't know – but if we are prepared to give up some of our desire for totalizing knowledge, if we are prepared to treat understanding as a process and a negotiation, rather than as a route to mastery and dominance, then there is much we can learn and understand from the wisdom of others.

That is not the only thing we will have to give up. As well as admitting that we cannot know everything, we will have to admit – and act on – the fact that we cannot be everywhere. Because, ultimately, one of the outcomes of the internet of animals must be a situation in which we are able to live alongside our non-human comrades, while leaving most of them alone. Contemporary human debates about privacy and the 'right to be forgotten' – our data expunged entirely from the databases of corporations and governments – must be applied to non-humans too. We must ensure that our desire to engage with animals – even with the best of intentions – does not further degrade their liberty and livelihoods. And when we have learned what we need to, we must remove the tags and the trackers and retreat. The logical conclusion of the development of wildlife corridors and protected areas – even mobile ones – is that there are places which must be left to non-humans, even in a more-than-human world.

This is not a new suggestion. It goes back at least to the beginning of the twentieth century in Western culture, to the founding of National Parks in Europe and America, and it is intrinsic to non-Western conceptions of our place among the species of the planet. But there is a growing awareness that it is now more urgent and must be much more extensive than a few scattered parks and sanctuaries. Indeed, there is a strong scientific and moral case that it should comprise at least half the entire Earth.

*

The campaign to designate half of Earth's surface a human-free natural reserve originates in the work of E. O. Wilson, one of the world's foremost evolutionary biologists, and the founder of sociobiology. Wilson's work is founded upon a deep understanding of the real value of biodiversity, and also upon a deeply moral appreciation of the natural world. His 'theory of island biogeography', developed since the 1960s, provided the theoretical underpinning for Pluie's practical teachings, by showing how all confined landscapes, including natural parks, inevitably lose species, however well they are protected and maintained. In 2014, Wilson and others launched the Half-Earth Project, to further propagate these ideas and to campaign for their implementation.

According to Wilson's widely accepted theory, changes in the size of a habitat results in a change in the number of species it can support – and this number is mathematically predictable and stable. As reserves expand in size, the number and diversity of species within them grows – but as they shrink, the diversity within them declines swiftly, immediately and often irreparably. When 90 per cent of wild habitat is gone, the number of species which can be supported is halved – and this is approximately the point we have reached, globally, in the present. But if we lose just 10 per cent of the remainder, most or all surviving species will disappear.

The answer, for Wilson and the Half-Earth Project, is a radical expansion of protected areas. If we can protect half the Earth from human interference and alteration, an almost unimaginable goal, we can hope to protect some 85 per cent of remaining species. Wilson calls this the 'safe zone' for life on Earth – and unless we start working on it now, the consequences for biodiversity, and thus for life on Earth as a whole, will be catastrophic.[18]

Central to the Half-Earth Project is working out which half of the Earth we need to protect – and technology is crucial to this determination. As we are coming to understand in the design and implementation of wildlife corridors, the best solution for humans and non-humans alike is not total apartheid, but a kind of cross-hatching: some wilderness expanses, human-free zones and set-aside ocean territories for sure, but also corridors and passageways, green bridges and untouched streams, threaded through and around areas of human habitation. Wilson calls these interconnected spaces 'Long Landscapes', and imagines a day in the not-too-distant future when 'you'll be so surrounded, so enveloped by connected corridors that you'll almost never not be in a national park, or at any rate in a landscape that leads to a national park.'[19]

In figuring out how we advance the Half-Earth Project, the internet of animals will be key, particularly if we understand it as not only a technological assemblage for generating data, but also as a political assembly for engaging with and asking questions of our more-than-human comrades. Half-Earth is a survival project for all of us. Bringing it into being will require not only listening to others, but actively ceding some of our privilege and power to them. This is the lesson, however much it is resisted, of every struggle for human liberation, and it is no less true of more-than-human progress. We have the science and technology to accomplish this vision. The only thing standing in our way is our own hubris, and the selfishness and ignorance of those who benefit the most from the present system and who therefore want to maintain the status quo.

Achieving the aims of a Half-Earth – and other projects which will be necessary to maintain life on this planet – cannot be accomplished without a struggle. No great change in human affairs has ever been achieved without one party giving up some or all of their power, and in the struggle for an equitable, more-than-human future, it is the whole of humanity that is on the side of power. But this position is ours to reject, and the rewards of doing so will be great and glorious, particularly if it is beauty and diversity, change and possibility, rather than power and domination, that we seek to embrace.

'One word: poetry. That's what the world has to offer us,' wrote Wilson in 2020. 'A whole series of mysteries, of possible discoveries, of phenomena, of unexpected events, and objects, and things, and living organisms and so on. An infinitude, almost, on this planet, waiting out there to be enjoyed. There's so many of this in the world waiting to be explored, and savored, and described.'[20] Wilson calls this hopeful urge 'biophilia': the belief that people have an innate affinity for other species. I call it 'solidarity', the word that for me expresses both a deep love for the Earth and everything in it, and a practical politics of mutual respect, aid and support.

The realization of the Half-Earth Project is one endpoint of the processes described in this book, but it is with the wildlife corridors that get us there that it is most closely aligned. For me, they embody the coming together of everything this book has been about: the complexity and variety of non-human intelligence, the subjecthood and agency of every being, the potentiality and politics of technology, and the wealth of knowledge and ideas we have to gain by opening ourselves to the more-than-human world with which we are inextricably and gloriously entangled. More than anything, they emphasize that movement is what we need, and it is the journey itself which sustains us in a constantly changing and surprising world. Means, not ends, are our immediate concern.

I confess, I'm surprised that one answer to the question, 'how do we stop technology separating us from the natural world' turns out to be sticking tiny digital sensors on everything. But perhaps I shouldn't be. After all, if we can tune military radars to observe the migration of birds, or turn spy satellites around to learn about the origins of the universe, then we can put the tools of surveillance to work to build a more-than-human parliament.

And perhaps that was never the real question. What I think I've come to understand, more deeply than ever before, is that the enemy is not technology itself, but rather inequality and centralization of power and knowledge, and that the answer to these threats are education, diversity and justice. You don't need artificial intelligence to work that out. You need actual intelligence. But more importantly, you need all the actual intelligences – every person, animal, plant and bug; every critter, every stone and every natural and unnatural system. You need a crab computer the size of the world. The problem is never technology itself; after all, remember, the computer is like the world.

I remain as excited as ever about the power and possibilities of computers and networks as I have ever been; I just abhor the structures of power, injustice, extractive industry and computational thinking in which they are currently embedded. But I hope I've shown, to some degree, that it doesn't need to be this way. There are always other ways of doing technology, just as there are other ways of doing intelligence and politics. Technology, after all, is what we can learn to do.

# Conclusion

## *Down on the Metal Farm*

It's May 2021 and I'm in Epirus again – just a couple of hours, on winding roads, from the lake where this story began. I've come back to Northern Greece, to the other side of the Pindus mountains, to see another kind of material extraction in progress. This time I'm not confronted by bulldozer tracks, boreholes and other marks of a rampant human greed and corporate AI. Rather, I'm standing in a small, gently sloping field, surrounded by neatly marked out plots of green, flowering plants. It looks like a vegetable garden, albeit one overrun with enthusiastic weeds, but in fact this is a carefully managed agricultural experiment. It's a small farm, but unlike any I've seen before. This farm grows metal.

In the 1990s, a number of mining firms funded research into a newly identified family of plants – in fact, several families – called hyperaccumulators. These plants had a very unusual property in common: they were capable of growing in soils rich in various kinds of metals, either naturally occurring or as a result of industrial pollution; soils which are toxic to most other kinds of life. What the research discovered was that these plants drew up the metal from the soil through their roots and stored it above ground in their shoots and leaves. In the process, they actually remediated the soil, cleaning it of the metals and making it more hospitable to other plants.[1]

The metal farm in the Pindus is one result of this research: an experiment to see if it's possible not only to repair damaged soils by planting hyperaccumulators, but to harvest metal just as we harvest other crops. It is part of a continent-wide experiment, with active plots in a number of countries across Europe – but it turns out that

Northern Greece and Albania are particularly well suited to this process, which is known as agromining.

Today is harvest day, as the plants are in full flower: a couple of local farmers shuffle on their haunches through the beds, cutting the plants off at the base and piling them onto tarps. Kostas, one of the researchers, breaks off a stem from one of the growing plants and presses the leaking sap onto a strip of blotting paper. The paper immediately stains purple-red, an indicator that the sap contains a high volume of nickel. Maria Konstantinou, the leader of the team, explains that these soils are known as serpentines for their blue-green colouring and are globally extremely rare: only the Balkans, parts of the Alps, Cuba and a few scattered locations in North America have any kind of extensive covering. Because they are rich in heavy metals – including chromium, iron, cobalt and nickel – most plants struggle to thrive in them. But the harsh conditions they create have also driven some plants to adapt to them in unique ways.

The Pindus plot is growing three types of flowering plant: *Alyssum murale*, *Leptoplax emarginata* and *Bornmuellera tymphaea*. *Murale* grows in low bushes topped by bunches of yellow flowers; *emarginata* is taller and spindlier, with clusters of green leaves and white petals shooting out at all angles; *tymphaea* straggles across the ground, covered in a dense layer of white blossom. Each of these plants is unique to the region: *murale* is native to Albania and Northern Greece, *emarginata* is found only in Greece, while *tymphaea* is found only on the slopes of the Pindus mountains (its name comes from Mount Tymfi, one of the highest peaks of the range).

Maria and her team from the International Hellenic University in Thessaloniki have been working this plot for six years, and this is their third harvest. Already, the results have surprised them. When the plants from Pindus are cropped and dried, they are transported to France, where they are incinerated. This produces both heat – which can be used for energy generation – and metal ash, from which the nickel can be extracted. The plot produces between 6 and 13 tonnes of biomass (dried plants) per hectare, depending on the crop, and each of those hectares produces between 80 and 150 kilogrammes of nickel. By comparison, a tonne of mined nickel ore contains about 1 or 2 per cent nickel, or 10 to 20 kilogrammes. The plants, by concentrating the metal in their leaves, become richer than the earth.

Research into hyperaccumulators is taking place around the world. In Malaysia, *Phyllantus balgooyi*, a shrub native to the equatorial rainforest, draws up so much nickel that its sap is bright green: it bleeds metal when cut. In New Caledonia, in the South Pacific, the rubber sap of *Pycnandra acuminata* contains as much as 25 per cent nickel by dry weight. And it's not just nickel which can be mined this way: Indian mustard, or *Brassica juncea*, forms an alloy of silver, gold and copper in its leaves; rapeseed removes lead, zinc and selenium from contaminated soils. Even lithium, a core requirement for the batteries in everything from mobile phones to electric vehicles, can be found in the leaves of *Apocynum venetum*, the sword-leaf dogbane, long known in traditional medicine for reducing hypertension. Many more such abilities remain to be discovered.[2]

Agromining is in its infancy, and its potential for altering the ways we extract all kinds of metals, as well as for repairing the damage we have done to the Earth, has yet to be fully realized. That being said, it will never come close to supplying the world's current metal needs: it is far less efficient, far slower and requires far more care than simply ripping into the earth with drills and explosives. Nor would we want it to succeed unduly, for in doing so it would require the kind of destructive, industrial-scale agriculture that is the unfortunate hallmark of soybeans, biofuels, palm oil and other once lauded miracle crops.

Nevertheless, hyperaccumulators point us towards an understanding and accommodation with non-human life which might in the long run be much more valuable than our particular material desires. What strikes me most about these plants is that they have evolved, over vast stretches of time, in particular locations, to address specific questions of survival in their own ways. They are knowledgeable about the soil and have developed ways to endure and thrive in it. Meanwhile, we have ignored them for most of that span – considered them as weeds and pests – until we realized that they were aligned with our own needs. When, eventually, we set out in search of other, less destructive ways of finding what we needed in the earth, we discovered that the plants had got there long before us. They are, if not more, then better evolved than us – if, that is, we consider true evolution to be a mutually prosperous, non-destructive alignment with the Earth.

Plants have learned to mine metals. Perhaps one day they will invent phones and computers too: they have all the materials they need. Or perhaps we will build something better together. In March 2021, researchers at MIT announced they had taught spinach to use email, which it used to warn them of explosive materials in the soil, leached from ageing landmines. The reality is somewhat more prosaic, but far more interesting. What they had done was engineer the spinach to change colour in the presence of certain chemicals, and then used cameras and computers to detect and transmit this information. Nonetheless, the experiment revealed the abilities of all kinds of plants to be, as one researcher put it, 'very good analytical chemists', and showed how to use technology to help them communicate their findings.[3]

At the Salk Institute for Biological Studies in California, scientists are engineering strains of mouse-eared cress – *Arabidopsis thaliana*, the same plant that showed us, when assailed by caterpillars, that plants could hear sound – to demonstrate another extraordinary ability. By transferring genes from cork trees into the cress, the researchers found that they could encourage it to grow much deeper, denser root structures, packed with suberin, an impermeable, cork-like polymer which stores huge amounts of carbon.[4] If they can repeat the feat with widely grown crops such as wheat, rice and cotton, then vast areas of arable land might be transformed into engines of carbon sequestration: pulling climate change-accelerating carbon out of the atmosphere and storing it safely underground for decades, even centuries.

This approach seems far more promising than the kind of megascale geo-engineering schemes proposed by global tech corporations: vast stacks of reactors and chimneys which pull carbon directly from the atmosphere, while producing ever more soon-to-be-obsolescent infrastructure and material waste.[5] Our common future demands less industrial hubris and more cooperation with existing, and deeply knowledgeable, biological systems.

For now, I am just happy to have found this field in Epirus, filled with gently waving, flower-laden, metal-rich weeds, dappled sunlight, the hum of bees and gently harvesting humans. I'm just sixty kilometres from the site where I first smelled petroleum on the breeze and learned of plans to use rapacious, corporate AI to crack open and despoil the land, but the journey from one place to another has encompassed gibbons, elephants, giant redwoods and slime moulds; neural networks, non-binary computers, satellites and self-driving cars; the I Ching, the music of John Cage and Sámi joikers, new forms of ancient governance, and herds of GPS-augmented antelope. The world is a computer made out of crabs, infinitely entangled at every level, and singing, full-throated, the song of its own becoming. The only way forward is together.

# Notes

## INTRODUCTION: MORE THAN HUMAN

1. C. Stambolis, G. Papamihalopoulos, K. Nikolaou, *A Strategy for Unlocking Greece's Hydrocarbon Potential: An IENE Project*, IENE, Open Forum, 10 June 2015, Athens; https://www.iene.gr/hc-exploration2015/articlefiles/session1/stambolis-nikolaou-papamichalopoulos.pdf.
2. Repsol sits at number forty-six on a list of the one hundred companies responsible for 71 per cent of global emissions. See Tess Riley, 'Just 100 Companies Responsible for 71% of Global Emissions, Study Says', *The Guardian*, 10 July 2017; https://www.theguardian.com/sustainable-business/2017/jul/10/100-fossil-fuel-companies-investors-responsible-71-global-emissions-cdp-study-climate-change.
3. IBM, 'IBM and Repsol Launch World's First Cognitive Technologies Collaboration for Oil Industry Applications', press release, 30 October 2014; https://www-03.ibm.com/press/us/en/pressrelease/45278.wss.
4. Repsol, 'Optimizing Hydrocarbon Exploration and Production Processes'; https://www.repsol.com/en/energy-and-innovation/a-better-world/pegasus-excalibur/index.cshtml.
5. Anjli Raval, 'Google and Repsol Team Up to Boost Oil Refinery Efficiency', *Financial Times*, 4 June 2018; https://www.ft.com/content/5711812c-670c-11e8-b6eb-4acfcfb08c11.
6. For Greenpeace's report, see *Oil in the Cloud: How Tech Companies are Helping Big Oil Profit from Climate Destruction*, Greenpeace, 19 May 2020; https://www.greenpeace.org/usa/reports/oil-in-the-cloud/. For Google's response, see Sam Shead, 'Google Plans to Stop Making A.I. Tools for Oil and Gas Firms', CNBC, 20 May 2020; https://www.cnbc.com/2020/05/20/google-ai-greenpeace-oil-gas.html.
7. Matt O'Brien, 'Employee Activism Isn't Stopping Big Tech's Pursuit of Big Oil', *USA Today*, 2 October 2019; https://eu.usatoday.com/story/

tech/2019/10/02/microsoft-amazon-google-oil-gas-partnerships/3839379002/.

8. Jordan Novet and Annie Palmer, 'Amazon salesperson's pitch to oil and gas: "Remember that we actually consume your products!"', CNBC, 20 May 2020; https://www.cnbc.com/2020/05/20/aws-salesman-pitch-to-oil-and-gas-we-actually-consume-your-products.html.
9. For an elaboration of the paperclip hypothesis, see Nick Bostrom, 'Ethical Issues in Advanced Artificial Intelligence', 2003; https://www.nickbostrom.com/ethics/ai.html.
10. Samuel Gibbs, 'Elon Musk: Regulate AI to Combat "Existential Threat" Before It's Too Late', *The Guardian*, 17 July 2017; https://www.theguardian.com/technology/2017/jul/17/elon-musk-regulation-ai-combat-existential-threat-tesla-spacex-ceo.
11. Nick Statt, 'Bill Gates is Worried about Artificial Intelligence Too', CNET, 28 January 2015; https://www.cnet.com/news/bill-gates-is-worried-about-artificial-intelligence-too/.
12. Sam Shead, 'DeepMind's Elusive Third Cofounder is the Man Making Sure that Machines Stay On Our Side', *Business Insider*, 26 January 2017; https://www.businessinsider.com/shane-legg-google-deepmind-third-cofounder-artificial-intelligence-2017-1.
13. Charlie Stross, 'Invaders from Mars', *Charlie's Diary*, 10 December 2010; http://www.antipope.org/charlie/blog-static/2010/12/invaders-from-mars.html.
14. Ernst Haeckel, *Generelle Morphologie der Organismen* (1866), translated into English by R. C. Stauffer (1957); cited in Robert C. Stauffer, 'Haeckel, Darwin, and Ecology', *The Quarterly Review of Biology*, 32 (2), June 1957, pp. 138–44; http://www.jstor.org/stable/2816117.
15. Charles Darwin, *On the Origin of Species by Means of Natural Selection, or the Preservation of Favoured Races in the Struggle for Life* (London: John Murray, 1859). In response to criticism of his work by the Church, Darwin added the words 'by the Creator' to subsequent editions, so that the final sentence read 'There is grandeur in this view of life, with its several powers, having been originally breathed by the Creator into a few forms or into one . . .'
16. John Muir, *My First Summer in the Sierra* (Boston: Houghton Mifflin, 1911). The final text is a version of a longer phrase from Muir's journal, dated 27 July 1869: 'When we try to pick out anything by itself we find that it is bound fast by a thousand invisible cords that cannot be broken, to everything in the universe.' (See Stephen Fox, *John Muir and His Legacy: The American Conservation Movement*, (Boston: Little, Brown, 1981).)

17. Ursula K. Le Guin, 'A Rant About "Technology"', 2004; http://ursulakleguinarchive.com/Note-Technology.html.
18. Rachel Carson, *Silent Spring* (New York: Houghton Mifflin, 1962), p. 5.
19. The quote is from a letter written by Blake to one of his patrons, the Reverend John Trusler, in the summer of 1799. Trusler had hired Blake to produce a series of moral artworks in the style of contemporary caricature. When Trusler criticized Blake's artistic vision as too 'imaginative', Blake let Trusler know in no uncertain terms what he thought of his aesthetic sense. 'If I am wrong, I am wrong in good company . . . What is grand is necessarily obscure to weak men. That which can be made explicit to the idiot is not worth my care.' (Collected in Alfred Kazin (ed.), *The Portable Blake* (London: Penguin Classics, 1979).)
20. Lynn Margulis, *The Symbiotic Planet: A New Look at Evolution* (Weidenfeld & Nicolson, 2013), p. 112.
21. From 'What is Reality?', collected in Alan Watts, *Become What You Are* (Boston, MA: Shambhala Publications Inc., 2003).
22. Eduardo Viveiros de Castro, 'The Transformation of Objects into Subjects in Amerindian Ontologies', quoted in David Harvey, *Animism* (London: C. Hurst & Co., 2017).
23. Churchill's original phrase, from a speech in the House of Commons in 1943, was 'We shape our buildings and afterward our buildings shape us.' McLuhan's colleague and friend John Culkin adapted and popularized the phrase in interviews with McLuhan. The whole saga can be followed athttps://quoteinvestigator.com/2016/06/26/shape/.

## 1. THINKING OTHERWISE

1. The learning software was derived from that released by Comma.ai, an open-source framework for developing self-driving cars. 'Open-source' means that anyone can review, edit and run the code themselves. To measure speed, position and turning angle, I wrote a smartphone app which is also open source. It's documented at https://github.com/stml/Austeer.
2. For an excellent account as well as critique of the *flâneur/se*, see Lauren Elkin, *Flâneuse: Women Walk the City* (London: Chatto & Windus, 2017). For Debord's theories, see Guy Debord, 'Theory of the *Dérive*', *Les Lèvres Nues*, 9 (November 1956).
3. The example of the tick is taken from Jakob von Uexküll and Georg Kriszat, *Streifzüge durch die Umwelten von Tieren und Menschen. Ein Bilderbuch unsichtbarer Welten* (Hamburg: Rowohlt, 1956). (English

edition: Jakob von Uexküll, *A Foray Into the Worlds of Animals and Humans: With a Theory of Meaning*, translated by Joseph D. O'Neil (Minneapolis /London: University of Minnesota Press, 2010).)

The elaboration given here is from Giorgio Agamben, *The Open: Man and Animal*, translated from Italian by Kevin Attell (Stanford, CA: Stanford University Press, 2004).

4. The creation of these images was supported by Nome Gallery, as part of *Failing to Distinguish Between a Tractor Trailer and the Bright White Sky,* an exhibition held in Berlin, April–July 2017. For more on the exhibition, visit https://nomegallery.com/exhibitions/failing-to-distinguish-between-a-tractor-trailer-and-the-bright-white-sky/.
5. A. M. Turing, 'Computing Machinery and Intelligence', *Mind*, 49 (1950), pp. 433–60.
6. H. F. Harlow, H. Uehling and A. H. Maslow, 'Comparative Behavior of Primates. I. Delayed Reaction Tests on Primates from the Lemur to the Orang-Outan', *Journal of Comparative Psychology*, 13(3) (1932), pp. 313–43. In the original study, all the apes have familiar as well as scientific names, some of dubious provenance. Alongside Charlotte, other participants included an orang-utan called Jiggs, and a number of macaques called Sourface, Sooty, Blackie and Kewpie (a baby doll) – as well as a quintet labelled merely Macaques I–V. The mandrills faired a little better; one pair, presumably partners, were called Socrates and Xanthippe.
7. Benjamin B. Beck, 'A Study of Problem Solving by Gibbons', *Behaviour*, 28(1–2), 1 January 1967, p. 95.
8. For this and many other accounts of experimental design in animal intelligence experiments, see Frans de Waal, *Are We Smart Enough To Know How Smart Animals Are?* (London: Granta, 2016).
9. Details of Jenny's life and her association with Darwin are drawn from 'Portrait of Jenny', Zoological Society of London, 1 June 2008; https://www.zsl.org/blogs/artefact-of-the-month/portrait-of-jenny.
10. For the history of Gallup's development of the mirror test, see Chelsea Wald, 'What Do Animals See in a Mirror?', *Nautilus*, May 2014; http://nautil.us/issue/13/symmetry/what-do-animals-see-in-a-mirror.
11. From Gordon G. Gallup, 'Chimpanzees: Self-Recognition', *Science*, 167(3914), 2 January 1970, pp. 86–7.
12. For dolphin tests, see K. Marten and S. Psarakos, 'Evidence of Self-Awareness in the Bottlenose Dolphin (Tursiops truncatus)', in S. T. Parker, R. W. Mitchell and M. L. Boccia (eds), *Self-Awareness in Animals and Humans: Developmental Perspectives* (Cambridge: Cambridge University Press, 1994), pp. 361–79. For orcas, see F. Delfour, K. Marten,

'Mirror Image Processing in Three Marine Mammal Species: Killer Whales (Orcinus orca), False Killer Whales (Pseudorca crassidens) and California Sea Lions (Zalophus californianus)', *Behavioural Processes*, 53(3), May 2001, pp. 181–90. For magpies, see H. Prior, A. Schwarz and O. Güntürkün, 'Mirror-Induced Behavior in the Magpie (Pica pica): Evidence of Self-Recognition', *PLoS Biology*, 6(8), August 2008, p. e202. For ants, see Marie-Claire Cammaerts Tricot and Roger Cammaerts, 'Are Ants (Hymenoptera, Formicidae) Capable of Self Recognition?', *Journal of Science*, 5(7), January 2015, pp. 521–32.

13. For 'dolphin porno tapes', see Chelsea Wald, 'What Do Animals See in a Mirror?', *Nautilus*, May 2014; http://nautil.us/issue/13/symmetry/what-do-animals-see-in-a-mirror. For the follow-up studies, see D. Reiss and L. Marino, 'Mirror Self-Recognition in the Bottlenose Dolphin: A Case of Cognitive Convergence', *Proceedings of the National Academy of Sciences*, 98(10), June 2001, pp. 5937–42.
14. D. J. Povinelli, 'Failure to Find Self-Recognition in Asian Elephants (Elephas maximus) in Contrast to Their Use of Mirror Cues to Discover Hidden Food', *Journal of Comparative Psychology*, 103(2), 1989, pp. 122–31; doi:10.1037/0735-7036.103.2.122.
15. J. M. Plotnik, F. B. M. de Waal and D. Reiss, 'Self-Recognition in an Asian Elephant', *Proceedings of the National Academy of Sciences of the USA*, 103(45), 7 November 2006, pp. 17053–7.
16. '2014 Ten Worst Zoos For Elephants', In Defense of Animals press release, 29 December 2016; https://www.idausa.org/assets/files/cam paign/Elephants/2014 Ten Worst Zoos for Elephants 2014- In Defense of Animals.pdf.
17. For the mirror test videos, see https://www.pnas.org/content/suppl/2006/10/26/0608062103.DC1. For accounts of her life, see Brad Hamilton, 'Happy the Elephant's Sad Life Alone at the Bronx Zoo', *New York Post*, 30September2012;https://nypost.com/2012/09/30/happy-the-elephants-sad-life-alone-at-the-bronx-zoo/.
18. For more on Happy's history, and the timeline of the Nonhuman Rights Project's case, see Chapter 9, as well as Happy's client page at the NhRP's website: https://www.nonhumanrights.org/client-happy/.
19. T. Suddendorf and E. Collier-Baker, 'The Evolution of Primate Visual Self-Recognition: Evidence of Absence in Lesser Apes', *Proceedings of the Royal Society B: Biological Sciences*, 276(1662), 7 May 2009, pp. 1671–7.
20. A. Z. Rajala, K. R. Reininger, K. M. Lancaster and L. C. Populin, 'Rhesus Monkeys (*Macaca mulatta*) Do Recognize Themselves in the Mirror:

Implications for the Evolution of Self-Recognition', *PLoS ONE* 5(9), 29 September 2010, e12865; https://doi.org/10.1371/journal.pone.0012865.

21. D. J. Shillito, G. G. Gallup Jr, and B. B. Beck, 'Factors Affecting Mirror Behavior in Western Lowland Gorillas, Gorilla gorilla', *Animal Behaviour*, 57(5), May 1999, pp. 999–1004.
22. F. G. P. Patterson and R. H. Cohn, 'Self-Recognition and Self-Awareness in Lowland Gorillas', in S. T. Parker, R. W. Mitchell and M. L. Boccia (eds), *Self-Awareness in Animals and Humans: Developmental Perspectives* (Cambridge: Cambridge University Press, 1994), pp. 273–90.
23. Melinda R. Allen, 'Mirror Self-Recognition in a Gorilla (Gorilla gorilla gorilla)', MSc Thesis, Florida International University, April 2007.
24. Patterson and Cohn, 'Self-Recognition and Self-Awareness in Lowland Gorillas', in Parker, Mitchell and Boccia (eds), *Self-Awareness in Animals and Humans: Developmental Perspectives*.
25. T. Broesch, T. Callaghan, J. Henrich, C. Murphy, and P. Rochat, 'Cultural Variations in Children's Mirror Self-Recognition', *Journal of Cross-Cultural Psychology*, 42(6), August 2011, pp. 1018–29.
26. For Inky's bid for freedom, see Dan Bilefsky, 'Inky the Octopus Escapes From a New Zealand Aquarium', *New York Times*, 13 April 2016; https://www.nytimes.com/2016/04/14/world/asia/inky-octopus-new-zealand-aquarium.html. For Sid's escapades, see Kathy Marks, 'Legging It: Evasive Octopus Who Has Been Allowed to Look for Love', *The Independent*, 14 February 2009; https://www.independent.co.uk/environment/nature/legging-it-evasive-octopus-who-has-been-allowed-to-look-for-love-1609168.html. For tales of octopus saboteurs, see Bob Pool, 'Did this Mollusk Open a Bivalve', *Los Angeles Times*, 27 February 2009; https://www.latimes.com/archives/la-xpm-2009-feb-27-me-octopus27-story.html. For Otto and his home redecoration attempts, see 'Otto the Octopus Wreaks Havoc', *The Telegraph*, 31 October 2008; https://www.telegraph.co.uk/news/newstopics/howaboutthat/3328480/Otto-the-octopus-wrecks-havoc.html.
27. For these and other anecdotes about cephalopod behaviour, see Peter Godfrey-Smith, *Other Minds: The Octopus, the Sea, and the Deep Origins of Consciousness* (New York: Farrar, Straus and Giroux, 2016). For octopus facial recognition, see R. C. Anderson, J. A. Mather, M. Q. Monette and S. R. M. Zimsen, 'Octopuses (*Enteroctopus dofleini*) Recognize Individual Humans', *Journal of Applied Animal Welfare Science*, 13(3), 18 June 2010, pp. 261–72; https://doi.org/10.1080/10888705.2010.483892.
28. Adrian Tchaikovsky, *Children of Time* (London: Tor Books, 2015).

29. This account of Smuts' studies, as well as her direct quotes, are drawn from Barbara Smuts, 'Encounters with Animal Minds', *Journal of Consciousness Studies* 8(5–7), January 2001, pp. 293–309.
30. These reflections are found in Jane Goodall, 'Primate Spirituality', in the *Encyclopedia of Religion and Nature* (London & New York: Continuum, 2005).

## 2. WOOD WIDE WEBS

1. For an overview of these processes, see Tom Reimchen, 'Salmon Nutrients, Nitrogen Isotopes and Coastal Forests', *Ecoforestry*, Fall 2001; https://web.uvic.ca/~reimlab/reimchen_ecoforestry.pdf.
2. Richard Brautigan, 'All Watched Over by Machines of Loving Grace', 1967.
3. Rafi Letzter, 'There Are Plants and Animals on the Moon Now (Because of China)', Space.com, 4 January 2019; https://www.space.com/42905-china-space-moon-plants-animals.html.
4. H. M. Appel and R. B. Cocroft, 'Plants Respond to Leaf Vibrations Caused by Insect Herbivore Chewing', *Oecologia*, 175, 2 July 2014, pp. 1257–66; https://doi.org/10.1007/s00442-014-2995-6.
5. For more on these open questions, see R. Lederer, 'The Mysteries of Eggs', *Ornithology: The Science of Birds*, 25 June 2019; https://ornithology.com/the-mysteries-of-eggs; and G. E. Hutchinson, 'The Paradox of the Plankton', *American Naturalist*, 95(882), May–June 1961, pp. 137–45.
6. For Monica Gagliano's account of this and subsequent experiments, see her book *Thus Spoke the Plant* (Berkeley, CA: North Atlantic Books, 2018). For the mimosa experiment in the scientific literature, see M. Gagliano, M. Renton, M. Depczynski et al., 'Experience Teaches Plants to Learn Faster and Forget Slower in Environments Where it Matters', *Oecologia*, 175(1), May 2014, pp. 63–72; https://doi.org/10.1007/s00442-013-2873-7.
7. Gagliano included repetition of many existing studies in her experiment; an overview of these can be found in Charles I. Abramson, Ana M. Chicas-Mosier, 'Learning in Plants: Lessons from *Mimosa pudica*', *Frontiers in Psychology*, 7(417), 31 March 2016; https://www.frontiersin.org/article/10.3389/fpsyg.2016.00417.
8. For research on plant smells and sniffing, see Anjel M. Helms, Consuelo M. De Moraes, John F. Tooker and Mark C. Mescher, 'Insect Odors and Plant Defense', *Proceedings of the National Academy of Sciences*, 110 (1), 2 January 2013, pp. 199–204; https://doi.org/10.1073/pnas.1218606110; and M. Mescher and C. De Moraes, 'Pass the Ammunition', *Nature*,

510, 11 June 2014, pp. 221–2; https://doi.org/10.1038/510221a. For the Dodder vine, see Justin B. Runyon, Mark C. Mescher, Consuelo M. De Moraes, 'Volatile Chemical Cues Guide Host Location and Host Selection by Parasitic Plants', *Science*, 313(5795), 29 September 2006, pp. 1964–7; https://doi.org/10.1126/science.1131371. For decision-making in plants, see M. Gruntman, D. Groß, M. Májeková et al., 'Decision-Making in Plants under Competition', *Nature Communications*, 8(2235), 21 December 2017; https://doi.org/10.1038/s41467-017-02147-2. For plant responses to aggression, see A. D. Zinn, D. Ward and K. P. Kirkman, 'Inducible Defences in *Acacia Sieberiana* in Response to Giraffe Browsing', *African Journal of Range and Forage Science*, 24(3), October 2007, pp. 123–9; DOI: 10.2989/AJRFS.2007.24.3.2.295, and M. J. Couvillon, H. Al Toufailia, T. M. Butterfield, F. Schrell, F. L. W. Ratnieks and R. Schürch, 'Caffeinated Forage Tricks Honeybees into Increasing Foraging and Recruitment Behaviors', *Current Biology*, 25(21), 2 November 2015, pp. 2815–18; DOI:10.1016/j.cub.2015.08.052. For proprioception, see O. Hamant and B. Moulia, 'How Do Plants Read Their Own Shapes?', *New Phytologist*, 212(2), October 2016, pp. 333–7; DOI: 10.1111/nph.14143. For plant families, see Richard Karban, Kaori Shiojiri, Satomi Ishizaki, William C. Wetzel and Richard Y. Evans, 'Kin Recognition Affects Plant Communication and Defence', *Proceedings of the Royal Society B: Biological Sciences*, 280(1756), 7 April 2013; http://doi.org/10.1098/rspb.2012.3062.

9. For an in-depth study of Pando, see Jeffry B. Mitton and Michael C. Grant, 'Genetic Variation and the Natural History of Quaking Aspen: The Ways in which Aspen Reproduces Underlie its Great Geographic Range, High Levels of Genetic Variability, and Persistence', *BioScience*, 46(1), January 1996, pp. 25–31; https://doi.org/10.2307/1312652.
10. P. C. Rogers and D. J. McAvoy, 'Mule Deer Impede Pando's Recovery: Implications for Aspen Resilience from a Single-Genotype Forest', *PLOS ONE*, 13(10), 17 October 2018, e0203619; DOI:10.1371/journal.pone.0203619.
11. Merlin Sheldrake, *Entangled Life: How Fungi Make Our Worlds, Change Our Minds and Shape Our Futures* (London: The Bodley Head, 2020). I am deeply indebted to Merlin for much of my understanding of fungal life, the origins of the Wood Wide Web, and many of the examples and explanations given here, although any errors in such understanding are my own.
12. For the mycorrhizal climate model, see Benjamin J. W. Mills, Sarah A. Batterman and Katie J. Field, 'Nutrient Acquisition by Symbiotic

Fungi Governs Palaeozoic Climate Transition', *Philosophical Transactions of the Royal Society B: Biological Sciences*, 373(1739), 5 February 2018; http://doi.org/10.1098/rstb.2016.0503. For the relationship between mycorrhiza and ancient climates, see K. Field, D. Cameron, J. Leake et al., 'Contrasting Arbuscular Mycorrhizal Responses of Vascular and Non-Vascular Plants to a Simulated Palaeozoic $CO_2$ Decline', *Nature Communications*, 3(835), 15 May 2012; https://doi.org/10.1038/ncomms1831.

13. For Simard's original paper, see S. Simard, D. Perry, M. Jones et al., 'Net Transfer of Carbon Between Ectomycorrhizal Tree Species in the Field', *Nature*, 388, 7 August 1997, pp. 579–82; https://doi.org/10.1038/41557. For Read's commentary, see D. Read, 'The Ties that Bind', *Nature*, 388, 7 August 1997, pp. 517–18; https://doi.org/10.1038/41426.
14. For more on Barabási's description of the internet, see A.-L. Barabási, 'The Physics of the Web', *Physics World*, 14(7), July 2001, pp. 33–8.

## 3. THE THICKET OF LIFE

1. Yes, my school had a cupboard full of radioactive sources. No, I don't know if this was sensible or even legal.
2. The symposium, entitled *Through Post-Atomic Eyes*, was held in October 2015, and convened by OCAD University and the University of Toronto. A catalogue of contributions, edited by Claudette Lauzon and John O'Brian, was published under the same title in 2020 (Montreal, QC: McGill-Queen's University Press, 2020). For my address on protest and personal data, see 'Big Data, No Thanks', Booktwo.org, 2 November 2015; http://booktwo.org/notebook/big-data-no-thanks/.
3. For more on Karen Barad's work, see *Meeting the Universe Halfway: Quantum Physics and the Entanglement of Matter and Meaning* (Durham, NC: Duke University Press, 2007), or her essay 'On Touching – The Inhuman That Therefore I Am', *Differences*, 23(3), 1 December 2012, pp. 206–23; DOI:10.1215/10407391-1892943
4. 'In Earth's Hottest Place, Life Has Been Found in Pure Acid', BBC Future, August 2017; https://www.bbc.com/future/article/20170803-in-earths-hottest-place-life-has-been-found-in-pure-acid.
5. F. Gómez, B. Cavalazzi, N. Rodríguez et al., 'Ultra-Small Microorganisms in the Polyextreme Conditions of the Dallol Volcano, Northern Afar, Ethiopia', *Scientific Reports*, 9(1), May 2019; DOI:10.1038/s41598-019-44440-8.

6. J. Belilla, D. Moreira, L. Jardillier, G. Reboul, K. Benzerara, J. M. López-García, . . . P. López-García, 'Hyperdiverse Archaea Near Life Limits at the Polyextreme Geothermal Dallol Area', *Nature Ecology & Evolution*, 3(11), October 2019, pp. 1552–61; DOI:10.1038/s41559-019-1005-0.
7. For lithotrophic bacteria, see J. L. Sanz, N. Rodríguez, E. E. Díaz and R. Amils, 'Methanogenesis in the Sediments of Rio Tinto, an Extreme Acidic River', *Environmental Microbiology*, 13(8), August 2011, pp. 2336–41; DOI:10.1111/j.1462-2920.2011.02504.x. For the microbial mats of the Andes, see Elizabeth K. Costello, Stephan R. P. Halloy, Sasha C. Reed, Preston Sowell and Steven K. Schmidt, 'Fumarole-Supported Islands of Biodiversity within a Hyperarid, High-Elevation Landscape on Socompa Volcano, Puna de Atacama, Andes', *Applied and Environmental Microbiology*, 75(3), February 2009, pp. 735–47; DOI: 10.1128/AEM.01469-08. For an exploration of deep-sea vents, see H. W. Jannasch and M. J. Mottl, 'Geomicrobiology of Deep-Sea Hydrothermal Vents', *Science*, 229(4715), 23 August 1985, pp. 717–25; DOI:10.1126/science.229.4715.717.
8. N. Patterson, D. J. Richter, S. Gnerre, E. S. Lander and D. Reich, 'Genetic Evidence for Complex Speciation of Humans and Chimpanzees', *Nature*, 441(7097), 29 June 2006, pp. 1103–1108.
9. A. Durvasula and S. Sankararaman, 'Recovering Signals of Ghost Archaic Introgression in African Populations', *Science Advances*, 6(7), 12 February 2020; DOI: 10.1126/sciadv.aax5097.
10. Jon Mooallem, 'Neanderthals Were People, Too', *The New York Times*, 11 January 2017; https://www.nytimes.com/2017/01/11/magazine/neanderthals-were-people-too.html.
11. Elizabeth Kolbert, 'Sleeping with the Enemy', *New Yorker*, 8 August 2011; https://www.newyorker.com/magazine/2011/08/15/sleeping-with-the-enemy.
12. Watch Ljuben Dimkaroski play the Divje Babe Bone Flute at https://www.youtube.com/watch?v=AZCWFcyxUhQ. For more information about the European Music Archaeology project, visit http://www.emaproject.eu. For the Adagio performance, see https://www.classicfm.com/discover-music/instruments/flute/worlds-oldest-instrument-neanderthal-flute/.
13. A. W. G. Pike, D. L. Hoffmann, M. García-Diez, P. B. Pettitt, J. Alcolea, R. De Balbin, . . . J. Zilhão, 'U-Series Dating of Paleolithic Art in 11 Caves in Spain', *Science*, 336(6087), 15 June 2012, pp. 1409–13; DOI:10.1126/science.1219957.

14. D. L. Hoffmann, C. D. Standish, M. García-Diez, P. B. Pettitt, J. A. Milton, J. Zilhão, . . . A. W. G. Pike, 'U-Th Dating of Carbonate Crusts Reveals Neandertal Origin of Iberian Cave Art', *Science*, 359(6378), 23 February 2018, pp. 912–15.
15. Ed Yong, 'A Shocking Find in a Neanderthal Cave in France', *The Atlantic*, 25 May 2016; https://www.theatlantic.com/science/archive/2016/05/the-astonishing-age-of-a-neanderthal-cave-construction-site/484070/.
16. J. Gresky, J. Haelm and L. Clare, 'Modified Human Crania From Göbekli Tepe Provide Evidence For a New Form of Neolithic Skull Cult', *Science Advances*, 3(6), June 2017, e1700564; DOI:10.1126/sciadv.1700564.
17. M. Krings, A. Stone, R. W. Schmitz, H. Krainitzki, M. Stoneking and S. Pääbo et al., 'Neandertal DNA Sequences and the Origin of Modern Humans', *Cell*, 90(1), 11 July 1997, pp. 19–30.
18. K. Prüfer, F. Racimo, N. Patterson, et al., 'The Complete Genome Sequence of a Neanderthal From the Altai Mountains', *Nature*, 505, 2 January 2014, pp. 43–9.
19. S. R. Browning, B. L. Browning, Y. Zhou, S. Tucci and J. M. Akey, 'Analysis of Human Sequence Data Reveals Two Pulses of Archaic Denisovan Admixture', *Cell*, 173(1), 22 March 2018, pp. 53–61.e9; DOI: 10.1016/j.cell.2018.02.031.
20. For Tibetan adaptations from Denisovans, see E. Huerta-Sánchez, X. Jin, Asan, et al., 'Altitude Adaptation in Tibetans Caused by Introgression of Denisovan-like DNA', *Nature*, 512(7513), 14 August 2014, pp. 194–7; DOI:10.1038/nature13408. For Greenland, see F. Racimo, D. Gokhman, M. Fumagalli et al., 'Archaic Adaptive Introgression in TBX15/WARS2', *Molecular Biology and Evolution*, 34(3), 1 March 2017, pp. 509–24; DOI:10.1093/molbev/msw283. For 'ghost' populations and migration into Africa, see F. A. Villanea and J. G. Schraiber, 'Multiple Episodes of Interbreeding Between Neanderthals and Modern Humans', *Nature Ecology & Evolution*, 3(1), 26 May 2019, pp. 39–44, and A. Gibbons, 'Prehistoric Eurasians Streamed into Africa, Genome Shows', *Science*, 350(6257), 9 October 2015, p. 149; DOI: 10.1126/science.350.6257.149.
21. V. Slon, F. Mafessoni, B. Vernot, et al., 'The Genome of the Offspring of a Neanderthal Mother and a Denisovan Father', *Nature*, 561(7721), September 2018, pp. 113–16; DOI: 10.1038/s41586-018-0455-x.
22. Matthew Warren, 'Mum's a Neanderthal, Dad's a Denisovan: First Discovery of an Ancient-Human Hybrid', *Nature*, 22 August 2018; https://www.nature.com/articles/d41586-018-06004-0.

23. For chimpanzee oxytocin levels, see R. M. Wittig, C. Crockford, T. Deschner, K. E. Langergraber, T. E. Ziegler and K. Zuberbuhler, 'Food Sharing is Linked to Urinary Oxytocin Levels and Bonding in Related and Unrelated Wild Chimpanzees', *Proceedings of the Royal Society B: Biology*, 281(1778), 7 March 2014, 20133096. For Bonobos, see T. Kano, 'Social Behavior Of Wild Pygmy Chimpanzees (*Pan paniscus*) of Wamba: A Preliminary Report', *Journal of Human Evolution*, 9(4), May 1980, pp. 243–54. For gorilla infant care, see S. Rosenbaum, L. Vigilant, C. W. Kuzawa and T. S. Stoinski, 'Caring for Infants is Associated with Increased Reproductive Success For Male Mountain Gorillas', *Scientific Reports*, 8(15223), 15 October 2018.
24. Y. Fernández-Jalvo, J. Carlos Díez, I. Cáceres and J. Russell, 'Human Cannibalism in the Early Pleistocene of Europe (Gran Dolina, Sierra de Atapuerca, Burgos, Spain)', *Journal of Human Evolution*, 37(3–4), September 1999, pp. 591–622.
25. Richardson explicitly discusses the coastline paradox in Lewis F. Richardson, 'The Problem of Contiguity: An Appendix to Statistics of Deadly Quarrels', *General Systems: Yearbook of the Society for the Advancement of General Systems Theory*, 6(139), 1961, pp. 139–87.
26. Mandelbrot built directly on Richardson's work, inspired by his work on coastlines. See B. Mandelbrot, 'How Long is the Coast of Britain? Statistical Self-Similarity and Fractional Dimension', *Science*, new series, 156(3775), 5 May 1967, pp. 636–8.
27. For Woese and Fox's original paper, see C. R. Woese and G. E. Fox, 'Phylogenetic Structure of the Prokaryotic Domain: The Primary Kingdoms', *Proceedings of the National Academy of Science, USA*, 74(11), November 1977, pp. 5088–90; DOI: 10.1073/pnas.74.11.5088. For more on the discovery of Archaea and HGT, see David Quammen, *The Tangled Tree: A Radical New History of Life* (London: William Collins, 2018).
28. S. Gilbert, J. Sapp, A. Tauber, with J. D. Thomson and S. C. Stearns (eds), 'A Symbiotic View of Life: We Have Never Been Individuals', *The Quarterly Review of Biology*, 87(4), December 2012, pp. 325–41; DOI:10.1086/668166.
29. For viral infection as a vector for DNA transmission in humans, see M. Varela, T. E. Spencer, M. Palmarini and F. Arnaud, 'Friendly Viruses: the Special Relationship between Endogenous Retroviruses and Their Host', *Annals of the New York Academy of Sciences*, 1178(1), October 2009, pp. 157–72; DOI:10.1111/j.1749-6632.2009.05002.x. For the development of the mammalian uterus, see S. Mi, X. Lee, X. Li, et al., 'Syncytin is a Captive Retroviral Envelope Protein Involved in Human

Placental Morphogenesis', *Nature*, 403, October 2000, pp. 785–9; https://doi.org/10.1038/35001608. For the – much debated – spirochete origin of sperm, see Lynn Margulis, *Origin of Eukaryotic Cells* (New Haven, CT: Yale University Press, 1970).

30. For more on the ghosts of symbiosis, see Connie Barlow, *The Ghosts of Evolution: Nonsensical Fruit, Missing Partners, and other Ecological Anachronisms* (New York: Basic Books, 2000).
31. For the connection between the microbiome and brain development, see J. Cryan and T. Dinan, 'Mind-Altering Microorganisms: The Impact of the Gut Microbiota on Brain and Behaviour', *Nature Reviews Neuroscience*, 13, October 2012, pp. 701–12; https://doi.org/10.1038/nrn3346. For the microbiome and illness in the elderly, see M. Claesson, I. Jeffery, S. Conde, et al., 'Gut Microbiota Composition Correlates With Diet and Health in the Elderly', *Nature*, 488(7410), August 2012, pp. 178–84; https://doi.org/10.1038/nature11319.
32. Based on genomic sequencing, *Rickettsia prowazekii*, the bacteria responsible for typhus, is the closest free-living relative of mitochondria. See S. G. E. Andersson, C. G. Kurland, A. Zomorodipour, J. O. Andersson, et al., 'The Genome Sequence of *Rickettsia prowazekii* and the Origin of Mitochondria', *Nature*, 396(6707), November 1998, pp. 133–40; DOI:10.1038/24094.
33. W. F. Doolittle, 'Phylogenetic Classification and the Universal Tree', *Science*, 284(5423), June 1999, pp. 2124–9; DOI:10.1126/science.284.5423.2134.
34. Lynn Margulis, *The Symbiotic Planet: A New Look At Evolution* (London: Weidenfeld & Nicolson, 1998).
35. For Facebook's fifty-one genders, see Brandon Griggs, 'Facebook Goes Beyond "Male" And "Female" With New Gender Options', CNN, 13 February 2014; https://edition.cnn.com/2014/02/13/tech/social-media/facebook-gender-custom/index.html. For the list of seventy-one, see Rhiannon Williams, 'Facebook's 71 Gender Options Come to UK Users', *The Telegraph*, 27 June 2014; https://www.telegraph.co.uk/technology/facebook/10930654/Facebooks-71-gender-options-come-to-UK-users.html. For Facebook's announcement of a free-form field for gender, see Facebook Diversity's post of 26 February 2015; https://www.facebook.com/facebookdiversity/posts/774221582674346.
36. Lynn Margulis and Dorion Sagan, *Origins of Sex: Three Billion Years of Genetic Recombination* (New Haven, CT: Yale University Press, 1986).

## 4. SEEING LIKE A PLANET

1. For the chelidonia festival, see Greek Wikipedia (Stilpon Kyriakidis), 'Language and folk culture of the modern Greeks', Association for the Distribution of Greek Letters, Athens, 1946. For Letnik, see 'Leto – Old European Culture', blog, 27 January 2017; http://oldeuropeanculture.blogspot.com/2017/01/leto.html. For reports of the 2020 swallow migration, see Tasos Kokkinidis, 'Thousands of Swallows and Other Birds Die in Greece After Migration', *Greek Reporter*, 9 April 2020; https://greece.greekreporter.com/2020/04/09/thousands-of-swallows-and-other-birds-die-in-greece-after-migration/, and Horatio Clare, 'Where Have All the Swallows Gone?', *Daily Mail*, 23 June 2020; https://www.dailymail.co.uk/news/article-8449275/Where-swallows-gone-Theyve-harbingers-summer-year-numbers-low.html.
2. E. Syrjämäki and T. Mustonen, *It is the Sámi who Own this Land – Sacred Landscapes and Oral Histories of the Jokkmokk Sámi* (Vaasa, Finland: Snowchange Cooperative, 2013).
3. Craig Welch, 'Half of All Species are on the Move – and We're Feeling It', *National Geographic*, 27 April 2017; https://www.nationalgeographic.com/news/2017/04/climate-change-species-migration-disease/.
4. For a comprehensive overview of the creation and implementation of global time, see Vanessa Ogle, *The Global Transformation of Time 1870–1950* (Cambridge, MA: Harvard University Press, 2015).
5. Tracy Kidder, *The Soul of a New Machine* (New York: Little, Brown and Company, 1981). The engineer comfortable with nanoseconds was Ed Rasala; the one who left for Vermont was Josh Rosen.
6. For a longer account of the Marsham Record, see I. D. Margary, 'The Marsham Phenological Record in Norfolk, 1736–1925, and Some Others', *Quarterly Journal of the Royal Meteorological Society*, 52(217), January 1926, pp. 27–54; DOI:10.1002/qj.49705221705. For analysis, see T. H. Sparks and P. D. Carey, 'The Responses of Species to Climate Over Two Centuries: An Analysis of the Marsham Phenological Record, 1736–1947', *Journal of Ecology*, 83(2), April 1995, p. 321; DOI:10.2307/2261570.
7. For more on the Marsham Record, phenology, caribou and technological time, see James Bridle, 'Phenological Mismatch', *e-flux* journal, June 2019; https://www.e-flux.com/architecture/becoming-digital/273079/phenological-mismatch/.
8. This calculation is taken from S. Loarie, P. Duffy, H. Hamilton, et al., 'The Velocity of Climate Change', *Nature*, 462, 24 December 2009, pp. 1052–5; https://doi.org/10.1038/nature08649.

9. For an excellent primer on Forest Migration, see L. F. Pitelka and the Plant Migration Workshop Group, 'Plant Migration and Climate Change', *American Scientist*, 85(5), September–October 1997, pp. 464–73.
10. S. Fei, J. M. Desprez, K. M. Potter, I. Jo, J. A. Knott and C. M. Oswalt, 'Divergence of Species Responses to Climate Change', *Science Advances*, 3(5), 17 May 2017, e1603055; DOI:10.1126/sciadv.1603055.
11. L. Kullman, 'Rapid recent range-margin rise of tree and shrub species in the Swedish Scandes', *Journal of Ecology*, 90(1), February 2002, pp. 68–77; DOI:10.1046/j.0022-0477.2001.00630.x
12. K. Tape, M. Sturm and C. Racine, 'The Evidence for Shrub Expansion in Northern Alaska and the Pan-Arctic', *Global Change Biology*, 12(4), April 2006, pp. 686–702; DOI:10.1111/j.1365-2486.2006.01128.x.
13. Clement Reid, *The Origin of the British Flora* (London: Dulau & Co., 1899).
14. R. G. Pearson, 'Climate Change and the Migration Capacity of Species', *Trends in Ecology & Evolution*, 21(3), March 2006, pp. 111–13; DOI: 10.1016/j.tree.2005.11.022.
15. K. D. Woods and M. B. Davis, 'Paleoecology of Range Limits: Beech in the Upper Peninsula of Michigan', *Ecology*, 70(3), June 1989, pp. 681–96; DOI:10.2307/1940219.
16. L. Kullman, 'Norway Spruce Present in the Scandes Mountains, Sweden at 8000 BP: New Light on Holocene Tree Spread', *Global Ecology and Biogeography Letters*, 5(2), March 1996, pp. 94–101; DOI:10.2307/2997447.
17. Schultz studies metabolic pathways in plants and animals, and draws attention to the way many chemical interactions occur along similar pathways in both kingdoms. See Jack C. Schultz, 'Plants are Just Very Slow Animals', March 2010; https://mospace.umsystem.edu/xmlui/handle/10355/6759.
18. Richard Powers's version was the first of this fable that I heard, but the story almost certainly originates as an original *Star Trek* episode, called 'Wink of an Eye'. It's also referenced by plant neurobiologist Stefano Mancuso, in an interview with Michael Pollan in 'The Intelligent Plant', *New Yorker*, 15 December 2013 (https://www.newyorker.com/magazine/2013/12/23/the-intelligent-plant), which is possibly where Powers heard it.
19. Charles Darwin, assisted by Francis Darwin, *The Power of Movement in Plants* (New York: D. Appleton and Company, 1887).
20. Letter from Charles Darwin to Alphonse de Candolle, 28 May 1880, in Francis Darwin (ed.), *The Life and Letters of Charles Darwin*, Vol. II (London: John Murray, 1887), p. 506.

21. For the military origins of the internet, see Yasha Levine, *Surveillance Valley: The Secret Military History of the Internet* (London: Icon Books, 2019). For a critique of Silicon Valley neoliberalism, see Richard Barbrook and Andy Cameron, 'The Californian Ideology', *Mute*, 1(3), 1 September 1995; http://www.metamute.org/editorial/articles/californian-ideology.
22. I. O. Buss, 'Bird Detection by Radar', *The Auk*, 63(3), 1 July 1946, pp. 315–18; https://doi.org/10.2307/4080116.
23. For an account of Varley and Lack's work, see A. D. Fox and P. D. L. Beasley, 'David Lack and the Birth of Radar Ornithology', *Archives of Natural History*, 37(2), October 2010, pp. 325–32. For the herring gull experiment, see D. Lack and G. C. Varley, 'Detection of Birds by Radar', *Nature*, 156(3963), 1945, p. 446.
24. For Lack's post-war experiments, see D. Lack, 'Migration across the North Sea Studied by Radar. Part 1. Survey through the Year', *Ibis*, 101, 1959, pp. 209–34 (and subsequent issues).
25. For examples of patterns, see https://birdcast.info/forecast/understanding-radar-and-birds-part-1/. For live images, visit https://radar.weather.gov/. For tracking individual birds, see B. Bruderer, T. Steuri and M. Baumgartner, 'Short-Range High-Precision Surveillance of Nocturnal Migration and Tracking of Single Targets', *Israel Journal of Zoology*, 41(3), 1995, pp. 207–20.
26. The story of NRO's offer to NASA is told in Dennis Overbye, 'Ex-Spy Telescope May Get New Identity as a Space Investigator', *New York Times*, 4 June 2012.
27. Joel Achenbach, 'NASA Gets Two Military Spy Telescopes for Astronomy', *Washington Post*, 4 June 2012.
28. Geoff Brumfiel, 'Trump Tweets Sensitive Surveillance Image of Iran', NPR, 30 August 2019.
29. For examples of these images, see 'Mapping the Mighty Mangrove', NASA, December 2019; https://landsat.gsfc.nasa.gov/mapping-the-mighty-mangrove/; M. Arekhi, A. Yesil, U. Y. Ozkan and F. B. Sanli, 'Detecting Treeline Dynamics in Response to Climate Warming Using Forest Stand Maps and Landsat Data in a Temperate Forest', *Forest Ecosystems*, 5 (23), 11 May 2018; https://doi.org/10.1186/s40663-018-0141-3; and G. Mancino, A. Nolè, F. Ripullone and A. Ferrara, 'Landsat TM Imagery and NDVI Differencing to Detect Vegetation Change: Assessing Natural Forest Expansion in Basilicata, Southern Italy', *iForest – Biogeosciences and Forestry*, 7(2), April 2014, pp. 75–84; https://doi.org/10.3832/ifor0909-007.

30. Jean Epstein, *Photogénie de l'impondérable* (Paris: Éditions Corymbe, 1935), quoted in Teresa Castro, 'The Mediated Plant', *e-flux Journal*, 102, September 2019; https://www.e-flux.com/journal/102/283819/the-mediated-plant/.

## 5. TALKING TO STRANGERS

1. A selection of artefacts from the exhibition, including many of those mentioned here, are recorded at the website of Il Paese di Cuccagna, I-DEA, Matera City of Culture 2019; https://idea.matera-basilicata2019.it/en/exhibition/il-paese-di-cuccagna.
2. For more spells and incantations, see Ernesto de Martino, *Magic: A Theory from the South*, translated by Dorothy Zinn (London: HAU Books, 2015), as well as de Martino's masterwork, *Il mondo magico*; published in English as *The World of Magic* (New York: Pyramid Communications, 1972).
3. 'Hymn to Malandrina', an edited selection of recordings of the calls from Viggianello, named after Francesco Caputo's most troublesome she-goat, is available at https://vimeo.com/368478999.
4. The article 'Meet the Greater Honeyguide, the Bird That Understands Humans' by Purbita Saha and Claire Spottiswoode, 22 August 2016, Audobon.org; https://www.audubon.org/news/meet-greater-honeyguide-bird-understands-humans, contains a recording of the *brrrrr-hm* call.
5. C. N. Spottiswoode, K. S. Begg and C. M. Begg, 'Reciprocal Signaling in Honeyguide–Human Mutualism', *Science*, 353(6297), July 2016, pp. 387–9; DOI:10.1126/science.aaf4885.
6. C. A. Zappes, A. Andriolo, P. C. Simões-Lopes and A. P. M. Di Beneditto, '"Human-Dolphin (*Tursiops Truncatus* Montagu, 1821) Cooperative Fishery" and its Influence on Cast Net Fishing Activities in Barra de Imbé/Tramandaí, Southern Brazil', *Ocean & Coastal Management*, 54(5), May 2011, pp. 427–32; DOI:10.1016/j.ocecoaman.2011.02.003.
7. B. M. Wood, H. Pontzer, D. A. Raichlen and F. W. Marlowe, 'Mutualism and Manipulation in Hadza–Honeyguide Interactions', *Evolution and Human Behavior*, 35(6), 1 November 2014, pp. 540–46; DOI:10.1016/j.evolhumbehav.2014.07.007.
8. For a recording of the Hadza whistle, see the film by Theresa Lucrisia-Bradley, *A Hadzabe Honey Hunt*, 2020; available at https://www.youtube.com/watch?v=ubfkz3p8t6c. For the Boran technique, see H. A. Isack and H.-U. Reyer, 'Honeyguides and Honey Gatherers: Interspecific

Communication in a Symbiotic Relationship', *Science*, 243(4896), 10 March 1989, pp. 1343–6; DOI:10.1126/science.243.4896.1343.

9. H. Kaplan, K. Hill, J. Lancaster and A. M. Hurtado, 'A Theory of Human Life History Evolution: Diet, Intelligence, and Longevity', *Evolutionary Anthropology*, 9(4), 16 August 2000, pp. 156–85; DOI:10.1002/1520-6505(2000)9:4<156::aid-evan5>3.0.co;2-7.
10. F. Max Müller, 'The Theoretical Stage, and the Origin of Language', Lecture 9 from *Lectures on the Science of Language* (1861; Ebook release, June 2010); https://www.gutenberg.org/files/32856/32856-pdf.pdf.
11. I first heard *cantu a tenòre*, and this explanation of their roles, in a public park in Cagliari, as related by one of the best-known contemporary groups, the Tenores di Bitti. To experience a little of the effect and the explanation, see the short film *Tenores di Bitti 'Mialinu Pira' a Belluno*; https://www.youtube.com/watch?v=mMddrMMqmoo.
12. Theodore Levin, *Where Rivers and Mountains Sing: Sound, Music, and Nomadism in Tuva and Beyond* (Bloomington, IN: Indiana University Press, 2006).
13. In recent decades, it has been observed that climate change is causing irrevocable breakdowns between species' evolved act of attuning to historic signals of seasonality and the actual availability of fodder and other needs (see James Bridle, 'Phenological Mismatch', *e-flux* journal, June 2019). It would be intriguing, if devastating, to discover if Tuvan songs contain an aural record of climatic conditions which no longer exist, just as the landscape painters of pre-twentieth century Europe documented unpolluted skies of a kind that none of us will ever know today.
14. Maurice Merleau-Ponty, *The Visible and the Invisible*, translated by Alphonso Lingis (Evanston, IL: Northwestern University Press, 1968). I am indebted to David Abram and *The Spell of the Sensuous: Perception and Language in a More-Than-Human World* (New York: Vintage, 1996) for his elucidation of Merleau-Ponty's work on language, as well as his thoughts on writing, the alphabet and phenomenology.
15. I am particularly grateful to Einar Sneve Martinussen for introducing me to the history of the Áltá action, and to Joar Nango's 'European Everything' crew for introducing me to joiking – and my apologies for my deeply reductive telling of the tradition. For more on the Áltá action and its cultural resonances, see the 'Let the River Flow. The Sovereign Will and the Making of a New Worldliness', published to coincide with an exhibition of the same name held at the Office for Contemporary Art Norway, Oslo, April–June 2018.

16. For these and other thoughts on the nature of written language and the alphabet, I am again indebted to David Abram's *The Spell of the Sensuous*.
17. These examples are from the linguist Sali Tagliamonte's book *Variationist Sociolinguistics: Change, Observation, Interpretation* (Hoboken, NJ: Wiley-Blackwell, 2011). The original sources are Ælfric, *c.* 1000, *Grammar*, xlviii. (Z) 279; Chaucer, *Prioress's Prologue and Tale* in CT. 5 (Harleian MS.); Shakespeare, *Much Ado About Nothing* (written 1598–9), Act IV, scene 1.
18. S. A. Tagliamonte and D. Denis, 'Linguistic Ruin? Lol! Instant Messaging and Teen Language', *American Speech*, 83(1), March 2008, pp. 3–34; DOI:10.1215/00031283-2008-001.
19. See, for example, Jennifer Lee, 'I Think, Therefore IM', *New York Times*, 19 September 2002; http://www.nytimes.com/2002/09/19/technology/circuits/19MESS.html.
20. In recounting anecdotes such as this, I can feel myself slipping back into technophobia and fears of a computationally mediated stupidity. At the same time, I find some amusement in Excel's dogged determination to relate the abstraction of genomic acronyms to terrestrial events, and a comfort in its repeated questions, over and over again, '*Is this what you really mean? Because, you know, humans mean a lot of different things.*'
21. Reuters, '"Master" and "Slave" Computer Labels Unacceptable, Officials Say', CNN.com, 26 November 2003; http://edition.cnn.com/2003/TECH/ptech/11/26/master.term.reut/.
22. Julia Horowitz, 'Twitter and J. P. Morgan are Removing "Master", "Slave" and "Blacklist" from Their Code', CNN Business, 3 July 2020; https://edition.cnn.com/2020/07/03/tech/twitter-jpmorgan-slave-master-coding/index.html.
23. Ron Eglash, 'Broken Metaphor: The Master–Slave Analogy in Technical Literature', *Technology and Culture*, 48(2), April 2007, pp. 360–69; DOI:10.1353/tech.2007.0066.
24. For figures on racial minorities in the tech industry, see Sam Dean and Johana Bhuiyan, 'Why are Black and Latino People Still Kept Out of the Tech Industry?', *Los Angeles Times*, 24 June 2020; https://www.latimes.com/business/technology/story/2020-06-24/tech-started-publicly-taking-lack-of-diversity-seriously-in-2014-why-has-so-little-changed-for-black-workers. For extensive accounts of racial and other bias within tech products, see Virginia Eubanks, *Automating Inequality: How High-Tech Tools Profile, Police, and Punish the Poor* (New York: St Martin's

Press, 2018) and Safiya Umoja Noble, *Algorithms of Oppression: How Search Engines Reinforce Racism* (New York: NYU Press, 2018).

25. The other two laws, first published in Clarke's *Profiles of the Future* (London: Victor Gollancz, 1962) are 'When a distinguished but elderly scientist states that something is possible, he is almost certainly right. When he states that something is impossible, he is very probably wrong' and 'The only way of discovering the limits of the possible is to venture a little way past them into the impossible.'
26. Even the act of naming something قلب produces interesting ructions in the infosphere: visiting the website for the project necessitates using Arabic characters rather than English ones, as it is located at https://nas.sr/قلب/ – an address made possible by multilingual standards which were only introduced to the web in 2008. That this long-delayed process was called 'internationalization' tells you all you need to know about the default, nationalistic assumptions of the internet.
27. 'Arabic Programming Language At Eyebeam: قلب Opens The World', AnimalNewYork.com, 24 January 2013; http://animalnewyork.com/2013/arabic-programming-language-at-eyebeam-قلب-opens-the-world/.
28. For examples of Piet programmes, see https://www.dangermouse.net/esoteric/piet.html. For Emojicode, visit https://www.emojicode.org/.
29. If you want to play with Brainfuck, an online composer is available at http://www.bf.doleczek.pl/.
30. For a full specification of Ook!, see https://www.dangermouse.net/esoteric/ook.html.
31. For marmoset understanding, see S. Verma, K. Prateek, K. Pandia, N. Dawalatabad, R. Landman, J. Sharma, M. Sur and H. A. Murthy, 'Discovering Language in Marmoset Vocalization', *Proceedings, Interspeech 2017*, 2426-2430; DOI: 10.21437/Interspeech.2017-842. To explore Google's archive of whale song, visit https://patternradio.withgoogle.com/. For prairie dog research, see Jeff Rice, 'Rodents' Talk Isn't Just "Cheep"', *Wired*, 6 October 2005; https://www.wired.com/2005/06/rodents-talk-isnt-just-cheep/.
32. For the story of Roah, see Konrad Lorenz, *King Solomon's Ring* (1952; Abingdon: Routledge, 2002), pp. 86–7 – but the whole book is wonderful.
33. This may be unfair to parrots, although Lorenz was writing before members of that species had participated in quite so many studies as they have today. In fact, African grey parrots have been shown to display cognitive competence, meaning they do use appropriate words in appropriate situations (see E. N. Colbert-White, H. C. Hall, D. M. Fragaszy, 'Variations in an African Grey Parrot's Speech Patterns Following Ignored and Denied

Requests', *Animal Cognition*, 19(3), May 2016, pp. 459–69; DOI:10.1007/s10071-015-0946-1). Alex, a grey parrot who worked with the animal psychologist Irene Pepperberg for more than thirty years, had a vocabulary of over a hundred words, could identify complex shapes and colours as well as abstract concepts, and responded to the emotional states of human researchers. His last words, which were the same ones he spoke every night to Pepperberg when she left the lab, were 'You be good, I love you. See you tomorrow.' Irene M. Pepperberg, *Alex & Me: How a Scientist and a Parrot Discovered a Hidden World of Animal Intelligence and Formed a Deep Bond in the Process* (London: HarperCollins, 2008). For a rather different take on parrot communication, seek out the music of Hatebeak, an American death metal band whose lead singer is a grey parrot called Waldo. In contrast to some of the zoomusicology portrayed earlier in this chapter, Hatebeak's music has been described as 'a jackhammer being ground in a compactor'.

34. The Dolphin Embassy is critically under-researched, but this excerpt from the catalogue for Ant Farm's 2004 retrospective at Berkeley Art Museum, along with links to further material is available at Greg.org: https://greg.org/archive/2010/06/01/cue-the-dolphin-embassy.html.
35. Lilly has a special place in most accounts of human–dolphin communication, and he is famous for the flooded laboratory he constructed, with Margaret Howe Lovatt, on the Caribbean island of St Thomas. There, they carried out a multi-year research programme on cetacean intelligence, which involved living in close contact with captive dolphins, as well as undertaking experiments involving sensory deprivation tanks (which Lilly invented), electrodes inserted into the dolphins' brains and forced injections of uncontrolled doses of LSD. While their work has been overshadowed by later, lurid tales of Lovatt's relationship with one of the dolphins, it revealed many aspects of cetacean physiognomy and ability which were previously unknown. None of that, however, justifies the coercion and abuse, however well-intentioned, which was part and parcel of their work.
36. Martín Abadi and David G. Andersen, 'Learning to Protect Communications with Adversarial Neural Cryptography', ArXiv.org, 2016; https://arxiv.org/abs/1610.06918.
37. For Facebook's own account of the experiments, see Mike Lewis, Denis Yarats, Devi Parikh and Dhruv Batra, 'Deal or No Deal? Training AI Bots to Negotiate', Facebook Engineering, 14 June 2017; https://engineering.fb.com/ml-applications/deal-or-no-deal-training-ai-bots-to-negotiate/, and Lewis et al., 'Deal or No Deal? End-to-End Learning for Negotiation

Dialogues', ArXiv.org, 16 June 2017; https://arxiv.org/pdf/1706.05125.pdf. For an account of the language variations produced, see Mark Wilson, 'AI is Inventing Languages Humans Can't Understand. Should We Stop It?', Fast Company, 14 July 2017; https://www.fastcompany.com/90132632/ai-is-inventing-its-own-perfect-languages-should-we-let-it.

38. For an overview of Ebonics and related Black idiolects, see John Russell Rickford and Russell John Rickford, *Spoken Soul: The Story of Black English* (New York: John Wiley, 2000). I'm indebted to Eva Barbarossa for her insights into computational language and the pointer towards pronoun forms.
39. 'The Author of the Acacia Seeds' (1974) is collected in Ursula K. Le Guin, *The Compass Rose* (London: Gollancz, 1983). Thanks to Matt Webb for bringing the story to my attention. It is at least partly responsible for most of this book.
40. Tyson Yunkaporta, *Sand Talk: How Indigenous Thinking Can Save the World* (Melbourne: Text Publishing, 2019). For an extract from the book concerning the liveliness of rocks, see 'Friday Essay: Lessons From Stone – Indigenous Thinking and the Law', The Conversation, 5 September 2019; https://theconversation.com/friday-essay-lessons-from-stone-indigenous-thinking-and-the-law-122617.
41. For an analysis of speech-like patterns in Instant Messaging, see K. Ferrara, H. Brunner and G. Whittemore, 'Interactive Written Discourse as an Emergent Register', *Written Communication*, 8(1), 1 January 1991, pp. 8–34. For research on Arabic IM use, see D. Palfreyman and M. al Khalil, '"A Funky Language for Teenzz to Use": Representing Gulf Arabic in Instant Messaging', *Journal of Computer-Mediated Communication*, 9(1), 2003, pp. 23–44.
42. Naomi S. Baron, 'Why Email Looks Like Speech: Proofreading Pedagogy and Public Face', in Jean Aitchison and Diana M. Lewis (eds), *New Media Language* (London: Routledge, 2003), pp. 85–94.

## 6. NON-BINARY MACHINES

1. Plato, *Apology*, 21a–d.
2. A. M. Turing, 'On Computable Numbers, With an Application to the *Entscheidungsproblem*' (1936), *Proceedings of the London Mathematical Society*, Series 2, 42, 1937, pp. 230–65; DOI:10.1112/plms/s2-42.1.230.
3. A. M. Turing, 'Systems of Logic Based on Ordinals', *Proceedings of the London Mathematical Society*, Series 2, 45, 1939, pp. 161–228.

4. B. Jack Copeland and Diane Proudfoot, 'Alan Turing's Forgotten Ideas in Computer Science', *Scientific American*, 280(4), April 1999, pp. 98–103; DOI: 10.1038/scientificamerican0499-98.
5. For a full account of the wiring and behaviour of the tortoises – including the Carroll quote – see W. Grey Walter, 'An Imitation of Life', *Scientific American*, May 1950. In the preparation of this chapter, I am particularly indebted to Andrew Pickering, whose extensive work on the British-based cyberneticians Stafford Beer, Gordon Pask, Grey Walter and Ross Ashby can be found in numerous papers, and is summarized in *The Cybernetic Brain: Sketches of Another Future* (Chicago, IL: University of Chicago Press, 2011).
6. Walter, 'An Imitation of Life'.
7. Ibid.
8. Ibid.
9. The origin of the term is in Norbert Wiener, *Cybernetics: Or Control and Communication in the Animal and the Machine* (Cambridge, MA: MIT Press, 1948).
10. For a description of the homeostat and its operation, see W. R. Ashby, 'Design for a Brain', *Electrical Engineering*, 20, December 1948, pp. 379–83.
11. Walter, 'An Imitation of Life'.
12. Ashby, 'Design for a Brain'.
13. For Andrew Pickering's take on cybernetic minds, see A. Pickering, 'Beyond Design: Cybernetics, Biological Computers and Hylozoism', *Synthese*, 168(3), June 2009, pp. 469–9; https://doi.org/10.1007/s11229-008-9446-z.
14. Sadly, the *U* doesn't stand for anything exciting like 'Universal' or 'Unknowable'. It refers to Beer's mathematical description of the U-machine's operation and its relation to the T-Machine, which handled inputs, and the V-machine, which handled outputs.
15. S. Beer, 'Towards the Automatic Factory', paper presented to a symposium at the University of Illinois in June 1960, reprinted in Roger Harnden and Allenna Leonard (eds), *How Many Grapes Went into the Wine? Stafford Beer on the Art and Science of Holistic Management* (New York: John Wiley, 1994), pp. 163–225.
16. A. Pickering, 'The Science of the Unknowable: Stafford Beer's Cybernetic Informatics', *Kybernetes*, 33(3/4), March 2004, pp. 499–521; https://doi.org/10.1108/03684920410523535.
17. There are not a lot of details available on Beer's pond experiments, which are briefly summarized in the edited collection of his work, *How*

*Many Grapes Went into the Wine? Stafford Beer on the Art and Science of Holistic Management*. I am indebted to Matt Webb for first bringing it to my attention and for 'putting a pond through business school'.

18. For slime mould's memory abilities, see T. Saigusa, A. Tero, T. Nakagaki and Y. Kuramoto, 'Amoebae Anticipate Periodic Events', *Physical Review Letters*, 100(1), 11 January 2008, no. 018101; https://doi.org/10.1103/PhysRevLett.100.018101.
19. A. Tero, S. Takagi, T. Saigusa, K. Ito, D. P. Bebber, M. D. Fricker and T. Nakagaki, 'Rules for Biologically Inspired Adaptive Network Design', *Science*, 327(5964), 22 January 2010, pp. 439–42; DOI:10.1126/science.1177894.
20. This is a generalization of Leonhard Euler's Seven Bridges of Königsberg problem, which we encountered in our discussion of fungi and network theory in Chapter 2.
21. L. Zhu, S.-J. Kim, M. Hara and M. Aono, 'Remarkable Problem-Solving Ability of Unicellular Amoeboid Organism and its Mechanism', *Royal Society Open Science*, 5(12), 19 December 2018, no. 180396; DOI: 10.1098/rsos.180396.
22. Actually, 840 times more, but come on, it's a slime mould. A four-city network has only 3 × 2 × 1/2 = 3 possible routes ((n–1)! / 2), but an eight-city network has 7 × 6 × 5 × 4 × 3 × 2 × 1/2 = 2,520 different possible routes.
23. For an overview of the history of memristor research, see 'The Memristor Revisited', *Nature Electronics*, 1(261), May 2018; https://doi.org/10.1038/s41928-018-0083-3.
24. For research into the memristic capabilities of slime moulds, see E. Gale, A. Adamatzky and B. de Lacy Costello, 'Slime Mould Memristors', *BioNanoScience*, 5(1), 1 March 2015, pp. 1–8. For plants and animals as memristors, see A. Adamatzky, S. L. Harding, V. Erokhin and R. Mayne et al., 'Computers from Plants We Never Made: Speculations', ArXiv:1702.08889 [cs.ET], 28 February 2017.
25. This theory of billiard ball circuits was published in E. Fredkin and T. Toffoli, 'Conservative Logic', *International Journal of Theoretical Physics*, 21(3–4), April 1982, pp. 219–53; DOI:10.1007/BF01857727.
26. For the crab computer, see Y.-P. Gunji, Y. Nishiyama and A. Adamatzky, 'Robust Soldier Crab Ball Gate', *Complex Systems*, 20(2), 15 June 2011; DOI:10.1063/1.3637777.
27. K. Nakajima, H. Hauser, T. Li and R. Pfeifer, 'Information Processing via Physical Soft Body', *Scientific Reports*, 5, 27 May 2015, no.10487; https://doi.org/10.1038/srep10487.

28. C. Fernando and S. Sojakka, 'Pattern Recognition in a Bucket', *Lecture Notes in Computer Science*, September 2003, pp. 588–97; DOI:10.1007/978-3-540-39432-7_63.
29. For more on Lukyanov and the Soviet hydraulic computers, see O. Solovieva, 'Water Machine Computers', *Science and Life*, 4, 2000; https://www.nkj.ru/archive/articles/7033/, and the archives of the Moscow Polytechnic Museum at http://web.archive.org/web/20120328115234/http://rus.polymus.ru/?h=relics&rel_id=9&mid=&aid=9&aid_prev=9.
30. For a fuller account of the Mississippi Basin model, see Kristi Cheramie, 'The Scale of Nature: Modeling the Mississippi River', *Places Journal*, March 2011; https://placesjournal.org/article/the-scale-of-nature-modeling-the-mississippi-river/.
31. For Rob Holmes's and the Dredge Research Collaborative's work on the Mississippi and other landscapes, see http://m.ammoth.us/.
32. For Zach Blas's work, see https://zachblas.info/works/queer-technologies/.
33. 'A Genderqueer Activist Explains What it Means to be Nonbinary on the Gender Spectrum', *Vox*, 15 June 2016; https://www.vox.com/2016/6/15/11906704/genderqueer-nonbinary-lgbtq.
34. For an introduction to open-source philosophy, see Lawrence Lessig, 'Open Code and Open Societies: Values of Internet Governance', Sibley Lecture at the University of Georgia, 16 February 1999; or, for a fictional account of its actual implementation, I recommend Cory Doctorow, *Walkaway* (New York: Macmillan, 2017). For examples of distributed processing initiatives, see https://setiathome.berkeley.edu and https://foldingathome.org/. The social networks Mastodon and Scuttlebutt, the Beaker web browser and Jitsi.org web conferencing are good examples of federated and peer-to-peer network projects.
35. For a description of the Optometrist Algorithm, see E. A. Baltz, E. Trask, M. Binderbauer, et al., 'Achievement of Sustained Net Plasma Heating in a Fusion Experiment with the Optometrist Algorithm', *Scientific Reports*, 7(6425), 25 July 2017; https://doi.org/10.1038/s41598-017-06645-7. For more on this kind of human and learning machine cooperation, see James Bridle, *New Dark Age: Technology and the End of the Future* (London: Verso, 2018), Chapter 4, 'Calculation'.
36. For a full description of the Cockroach Controlled Mobile Robot, see Garnet Hertz, *Cockroach Controlled Mobile Robot: Control and Communication in the Animal and the Machine*, video, Concept Lab, 2008; http://www.conceptlab.com/roachbot/.
37. For more on Project Cybersyn, see Eden Medina, *Cybernetic Revolutionaries: Technology and Politics in Allende's Chile* (Cambridge, MA:

MIT Press, 2011). Beer's Manchester Business School lecture of 1974 is available at https://www.youtube.com/watch?v=e_bXlEvygHg.

## 7. GETTING RANDOM

1. For more information about the role of the *klepsydra* and related objects, see the dead-media historian Julian Dibbell's 'Info Tech of Ancient Athenian Democracy', available at http://www.alamut.com/subj/artiface/deadMedia/agoraMuseum.html.
2. See Kerry Tomlinson, 'How Random are "Random" Lottery Numbers?', Archer Intelligence, 30 March 2016; https://archerint.com/how-random-are-random-lottery-numbers/.
3. This is not to say that people haven't tried. In 1980, a TV host, a studio technician and a lottery official successfully rigged a balls-in-the-bucket machine belonging to the Pennsylvania Lottery, by ejecting latex paint into some of the balls to weigh them down, taking away $1.2 million (before being caught). Some lotteries now use X-ray machines and sophisticated weighing scales to test the balls before draws.
4. For a timeline of ERNIE's development, see 'Meet Ernie', National Savings and Investments; https://www.nsandi.com/ernie.
5. For the origin of Lavarand, see 'Welcome to Lavarand!', Silicon Graphics, archived at https://web.archive.org/web/19971210213248/http://lavarand.sgi.com/. For Cloudflare's implementation, see Joshua Liebow-Feeser, 'Randomness 101: LavaRand in Production', Cloudflare, 6 November 2017; https://blog.cloudflare.com/randomness-101-lavarand-in-production/. For Hotbits, see http://www.fourmilab.ch/hotbits/; and for Random.org, see https://www.random.org.
6. For this and other details of the development of Monte Carlo, as well as the history of John and Klári von Neumann, see George Dyson, *Turing's Cathedral: The Origins of the Digital Universe* (London: Allen Lane, 2012).
7. Marcel Duchamp, who we saw at work in an earlier chapter, was also a Monte Carlo veteran. He too had a 'system' which he employed at the casino, which involved throwing dice to decide where to bet at roulette. He claimed it was successful – although excruciatingly slow – and in 1924 produced a series of prints, called the Monte Carlo Bonds, which were simultaneously conceptual artworks and legal documents, bearing the value of his winnings.
8. John von Neumann, 'Various Techniques Used in Connection with Random Digits' in *Proceedings of a Symposium* held 29, 30 June and 1 July

1949, in Los Angeles, California, under the sponsorship of the RAND Corporation and the National Bureau of Standards, with the cooperation of the Oak Ridge National Laboratory; also published in A. S. Householder, G. E. Forsythe and H. H. Germond (eds), *Monte Carlo Method* (Washington DC: US Government Printing Office, 1951), National Bureau of Standards Applied Mathematics Series 12.

9. RAND Corporation, *A Million Random Digits with 100,000 Normal Deviates* (Glencoe, IL: Free Press, 1955; revised pbk edn Santa Monica, CA: RAND Corporation, 2001); https://www.rand.org/pubs/monograph_reports/MR1418.html. For details on the process involved in generating the numbers, see George W. Brown, 'History of RAND's random digits – Summary', in Householder, Forsythe and Germond (eds), *Monte Carlo Method*, pp. 31–2; https://www.rand.org/content/dam/rand/pubs/papers/2008/P113.pdf.
10. These and subsequent quotes are drawn from Kenneth Silverman, *Begin Again: A Biography of John Cage* (New York: Alfred Knopf, 2010; pbk, Evanston, IL: Northwestern University Press, 2012). I am indebted to this work and to Cage's own writings, particularly the collection *Silence: Lectures and Writings* (London: Marion Boyars, 1994).
11. Hiller instituted a pseudo-random subroutine to generate ICHING's random integers, called ML3DST. This involved combining algorithmically generated fractions which, while not quite satisfying our desire for truly random numbers, was extensively tested to ensure it was 'random enough' for the application. Hiller discusses ICHING and ML3DST in L. Hiller, 'Programming the I-Ching Oracle', *Computer Studies in the Humanities and Verbal Behavior*, 3(3), October 1970, pp. 130–43. I'm indebted to Tiffany Funk for her discussion and assistance with understanding this process.
12. For more on SAM and the Cybernetic Serendipity exhibition, see Jasia Reichardt (ed.), *Cybernetic Serendipity – The Computer and the Arts*, a Studio International special issue (London: Studio International Foundation, 1968).
13. For more details on the origin and design of these programs, see S. Husarik, 'John Cage and LeJaren Hiller: HPSCHD, 1969', *American Music*, 1(2), Summer 1983, pp. 1–21; DOI: 10.2307/3051496; Tiffany Funk, 'Zen and the Art of Software Performance: John Cage and Lejaren A. Hiller Jr.'s HPSCHD (1967–1969)', Thesis, University of Illinois at Chicago, 2016.
14. For Leibniz's 'Explanation of Binary Arithmetic', see 'Explication de l'Arithmétique Binaire', in C. I Gerhardt (ed.), *Die mathematische Schriften von Gottfried Wilhelm Leibniz* ('Mathematical Works'), Vol. 7,

pp. 223–7. An English translation is available at http://www.leibniz-translations.com/binary.htm. For his religious beliefs and his understanding of the I Ching, see J. A. Ryan, 'Leibniz' Binary System and Shao Yong's *Yijing*', *Philosophy East and West*, 46(1), January 1996, p. 59; DOI: 10.2307/1399337. For the marble computer, see Ravi P. Agarwal and Syamal K. Sen, *Creators of Mathematical and Computational Sciences* (New York: Springer, 2014).

15. For the Grants' studies in the Galapagos, see Peter R. Grant and B. Rosemary Grant, *How and Why Species Multiply: The Radiation of Darwin's Finches* (Princeton, NJ: Princeton University Press, 2011).
16. For John Tyler Bonner's work on randomness in evolution, with particular reference to the size and complexity of organisms, see his book *Randomness in Evolution* (Princeton, NJ: Princeton University Press, 2013). For a strong defence of randomness against the dominance of natural selection in evolutionary theory, see M. Lynch, 'The Frailty of Adaptive Hypotheses for the Origins of Organismal Complexity', *Proceedings of the National Academy of Sciences, USA*, 104 (Suppl 1), 15 May 2007, pp. 8597–604; DOI:10.1073/pnas.0702207104.
17. For an account of Haeckel's travels, investigations and art, see Andrea Wulf, *The Invention of Nature: The Adventures of Alexander von Humboldt, the Lost Hero of Science* (London: John Murray, 2016).
18. Mary Minihan, 'Was Citizens' Assembly best way to deal with abortion question?', *Irish Times*, 29 April 2017.
19. For an argument for sortition and assemblies in the political process, see David Van Reybrouck, *Against Elections: The Case for Democracy*, translated by Liz Waters (London: Bodley Head, 2016). For an overview of the scientific research into cognitive diversity, see H. Landemore, 'Deliberation, Cognitive Diversity, and Democratic Inclusiveness: An Epistemic Argument for the Random Selection of Representatives', *Synthese*, 190, May 2013, pp. 1209–31; https://doi.org/10.1007/s11229-012-0062-6; and L. Hong and S. E. Page, 'Groups of Diverse Problem Solvers Can Outperform Groups of High-Ability Problem Solvers', *Proceedings of the National Academy of Sciences, USA*, 101(46), 16 November 2004, pp. 16385–9; DOI: 10.1073/pnas.0403723101.
20. For examples of Tamil sortition, see 'All eyes on Kudavolai method of elections', *The Hindu*, 25 September 2010; https://www.thehindu.com/news/cities/chennai/All-eyes-on-Kudavolai-method-of-elections/article16046086.ece. For the Iroquois Confederacy, see Sergia C. Coffey, 'The Influence of the Culture and Ideas of the Iroquois Confederacy on

European Economic Thought', paper presented at the Eastern Economic Association Conference, Washington DC, 2016.

## 8. SOLIDARITY

1. For a longer account of Topsy's life and death, see Michael Daly, *Topsy: The Startling Story of the Crooked-Tailed Elephant, P. T. Barnum, and the American Wizard, Thomas Edison* (New York: Atlantic Monthly Press, 2013).
2. For the French animal trials and the history of Bartholomew Chassenée, see Jeffrey St Clair, 'Let Us Now Praise Infamous Animals', Counterpunch.org, 3 August 2018. For an in-depth history of animal trials in medieval Europe and beyond, see E. P. Evans, *The Criminal Prosecution and Capital Punishment of Animals* (Clark, NJ: The Lawbook Exchange, 2009, reprint edn).
3. Jason Hribal, *Fear of the Animal Planet: The Hidden History of Animal Resistance* (Chico, CA: AK Press, 2011).
4. Peter Kropotkin, *Mutual Aid: A Factor of Evolution* (1902; London: Freedom Press, 2009, pbk edn).
5. For research on red deer democracy, see L. Conradt and T. J. Roper, 'Democracy in Animals: The Evolution of Shared Group Decisions', *Proceedings of the Royal Society B: Biological Sciences*, 274(1623), 22 September 2007; http://doi.org/10.1098/rspb.2007.0186. For the buffalo, see D. Wilson, 'Altruism and Organism: Disentangling the Themes of Multilevel Selection Theory', *The American Naturalist*, 150(S1), July 1997, pp. S122–S134; DOI:10.1086/286053. For pigeons with GPS backpacks, see M. Nagy, Z. Ákos, D. Biro and T. Vicsek, 'Hierarchical Group Dynamics in Pigeon Flocks', *Nature*, 464(7290), 8 April 2010, pp. 890–93; https://doi.org/10.1038/nature08891; and for cockroaches, see J.-M. Amé, J. Halloy, C. Rivault, C. Detrain, J. L. Deneubourg, 'Collegial Decision Making Based on Social Amplification Leads to Optimal Group Formation', *Proceedings of the National Academy of Sciences, USA*, 103(15), 11 April 2006, pp. 5835–40; DOI: 10.1073/pnas.0507877103.
6. For more on Frisch, Lindauer, the waggle dance and its political implications, see Thomas D. Seeley, *Honeybee Democracy* (Princeton, NJ: Princeton University Press, 2010), to which work I am deeply indebted.
7. When Hofstadter wrote the book, many of the assumptions he makes about bee swarms and ant colonies were still conjectures, but they have

been confirmed by subsequent studies. For the original, see Douglas R. Hofstadter, *Gödel, Escher, Bach: An Eternal Golden Braid* (New York: Basic Books, 1979). For an update, see J. A. R. Marshall, R. Bogacz, A. Dornhaus, R. Planqué, T. Kovacs and N. R. Franks, 'On Optimal Decision-Making in Brains and Social Insect Colonies', *Journal of the Royal Society, Interface*, 6(40), 6 November 2009; http://doi.org/10.1098/rsif.2008.0511.

8. For a full explanation of BeeAdHoc, see H. Wedde, M. Farooq, T. Pannenbaecker, B. Vogel, C. Mueller, J. Meth and R. Jeruschkat, 'BeeAdHoc: An Energy Efficient Routing Algorithm for Mobile Ad Hoc Networks Inspired by Bee Behavior', paper for the Genetic and Evolutionary Computation Conference (GECCO), 25–29 June 2005, Washington DC, pp. 153–60; DOI: 10.1145/1068009.1068034.
9. Find out more about the Nonhuman Rights Project at https://www.nonhumanrights.org/who-we-are/.
10. A full transcript of the arguments before the court, *In the Matter of: Nonhuman Rights Project* v. *James Breheny, et al.*, Supreme Court of the State of New York, 23 September 2019, can be found at https://www.nonhumanrights.org/content/uploads/92319-Happy-oral-arguments-corrected-transcript.pdf.
11. For Judge Truitt's opinion, see the decision of the Supreme Court of the State of New York in the case of *The Nonhuman Rights Project on behalf of Happy* v. *James Breheny, et al.*, 18 February 2020; https://www.nonhumanrights.org/content/uploads/HappyFeb182020.pdf.
12. The Uttarakhand quote is from the NhRP's legal arguments; for more on the story, see Saptarshi Ray, 'Animals Accorded Same Rights as Humans in Indian State', *The Telegraph*, 5 July 2018; https://www.telegraph.co.uk/news/2018/07/05/animals-accorded-rights-humans-indian-national-park/.
13. For the case of Cecilia, see Merritt Clifton, 'Argentinian Court Grants Zoo Chimp a Writ of Habeas Corpus', Animals 24-7, 8 November 2016; https://www.animals24-7.org/2016/11/08/argentinian-court-grants-zoo-chimp-a-writ-of-habeas-corpus/. For the case of Chucho, the bear in Colombia, see 'Colombia's Constitutional Court denies Habeas Corpus for Andean Bear', *The City Paper*, Bogota, 23 January 2020; https://thecitypaperbogota.com/news/colombias-constitutional-court-denies-habeas-corpus-for-andean-bear/23781.
14. For more on India's declaration of river rights, see 'Could making the Ganges a "person" save India's holiest river?', BBC News, 5 April 2017; https://www.bbc.com/news/world-asia-india-39488527.

15. For Ecuador's constitution, see Mihnea Tanasescu, 'The Rights of Nature in Ecuador: The Making of an Idea', *International Journal of Environmental Studies*, 70(6), December 2013, pp. 846–61; DOI:10.1080/00207233.2013.845715. For Colombia and the Amazon, see Anastasia Moloney, 'Colombia's Top Court Orders Government to Protect Amazon Forest in Landmark Case', Reuters, 6 April 2018; https://www.reuters.com/article/us-colombia-deforestation-amazon-IDUSKCN1HD21Y. For New Zealand's Whanganui, see Jeremy Lurgio, 'Saving the Whanganui: Can Personhood Rescue a River?', *The Guardian*, 29 November 2019; https://www.theguardian.com/world/2019/nov/30/saving-the-whanganui-can-personhood-rescue-a-river.
16. Dr Erin O'Donnell, a senior fellow at the University of Melbourne law school, quoted in the *Guardian* article, 'Saving the Whanganui', ibid.
17. References to Sophia's appearance, engineering and legal status are too numerous and dispiriting to single out, but are available from your favourite search engine. Please consider using DuckDuckGo, which is designed to protect your privacy, or Ecosia, which uses its ad revenue to plant trees, if you want to pursue further information on this matter.
18. For the European Parliament resolution, see the study, 'European Civil Law Rules in Robotics', Directorate-General for Internal Policies of the Union (European Parliament), December 2016; https://op.europa.eu/en/publication-detail/-/publication/19ea0f1c-9ab0-11e6-868c-01aa75ed71a1/language-en. For the open letter, see http://www.robotics-openletter.eu/.
19. The example of the Telepaths is drawn from Sue Donaldson and Will Kymlicka, *Zoopolis: A Political Theory of Animal Rights* (Oxford: Oxford University Press, 2011), who in turn based it on a realization by Michael A. Fox, in his book, *The Case for Animal Experimentation: An Evolutionary and Ethical Perspective* (Oakland, CA: University of California Press, 1986). That book is often cited as a sophisticated defence of the right of human beings to use animals for their benefit, but when Fox realized that the same arguments could be used by an alien species to enslave humans, he repudiated it and moved to a position supporting animal rights.
20. Calls to develop 'friendly' AI originate with the Machine Intelligence Research Institute, co-founded by Eliezer Yudkowsky, who also created the LessWrong community, but have been championed by some of the founders of academic and practical AI studies. See, for example, Stuart Russell and Peter Norvig's *Artificial Intelligence: A Modern Approach*,

3rd edn (Harlow: Pearson Education, 2016), the standard textbook on the subject, which cites Yudkowsky's concerns about AI safety.

21. The Trolley problem was first given that name by the moral philosopher Judith Jarvis Thomson in 'Killing, Letting Die, and the Trolley Problem', *The Monist*, 59 (2), April 1976, pp. 204–17. Her conclusion, from a number of examples, was that 'there are circumstances in which – even if it is true that killing is worse than letting die – one may choose to kill instead of letting die'. The Moral Machine can be found at https://www.moralmachine.net.
22. For discussion of Facebook's ethical policy, see Anna Lauren Hoffmann, 'Facebook Has a New Process For Discussing Ethics. But Is It Ethical?', *The Guardian*, 17 June 2016; https://www.theguardian.com/technology/2016/jun/17/facebook-ethics-but-is-it-ethical. For Google's failed Advanced Technology External Advisory Council, see Jane Wakefield, 'Google's Ethics Board Shut Down', BBC News, 5 April 2019; https://www.bbc.com/news/technology-47825833.
23. Cade Metz and Daisuke Wakabayashi, 'Google Researcher Says She Was Fired Over Paper Highlighting Bias in A.I.', *The New York Times*, 3 December 2020.
24. For more on the 'non-identity' of plants, and the ways in which they disrupt our social and political thinking, see Michael Marder, *Plant-Thinking: A Philosophy of Vegetal Life* (New York: Columbia University Press, 2013).
25. Henri Bergson, *Creative Evolution*, translated by Arthur Mitchell (New York: Henry Holt and Company, 1911).
26. I am grateful to Anab Jain, and her essay-lecture 'Calling for a More-Than-Human Politics', 19 February 2020, for this taxonomy of care. See https://superflux.in/index.php/calling-for-a-more-than-human-politics/ for the full text.

## 9. THE INTERNET OF ANIMALS

1. For an overview of animal tracking techniques, see L. David Mech, *Handbook of Animal Radio-Tracking* (Minneapolis, MN: University of Minnesota Press, 1983) and L. Mech and Shannon Barber-Meyer, 'A Critique of Wildlife Radio-Tracking and its Use in National Parks', US National Park Service Report, 2002, pp. 1–78. For Cochran and Lord's system, see W. W. Cochran and R. D. Lord, 'A Radio-Tracking System for Wild Animals', *Journal of Wildlife Management*, 27(1), January 1963, pp. 9–24; DOI:10.2307/3797775.

2. A good overview of the history of Argos can be found in Rebecca Morelle, 'Argos: Keeping Track of the Planet', BBC News, 7 June 2007; http://news.bbc.co.uk/1/hi/sci/tech/6701221.stm.
3. The story of Pluie is recounted in a number of places, among them Cornelia Dean, 'Wandering Wolf Inspires Project', *The New York Times*, 23 May 2006; https://www.nytimes.com/2006/05/23/science/earth/23wolf.html; 'Pluie the Wolf', 8 November 2017, National Park Service; https://www.nps.gov/glac/learn/education/pluie-the-wolf.htm; and 'Pluie, the Wolf Who Inspired Carnivore Recovery Across the West', *Wild Animals* podcast, series 1, episode 1, North Carolina Museum of Natural Sciences; https://naturalsciences.org/research-collections/wild-animals-podcast.
4. For the history and formation of Y2Y, as well as information about its current activities and accomplishments, see https://y2y.net/about/vision-mission/history/. For a broader history and definition of wildlife corridors, see D. K. Rosenberg, B. R. Noon and E. C. Meslow, 'Towards a Definition of Biological Corridor', in J. A. Bissonette and P. R. Krausman (eds), *Integrating People and Wildlife For a Sustainable Future*, International Wildlife Management Congress (Bethesda, MD; The Wildlife Society, 1995).
5. For *The West Wing* episode, see series 1, episode 5, 'The Crackpots and These Women'.
6. The Sredneussuriisky Wildlife Refuge was established in 2012; see 'New Corridor Links Amur Tiger Habitats in Russia and China', World Wide Fund for Nature, 19 October 2012; https://wwf.panda.org/?206504/new-corridor-links-amur-tiger-habitats-in-russia-and-china. For the Natuurbrug Zanderij Crailoo and other Dutch 'ecoducts', see https://www.atlasobscura.com/places/natuurbrug-zanderij-crailoo/. For Tony and the Ngare Ndare corridor, see Maurice O. Nyaligu and Susie Weeks, 'An Elephant Corridor in a Fragmented Conservation Landscape: Preventing the Isolation of Mount Kenya National Park and National Reserve', *Parks: The International Journal of Protected Areas and Conservation*, 19(1), March 2013, pp. 91–101.
7. For the European Green Belt, see https://www.europeangreenbelt.org; and Tony Paterson, 'From Iron Curtain to Green Belt: How New Life Came to the Death Strip', *The Independent*, 17 May 2009; https://www.independent.co.uk/environment/nature/from-iron-curtain-to-green-belt-how-new-life-came-to-the-death-strip-1686294.html. For the Korean DMZ, see Claire Harbage, 'In Korean DMZ, Wildlife Thrives. Some Conservationists Worry Peace Could Disrupt It', NPR, 20 April 2019; https://www.npr.org/2019/04/20/710054899/in-korean-dmz-wildlife-

thrives-some-conservationists-worry-peace-could-disrupt-i. For the wildlife of Pripyat, see 'The Chernobyl Exclusion Zone is Arguably a Nature Preserve', BBC News, April 2016; https://web.archive.org/web/20200315114228/http://www.bbc.com/earth/story/20160421-the-chernobyl-exclusion-zone-is-arguably-a-nature-reserve.

8. For Oslo's bee highway, see Agence France-Presse, 'Oslo Creates World's First "Highway" to Protect Endangered Bees', *The Guardian*, 25 June 2015; https://www.theguardian.com/environment/2015/jun/25/oslo-creates-worlds-first-highway-to-protect-endangered-bees; and for the managing association, ByBi, see https://bybi.no/. For the Norwegian Trekking Association, see https://english.dnt.no.
9. For the impact of the US border wall on wildlife, see Samuel Gilbert, '"An Incredible Scar": The Harsh Toll of Trump's 400-Mile Wall Through National Parks', *The Guardian*, 31 October 2020; https://www.theguardian.com/environment/2020/oct/31/trump-border-wall-wilderness-wildlife-impact; and Samuel Gilbert, 'Trump's Border Wall Construction Threatens Survival of Jaguars in the US', *The Guardian*, 1 December 2020; https://www.theguardian.com/environment/2020/dec/01/trump-border-wall-threatens-jaguars-revival-arizona-sky-islands.
10. For an account of wildlife under threat from the apartheid wall, see Miriam Deprez, 'Even Animals Are Divided by Israel's Wall and Occupation Threats to the Local Environment', *Middle East Monitor*, 20 August 2018; https://www.middleeastmonitor.com/20180820-even-animals-are-divided-by-israels-wall-and-occupation-threats-to-the-local-environment/; and Vanessa O'Brien, 'Israeli Army Opens West Bank Barrier For Animals', *DW News*, 2 November 2012; https://www.dw.com/en/israeli-army-opens-west-bank-barrier-for-animals/a-16351700.
11. For Thomas Wilner's recollection of the case, see 'The Rule of Law Oral History Project: The Reminiscences of Thomas B. Wilner', Columbia University, 2010; http://www.columbia.edu/cu/libraries/inside/ccoh_assets/ccoh_8626509_transcript.pdf . I'm grateful to David Birkin for first alerting me to the story. In addition to the non-human iguanas, a non-human corporation also spoke for the detainees, in the form of Hobby Lobby, a US corporation which had been granted religious liberties denied to Guantánamo detainees; see Philip J. Victor, 'Gitmo Detainees' Lawyers Invoke Hobby Lobby Decision in Court Filing', *Al-Jazeera America*, 5 July 2014; http://america.aljazeera.com/articles/2014/7/5/hobby-lobby-guantanamo.html.
12. For a history of wildlife and I-80, as well as an overview of climate and other pressures on migration, see Ben Guarino, 'Safe Passages', *Washington*

*Post*, 18 March 2020; https://www.washingtonpost.com/graphics/2020/climate-solutions/wyoming-wildlife-corridor/.

13. For maps and data from the Wyoming Migration Initiative, visit https://migrationinitiative.org.
14. For more details on ICARUS, see Andrew Curry, 'The Internet of Animals That Could Help to Save Vanishing Wildlife', *Nature*, 16 October 2018; https://www.nature.com/articles/d41586-018-07036-2, and the ICARUS homepage at https://www.icarus.mpg.de/en.
15. For Professor Maxwell's research, see 'Mobile Protected Areas Needed to Protect Biodiversity in the High Seas', *Inside Ecology*, 17 January 2020; https://insideecology.com/2020/01/17/mobile-protected-areas-needed-to-protect-biodiversity-in-the-high-seas/.
16. Ibid. For the proposal for Mobile Marine Protection Areas, see Sara M. Maxwell, Kristina M. Gjerde, Melinda G. Conners and Larry B. Crowder, 'Mobile Protected Areas for Biodiversity on the High Seas', *Science*, 367(6475), 17 January 2020, pp. 252–4.
17. For ICARUS's work in Etna and L'Aquila see University of Konstanz, 'The Sixth Sense of Animals: An Early Warning System For Earthquakes?', *Phys.org News*, 3 July 2020; https://phys.org/news/2020-07-sixth-animals-early-earthquakcs.html; and *Animals' Early Warning System*, ICARUS Report; https://www.icarus.mpg.de/28810/animals-warning-sensors.
18. For details on the Half-Earth Project's plans and calculations, see https://www.half-earthproject.org/. For more on E. O. Wilson and his work, see Tony Hiss, 'Can the World Really Set Aside Half of the Planet for Wildlife?', *Smithsonian Magazine*, September 2014; https://www.smithsonianmag.com/science-nature/can-world-really-set-aside-half-planet-wildlife-180952379/?no-ist.
19. Hiss, 'Can the World Really Set Aside Half of the Planet for Wildlife?'.
20. See E. O. Wilson's Introduction to Half-Earth Day 2020 at https://www.half-earthproject.org/half-earth-day-2020-highlights/.

## CONCLUSION: DOWN ON THE METAL FARM

1. For a history of agromining research, see Rufus L. Chaney, Alan J. M. Baker and Jean Louis Morel, 'The Long Road to Developing Agromining/Phytomining', in Antony van der Ent, Alan J. M. Baker, Guillaume Echevarria, Marie-Odile Simonnot and Jean Louis Morel (eds), *Agromining: Farming for Metals* (New York: Springer, 2021), pp. 1–22.
2. Dyna Rochmyaningsih, 'The rare plants that "bleed" nickel', BBC Future Planet, 26 August 2020; https://www.bbc.com/future/article/20200825-

indonesia-the-plants-that-mine-poisonous-metals; D. L. Callahan, U. Roessner, V. Dumontet, A. M. De Livera, A. Doronila, A. J. Baker and S. D. Kolev, 'Elemental and Metabolite Profiling of Nickel Hyperaccumulators from New Caledonia', *Phytochemistry*, 81, September 2012, pp. 80–89; DOI: 10.1016/j.phytochem.2012.06.010; L. E. Bennett, J. L. Burkhead, K. L. Hale, N. Terry, M. Pilon and E. A. H. Pilon-Smits, 'Analysis of Transgenic Indian Mustard Plants for Phytoremediation of Metal-Contaminated Mine Tailings', *Journal of Environmental Quality*, 32 (2), March–April 2003, pp. 432–40; DOI:10.2134/jeq2003.0432; Joseph L. Fiegl, Bryan P. McDonnell, Jill A. Kostel, Mary E. Finster and Dr Kimberly Gray, *A Resource Guide: The Phytoremediation of Lead to Urban, Residential Soils*, website adapted from report by Northwestern University (Evanston, IL: Northwestern University, 2011); https://web.archive.org/web/20110224034628/http://www.civil.northwestern.edu/ehe/html_kag/kimweb/MEOP/INDEX.HTM; Li Jiang, Lei Wang, Shu-Yong Mu and Chang-Yan Tian, '*Apocynum venetum*: A Newly Found Lithium Accumulator', *Flora: Morphology, Distribution, Functional Ecology of Plants*, 209 (5–6), June 2014; https://doi.org/10.1016/j.flora.2014.03.007.

3. Marthe de Ferrer, 'Scientists Have Taught Spinach to Send Emails and it Could Warn us About Climate Change', Euronews, 18 March 2021; https://www.euronews.com/green/2021/02/01/scientists-have-taught-spinach-to-send-emails-and-it-could-warn-us-about-climate-change.
4. Emily Dreyfuss, 'The Plan to Grab the World's Carbon With Supercharged Plants', *Wired*, 26 April 2019; https://www.wired.com/story/the-plan-to-grab-the-worlds-carbon-with-supercharged-plants/.
5. For an example of such a scheme, see John Vidal, 'How Bill Gates Aims to Clean Up the Planet', *The Observer*, 4 February 2018; https://www.theguardian.com/environment/2018/feb/04/carbon-emissions-negative-emissions-technologies-capture-storage-bill-gates.

# *Bibliography*

*The following works, not all of which feature explicitly in the text, were nevertheless crucial to developing my thinking, and any understanding gained thereof. They may be considered honoured ancestors of the present work, and as fertile ground for further exploration.* JB


Edward Abbey, *The Monkey Wrench Gang*, Penguin Classics, 2004

David Abrams, *The Spell of the Sensuous: Perception and Language in a More-Than-Human World*, Vintage Books, 1997

Andrew Adamatzky and Louis-Jose Lestocart (eds.), *Thoughts on Unconventional Computing*, Luniver Press, 2021

J. A. Baker, *The Peregrine*, HarperCollins, 1967

Karen Barad, *Meeting the Universe Halfway*, Duke University Press, 2007

Jane Bennett, *Vibrant Matter: A Political Ecology of Things*, Duke University Press Books, 2010

Hans Blohm, Stafford Beer and David Suzuki, *Pebbles to Computers: The Thread*, Oxford University Press, 1987

John Tyler Bonner, *Randomness in Evolution*, Princeton University Press, 2013

James Bradley, *Clade*, Titan Books, 2017

Federico Campagno, *Technic and Magic*, Bloomsbury Academic, 2018

Rachel Carson, *Silent Spring*, Penguin Classics, 2000

Eduardo Viveiros de Castro, *Cannibal Metaphysics*, University Of Minnesota Press, 2017

Emanuele Coccia, *The Life of Plants: A Metaphysics of Mixture*, Polity, 2018

Susan Curran and Gordon Park, *Micro Man*, MacMillan, 1982

Ernesto De Martino, *Magic: A Theory from the South*, HAU, 2015

T. J. Demos, *Decolonizing Nature: Contemporary Art and the Politics of Ecology*, Sternberg Press, 2016

Eva Diaz, *The Experimenters: Chance and Design at Black Mountain College*, University of Chicago Press, 2015

Susan Donaldson and Will Kymlicka, *Zoopolis: A Political Theory of Animal Rights*, Oxford University Press, 2011

George Dyson, *Darwin among the Machines*, Perseus, 1997

George Dyson, *Turing's Cathedral: The Origins of the Digital Universe*, Allen Lane, 2012

Claire Evans, *Broad Band*, Portfolio, 2018

Charles Foster, *Being a Beast*, Profile Books, 2016

Monica Gagliano, *Thus Spake the Plant*, North Atlantic Books, 2018

Amitav Ghosh, *The Great Derangement: Climate Change and the Unthinkable*, University of Chicago Press, 2016

Peter Godfrey-Smith, *Other Minds: The Octopus and the Evolution of Intelligent Life*, William Collins, 2017

Alexis Pauline Gumbs, *Undrowned: Black Feminist Lessons from Marine Mammals*, AK Press, 2021

Donna J. Haraway, *When Species Meet*, University of Minnesota Press, 2007

Donna J. Haraway, *Staying with the Trouble: Making Kin in the Chthulucene*, Duke University Press, 2015

Roger Harnden and Allenna Leonard (eds.), *How Many Grapes Went Into the Wine: Stafford Beer on the Art and Science of Holistic Management*, John Wiley & Sons, 1994

Graham Harvey, *Animism: Respecting the Living World*, C Hurst & Co, 2017

Stefan Helmreich, *Alien Ocean: Anthropological Voyages in Microbial Seas*, University of California Press, 2008

Jason Hribal, *Fear of the Animal Planet: The Hidden History of Animal Resistance*, AK Press, 2011

Aldous Huxley, *Island*, Harper, 1962

Humphrey Jennings, *Pandaemonium 1660–1886: The Coming of the Machine as Seen by Contemporary Observers*, The Free Press, 1986

Tracy Kidder, *The Soul of a New Machine*, Little Brown & Co, 1981

Robin Wall Kimmerer, *Braiding Sweetgrass: Indigenous Wisdom, Scientific Knowledge and the Teachings of Plants*, Penguin, 2020

Barbara Kingsolver, *Flight Behaviour*, Faber & Faber, 2012

Eduardo Kohn, *How Forests Think: Toward an Anthropology Beyond the Human*, California University Press, 2013

Peter Kropotkin, *Mutual Aid: A Factor of Evolution*, Freedom Press, 2009

Bruno Latour, *Down to Earth: Politics in the New Climatic Regime*, Polity Prss, 2018

Ursula K. Le Guin, *The Dispossessed*, Harper & Row, 1974

Konrad Lorenz, *King Solomons Ring*, Routledge, 2002

Robert Macfarlane, *Landmarks*, Hamish Hamilton, 2015
Andreas Malm, *Fossil Capital: The Rise of Steam-Power and the Roots of Global Warming*, Verso, 2015
Andreas Malm, *How to Blow Up a Pipeline: Learning to Fight in a World on Fire*, Verso, 2021
Michael Marder, *Plant-Thinking: A Philosophy of Vegetal Life*, Columbia University Press, 2013
Lyn Margulis, *The Symbiotic Planet: A New Look at Evolution*, Weidenfeld & Nicolson, 1998
Shannon Mattern, *Code and Clay, Data and Dirt: Five Thousand Years of Urban Media*, University of Minnesota Press, 2017
Pamela McCorduck, *Machines Who Think*, W. H. Freeman & Co, 1979
Eden Medina, *Cybernetic Revolutionaries: Technology and Politics in Allende's Chile*, MIT Press, 2014
Eva Meijer, *Animal Languages: The Secret Conversations of the Living World*, John Murray, 2019
Eva Meijer, *When Animals Speak: Toward an Interspecies Democracy*, NYU Press, 2019
Timothy Morton, *Humankind: Solidarity with Non-Human People*, Verso, 2019
Daniel Oberhaus, *Extraterrestrial Languages*, MIT Press, 2019
Mark O'Connell, *To Be a Machine*, Doubleday, 2017
Charles Patterson, *Eternal Treblinka: Our Treatment of Animals and the Holocaust*, Lantern Books, 2004
Fred Pearce, *The New Wild: Why Invasive Species Will be Nature's Salvation*, Beacon Press, 2015
Andrew Pickering, *The Cybernetic Brain*, University of Chicago Press, 2011
Richard Powers, *The Overstory*, William Heinemann, 2018
David Quammen, *The Tangled Tree: A Radical New History of Life*, Simon & Schuster, 2019
David van Reybrouck, *Against Elections: The Case for Democracy*, Random House, 2016
Kim Stanley Robinson, *The Ministry for the Future*, Orbit, 2021
Susan Schuppli, *Material Witness: Media, Forensics, Evidence*, MIT Press, 2020
James C. Scott, *Seeing Like a State: How Certain Schemes to Improve the Human Condition Have Failed*, Yale University Press, 1998
James C. Scott, *Against the Grain: A Deep History of the Earliest States*, Yale University Press, 2017
Thomas D. Seeley, *Honeybee Democracy*, Princeton University Press, 2010

Kenneth Silverman, *Begin Again: A Biography of John Cage*, Random House, 2010

Suzanne Simard, *Finding the Mother Tree: Uncovering the Wisdom and Intelligence of the Forest*, Allen Lane, 2021

Merlin Sheldrake, *Entangled Life*, Bodley Head, 2020

Adrian Tchaikovsky, *Children of Time*, Orbit, 2018

Anna Lowenhaupt Tsing, *The Mushroom at the End of the World: On the Possibility of Life in Capitalist Ruins*, Princeton University Press, 2015

Frans de Waal, *Are We Smart Enough to Know How Smart Animals Are?*, Granta Books, 2016

Xiaowei Wang, *Blockchain Chicken Farm*, FSG, 2020

Peter Wohlleben, *The Hidden Life of Trees*, William Collins, 2017

Andrea Wulf, *The Invention of Nature*, John Murray, 2015

Tyson Yunkaporta, *Sand Talk: How Indigenous Thinking Can Save the World*, HarperOne, 2020

# *Acknowledgements*

Many friends and acquaintances have been part of the discussions and ideas that went into this book; many more than I can name. Thank you to Lawrence Abu Hamdan, Nick Aikens, Tom Armitage, Eva Barbarossa, Zach Blas, Ingrid Burrington, Eduardo Cassina, Mat Dryhurst, Ben Eastham, Marco Ferrari, Kyriaki Goni, Tom Gordon-Martin, Holly Herndon, Samuel Hertz, Anab Jain, Julia Kaganskiy, Claudette Lauzon, Charlie Loyd, Jacob Moe, Julian Oliver, Matthew Plummer-Fernandez, Sascha Pohflepp (we miss you so much), Angelika Sgouros, Merlin Sheldrake, William Skeaping, Kevin Slavin, Zachos Varfis, Ben Vickers, Chris Woebken and Lydia Xynogala. Thanks are also due to Andrew Adamatzky, Alexandra Elbakyan, Maria Konstantinou, Susan Simard and Martin Wikelski for their expertise and generosity.

I am particularly grateful to Aslak Aamot Kjærulff, Einar Sneve Martinussen, Katherine Brydan, Rob Faure-Walker, Lucia Pietroiusti, Jonas Staal and Stefanos Levidis, who read deeply into the work and offered their thoughts and suggestions.

Thanks to Telis and Rosemarie in Aegina for all their kindness and hospitality in the strangest of years. Thanks to everyone at Akrogialia and Aiákeion. Thanks to Lila, Leonidas, Kostas and everyone in Zagori. Thanks to Joseph Grima, Chiara Siravo, Elisa Giuliano, Martha Schwindling, Antonio Elettrico and everyone in Matera. Thanks to the Infrastructure Club, again, and to everyone in the Littoral Group.

Thanks to everyone who's asked me to speak, and listened to me, in the last few years – and to Deborah Rey-Burns for all her work. Thank you to Luca Barbeni at Nome, and Janez Janša and Marcela Okretič at Aksioma, and Davor Mišković and Ivana Katic at Drugo More.

Thank you to Clemancy and Howard Gordon-Martin, for Howard's End (and much else); thanks to John Bridle for everything.

Thanks to my agent Antony Topping, and my editor at Penguin, Thomas Penn, for their enthusiasm and support for this book, and their thoughtful notes and comments. Thanks to Eric Chinski at FSG for also believing in it. Thanks to Jane Robertson for her editing skills and to Eva Hodgkin for wrangling the image rights. Thanks to all the other people who have put their time, effort and skills into this publication.

Thanks to the octopus I met this morning, who winked at me from beneath a rock, and to the three eagles who wheeled, shrieking, overhead, just as I wrote these words, for reminding me of my debt to the world and all the beings in it.

Thanks, above all, to Navine and Zephyr for being the greatest inspiration and deepest wellspring of love that anyone could imagine. Thank you.

Athens – London – Aegina
July 2019–September 2021

# *Index*